AAPS Advances in the Pharmaceutical Sciences Series

Volume 52

Series Editor

Yvonne Perrie, Strathclyde Institute of Pharmacy and Biomedical Sciences
University of Strathclyde
Glasgow, UK

The AAPS Advances in the Pharmaceutical Sciences Series, published in partnership with the American Association of Pharmaceutical Scientists, is designed to deliver volumes authored by opinion leaders and authorities from around the globe, addressing innovations in drug research and development, and best practice for scientists and industry professionals in the pharma and biotech industries. Indexed in Reaxys

SCOPUS Chemical Abstracts Service (CAS) SCImago

EMBASE

Hovhannes J. Gukasyan • Ashutosh Kulkarni
Editors

Antibody Drug Conjugates

Design, Discovery, and Development

Springer

aaps

Editors
Hovhannes J. Gukasyan
Pharmacology and Pharmaceutical Sciences, Regulatory and Quality Sciences
USC Alfred E Mann School of Pharmacy and Pharmaceutical Sciences (United States)
Los Angeles, CA, USA

Ashutosh Kulkarni
Early Portfolio Strategy and Operations
Allergan Aesthetics, an AbbVie Company (United States)
Irvine, CA, USA

ISSN 2210-7371 ISSN 2210-738X (electronic)
AAPS Advances in the Pharmaceutical Sciences Series
ISBN 978-3-031-89318-6 ISBN 978-3-031-89319-3 (eBook)
https://doi.org/10.1007/978-3-031-89319-3

Editorial Contact: Charlotte Nunes

This Springer imprint is published by the registered company Springer Nature Switzerland AG
The registered company address is: Gewerbestrasse 11, 6330 Cham, Switzerland

Contents

About the Editor

Hovhannes J. Gukasyan (Hovik) obtained his PhD from University of Southern California (USC), School of Pharmacy, Department of Pharmaceutical Sciences in 2004. Twenty-two years in pharma industry, he is currently the technology and CMC lead consultant for several small- to medium-sized biotech firms. He supported >80 "Pipeline" programs during his industry career, at Anadys Pharmaceuticals Inc, Pfizer Inc, and Allergan plc (now part of AbbVie Aesthetics), with successful examples marketed as Diquas®, Vyzulta®, Lorbrena®, as well as supported axitinib, sunitinib, crizotinib, ubrogepant, with several more in current clinical trials such as cenicriviroc and epigenetic modulator PF-06821497, also known as mevrometostat. His contributions include enabling development of new chemical entities (NCE), new biological entities (NBE), nonbiological complex drug products (NBCD, i.e., nanotechnology) using molecular (bio)pharmaceutics, with modeling and simulation techniques. He joined the University of Southern California (USC) Alfred E. Mann School of Pharmacy as Associate Professor of Pharmacology and Pharmaceutical Sciences in June 2021 and directs the Management of Drug Development program in the Regulatory and Quality Sciences department.

He has published 48 peer-reviewed (co) authored papers, 2 book chapters, and editing a textbook Molecules-to-Market: Case Studies on Drug Development. He is a "PhRMA Health Champion Fellow," and a "Pfizer Global R&D W.E. Upjohn Prize" award winner. He is a member of "Association for Research in Vision and Ophthalmology (ARVO)," "American Association of Pharmaceutical Scientists (AAPS)," "American Chemical Society (ACS)," and "Tear Film & Ocular Surface Society (TFOS)," and a reviewer of journals like *Investigative Ophthalmology and Visual Science*, *Life Sciences*, *Neuroscience Methods*, *Biochemical Pharmacology*, *Pharmaceutical Research*, and *Journal of Ocular Pharmacology and Therapeutics*.

Ashutosh Kulkarni, PhD, DABT, MBA received his bachelor's degree in Pharmaceutical Sciences from the University of Mumbai and his PhD in Pharmaceutical Sciences from the University of Southern California, School of Pharmacy under the supervision of Dr. Vincent H. L. Lee. He has worked in several pharmaceutical companies over the past 20+ years spanning roles in drug discovery and development across multiple therapeutic areas including Ophthalmology, Dermatology, Neuroscience, Cardiovascular and metabolic diseases, Oncology, and Immunology. He is currently a Project Director focusing on early stage portfolio in the Neurotoxin space at Allergan Aesthetics, an AbbVie company. During his 20+ year career, he has contributed to a number of IND and NDA submissions and was a key leader driving nonclinical as well as clinical pharmacology strategies across a broad portfolio of projects. The projects spanned multiple therapeutic modalities including small molecules and large molecule biologics including ADCs.

He has published about 20 peer-reviewed publications, 2 book chapters, and 25+ abstracts. Additionally, he serves as an adjunct professor at the USC School of Pharmacy and has been teaching there in the Department of Pharmacology and Pharmaceutical Sciences and Department of Regulatory and Quality Sciences for the past 15+ years.

Commercialized ADCs: A Market Snapshot for Exemplification

Ajmeeta Sangtani and Valentina Vozella

Abstract ADCs enable targeted delivery of cytotoxic payloads to cancer cells while minimizing systemic side effects. This review covers the various components of ADCs—the monoclonal antibody, linker, and cytotoxic drug payload—and the role each plays in the mechanism of action. It summarizes the 7 ADCs currently approved by FDA for cancer treatment, including their applications, pharmacokinetics, clinical efficacy and safety based on trial data. With more than 60 ADCs in clinical development, ongoing research continues to explore new disease indications, combination therapies, next-generation modifications to overcome limitations of early generation ADCs. Overall, ADCs represent an innovative therapeutic approach, using engineered antibodies to precisely target tumors with potent chemotherapies and expand the arsenal of oncology treatments.

Keywords Antibody-drug conjugates (ADCs) · Targeted cancer therapies · Monoclonal antibodies · Cytotoxic payloads · FDA-approved oncology treatments

Introduction

Currently, the main challenges facing chemotherapy are non-targeted delivery resulting in systemic toxicity and the development of drug resistance, both of which narrow the therapeutic window and limit the efficacy of the drug (Khongorzul et al. 2020). Antibody-drug conjugates (ADCs) are composed of monoclonal antibodies that target specific antigens overexpressed on cancer cells linked to a cytotoxic payload by a cleavable linker (Khongorzul et al. 2020). ADCs have advantages over traditional chemotherapeutics as they are a type of targeted delivery which

A. Sangtani
Kyfora Bio, Warrington, PA, USA

V. Vozella (✉)
Bausch + Lomb, Irvine, CA, USA

H. J. Gukasyan, A. Kulkarni (eds.), *Antibody Drug Conjugates*, AAPS Advances in the Pharmaceutical Sciences Series 52, https://doi.org/10.1007/978-3-031-89319-3_1

facilitates higher intratumoral drug concentration, thus sparing non-targeted tissues from chemotherapeutic exposure, lowering the toxic side effects. A version of them were first proposed by Nobel laureate Paul Ehrlich in 1900, who first theorized about antibody targeted delivery through the "magic bullet" concept (Strebhardt and Ullrich 2008). The goal of the "magic bullet" was delivery of an appended toxin payload to the intended structural target to reduce the likelihood of adverse effects on healthy cells. ADCs, similarly, facilitate selective delivery of highly potent cytotoxic payloads to tumor cells while evading healthy cells, thus improving the therapeutic window (Khongorzul et al. 2020).

ADCs function by binding to the specific antigen expressed on the cancer cell surface, forming an antigen-antibody complex (Birrer et al. 2019; Alley et al. 2010; Sievers and Senter 2013). The complex is then internalized via receptor-mediated endocytosis. After uptake into the cells, the labile-linker is cleaved, releasing the drug into the cytoplasm. The highly cytotoxic drug can then facilitate cell apoptosis resulting in eventual cell death. Certain determinants influence the clinical activity and tolerability of the ADC including the antigen target, the cytotoxic agent, and the cleavable linker binding the two. In terms of the antigen target, care needs to be taken in selecting an antigen that is sufficiently overexpressed in cancer cells to enhance the directed therapy nature of the ADC while minimizing uptake in normal cells. The cytotoxic payload, which can induce apoptosis and cell death through a variety of avenues including microtubule destabilization, DNA interference, or cell signaling disruption, needs to be stable in systemic circulation as well as within the acidic microenvironment of the tumor and in lysosomes where the drug is often cleaved from the conjugate. Similarly, the labile linker needs to be stable in systemic circulation, but cleavable in response to environmental changes (redox potential, pH) or lysosomal enzymes (Alley et al. 2010).

While the number of investigational ADCs in human trials have tripled over the last five years and there are more than 60 distinct ADCs under clinical evaluation, only seven ADCs have been approved by the FDA thus far (Birrer et al. 2019). This chapter provides an overview of the approved ADCs in chronological order.

Mylotarg (Fig. 1)

Acute myeloid leukemia (AML) is the most prevalent type of acute leukemia in adults and results from a mutation in the genes involved in hematopoietic differentiation or through large chromosomal translocation resulting in poorly differentiated myeloid cells (De Kouchkovsky and Abdul-Hay 2016). Currently, the standard of care for patients with AML is a '7 + 3' regimen of induction therapy (7 days of cytarabine with 3 days of an anthracycline) followed by consolidation therapy (if there is a favorable response) (De Kouchkovsky and Abdul-Hay 2016). Although prognosis for younger patients has improved due to advances in treatment (67% five year survival), prognosis for the elderly (60+), who account for the majority of new cases, still remains poor due to their unfavorable response to chemotherapy and

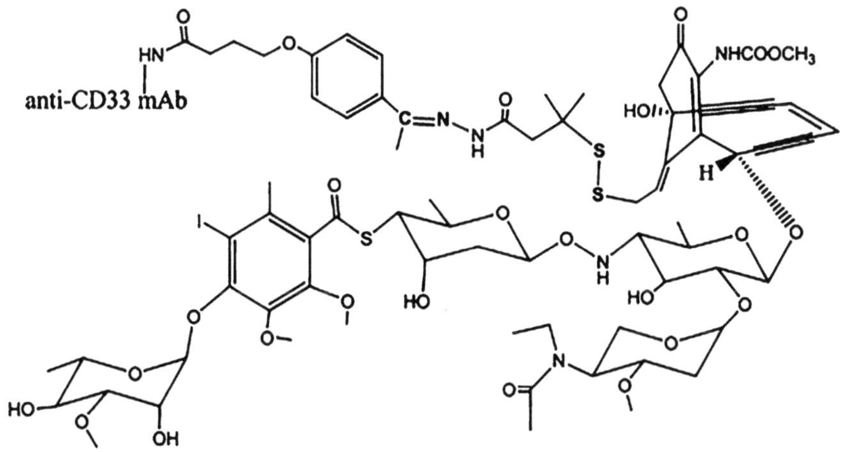

Fig. 1 Mylotarg structure

their susceptibility to treatment-related toxicities (27% five year survival) (De Kouchkovsky and Abdul-Hay 2016).

The first ADC to receive approval by the FDA was Mylotarg (Gemtuzumab ozogamicin), manufactured by Pfizer/Wyeth-Ayerst Laboratories (Mylotarg (Gemtuzumab Ozogamicin) 2020). It is composed of the anti-CD33 humanized monoclonal antibody, a selectively stable linker, and the chemotherapeutic calicheamicin and targets CD33. Cluster of differentiation-33 (CD33) is a sialic acid-dependent antigen expressed on (1) immature cells of myelomonocytic lineage including myelomonocytic progenitors, monocytes, and myeloid dendritic cells and (2) leukemic myeloblasts, but is not expressed on normal hematopoietic progenitors (Sievers and Linenberger 2001). CD33 is known to play a role in hematopoiesis, although the exact function is still being investigated. Additionally, it is present on leukemic blast cells in over 80% of patients with AML, and after antibody binding it accumulates in the marrow and internalizes into the blasts (Hauswirth et al. 2007; Scheinberg et al. 1991). Due to this accumulation and internalization following antibody binding, CD33 is an excellent target for antibody-drug conjugate chemotherapy in patients with AML (Sievers and Linenberger 2001).

In 2000, Mylotarg was first approved by the U.S. FDA under the accelerated approval regulation for first-relapse CD33+ AML patients who were 60 years or older and who were not a candidate for cytotoxic chemotherapy (Bross et al. 2001). Approval was based on three phase II studies comparing Mylotarg to standard chemotherapy. However, on June 21, 2010 Pfizer voluntarily pulled Mylotarg from the market due to a subsequent phase III trial (S0106) that showed no clinical benefit (no clinically significant improvement in complete response (CR) rate, disease-free survival, nor overall survival) and a higher death rate in the treatment arm compared to the control (Petersdorf et al. 2013; Selby et al. 2019). Although pulled from the market in the US, Mylotarg remained on the market in Japan and was available to patients in other countries, including the US, under the compassionate use program, which is designed to allow patients access to unauthorized medicines when they are not a candidate for clinical trials nor for currently marketed medicines (Selby et al. 2019). In September 2017, over 7 years after pulling it from the market, Mylotarg was reapproved by the FDA for the treatment of adults with newly diagnosed CD33+ AML (based on the Phase 3 ALFA-0702 clinical trial and the Phase 3 AML-19 clinical trial) and for the treatment of patients age 2+ with relapsed or refractory CD33+ AML (based on the Phase 2 MyloFrance-1 clinical trial). This new approval was based on a lower recommended dose, a different dosing schedule, in combination with chemotherapy or alone, and in a new patient population. Mylotarg also received orphan drug designation which provides funding to encourage the development of drugs for the treatment of rare diseases (Bross et al. 2001).

The recombinant humanized monoclonal antibody component of Mylotarg, gemtuzumab, is a CD33-targeting immunoglobulin G4 (IgG4 κ antibody hP67.6) (Stasi 2008; Giles et al. 2003). It is produced by CDR-grafting of an anti-CD33 murine antibody (hP67.6). The resulting antibody is a genetically engineered human IgG4 κ antibody that contains sequences from the murine antibody in the antigen binding domain. The linker connecting the antibody and drug is an acid-labile bifunctional linker (4-(4′-acetylphenoxy butanoic acid) acetyl butyrate). While stable at physiological pH, the linker hydrolyzes in low pH environments such as that found in late endosomes (5.0–6.5) and lysosomes (4.5–5.0), allowing the drug to separate from the complex. The drug in the conjugate, calicheamicin, is a semi-synthetic derivative of N-acetyl-γ-calicheamicin dimethyl hydrazine dichloride (Ikemoto et al. 1995). Naturally produced by the bacterium *Micromonospora echinospora ssp. Calichenis,* calicheamicin is a potent anti-tumor agent first identified by its ability to damage DNA in screen tests. When free in serum, the drug has a narrow therapeutic window and delayed toxicity that is 10^3–10^4 times more toxic to cells (including normal cells) than other anticancer agents currently on the market, such as doxorubicin. Additionally, it is known to be very toxic to both gram-positive and gram-negative bacteria. Thus, to increase the therapeutic effect and decrease the toxic side effects, delivery of this highly cytotoxic agent needs to be controlled to target specific cells through conjugation with an antibody. The ADC has a drug to antibody ratio (DAR) of 2 to 3 molecules.

The antibody component of Mylotarg has minimal therapeutic activity as it poorly fixes the complement, thus hindering complement-mediated antibody-dependent apoptosis. Thus, the therapeutic effect of Mylotarg is due solely to the appended drug, while the antibody plays a role in cell targeting and uptake. After binding to CD33 antigens on leukemic blasts, the complex is endocytosed into cells. Following endosomal maturation, lysosomal fusion, and acidification, the acid-labile linker is then cleaved and the calicheamicin moiety is released from the ADC complex. Calicheamicin then binds to the minor groove of the DNA double-helix in a sequence specific manner and undergoes a spontaneous reaction with glutathione in a Bergman-like reaction to activate and form a highly reactive benzenoid diradical which leads to scission of both DNA strands and eventual apoptosis (Ikemoto et al. 1995).

The pharmacokinetics of Mylotarg was characterized using a 2-compartment model with linear and time-dependent clearance components (Mylotarg (Gemtuzumab Ozogamicin) 2020; Sievers et al. 2001). The half-life ($t_{1/2}$) for the complex was found to be about 60 h after the first dose and 90 h after the second and the clearance from plasma was ~.35 L/hr. and .15 L/hr. after the first and second dose, respectively. The total volume of distribution of the complex in human was found to be 21.4 L, and calicheamicin alone was 97% bound to human plasma proteins in vitro. The plasma C_{max} for the gemtuzumab ozogamicin complex was found to be 3.0 mg/mL after the first dose and 3.6 mg/mL after the second. Calicheamicin is metabolized via a nonenzymatic reduction and serum levels in clinical trials were too low to quantify. Age, sex, race, mild or moderate renal impairment, and mild hepatic impairment showed no clinically significant effect on the pharmacokinetics of gemtuzumab ozogamicin, however, severe renal impairment and moderate or severe hepatic impairment were not tested.

As with any chemotherapy treatment, Mylotarg is known to be associated with certain adverse events including but not limited to hepatotoxicity, infusion-related reactions, hemorrhage, QT interval prolongation, and embryo-fetal toxicity (Mylotarg (Gemtuzumab Ozogamicin) 2020; Breccia and Lo-Coco 2011). One of the most severe adverse events associated with Mylotarg is veno-occlusive liver disease (VOD), which can be fatal. Analysis across the trials revealed that the greatest risk factors for VOD after Mylotarg were patients who received higher doses of the drug as a monotherapy, patients with moderate or severe hepatic impairment, patients who received human stem cell transplants (HSCT) before or after Mylotarg. Additionally, as with any intravenous (IV) therapy, infusion-related reactions are a life-threatening or even fatal risk, thus vital signs should be monitored frequently during infusion. Mylotarg is a myelosuppressive and can also cause severe hemorrhage, thus blood counts should be assessed prior to and after treatment. Similar to other drugs containing calicheamicin, Mylotarg can also cause QT interval prolongation, especially in patients with a predisposition for QTc prolongation or who are on medicinal products known to prolong QT. Additionally, as with many chemotherapies, Mylotarg can cause embryo-fetal toxicity and care should be taken with females of reproductive potential and males with female partners of reproductive potential. Finally, in subgroup studies of clinical trials, patients with adverse-risk cytogenetics did not have improved event-free survival with Mylotarg compared to control, thus individual risk-benefit should be analyzed. The maximum tolerated dose was 8.4 mg/kg.

As mentioned above, approval for Mylotarg in 2017 was based on 3 clinical trials: ALFA-0702, AML-19, and MyloFrance-1. In ALFA-0701, a Phase 3, multicenter, randomized, open-label study of 271 patients with newly diagnosed AML, patients were given either Mylotarg + chemotherapy or chemotherapy alone in the induction phase. After evaluation of response in the induction phase patients were again given Mylotarg + chemotherapy or chemotherapy alone in the consolidation phase. The primary endpoint for the study was event free survival (EFS) and results showed a 7.8 month increase in EFS for the Mylotarg arm compared to the control arm. Additionally, EFS probability at 2 and 3 years was more than double for patients in the Mylotarg arm compared to chemotherapy alone (Jen et al. 2018). AML-19 was a single-agent evaluation of Mylotarg in a Phase 3, randomized, open-label study of 237 patients with newly diagnosed CD33+ AML. Patients either received best supportive care (BSC) or Mylotarg in the induction phase, and same in the continuation phase for patients with no evidence of disease progression or significant toxicities. It is to be noted, that best supportive care was used as the control due to the fact that enrolled patients were either greater than 75 years of age, or 61 to 75 and unwilling to receive intensive chemotherapy. The primary endpoint was overall survival and Mylotarg demonstrated a 31% reduction in the risk of death compared to BSC. Additionally, the median disease-free survival was 5.3 months, and 37% of patients treated with Mylotarg achieved complete response (CR) or complete response with incomplete hematological recovery (CRi) (Amadori et al. 2016). In MyloFrance-1, a Phase 2, single arm, open-label study of 57 adults after first relapse CD33+ AML, patients were given Mylotarg alone during the induction phase. After evaluation of response, patients were given chemotherapy (Cytarabine) during the consolidation phase and efficacy was determined based on CR rate at day 43. At day 43, 33% of patients achieved remission, 26 patients achieved CR and the median overall survival was 8.4 months (Taksin et al. 2007).

Due to the CD33 antibody fragment of Mylotarg, the ADC is only effective against immature cells of myelomonocytic lineage and leukemic myeloblasts as they most abundantly express the CD33 antigen on their surface. However, the drug is still in clinical trials for a variety of related cancers and, as of January 2020, is being investigated in >30 clinical trials, including Phase IV trials, according to clinicaltrials.gov.

ADCETRIS (Fig. 2)

Hodgkin lymphoma (HL) and non-Hodgkin lymphoma (nHL) are uncommon malignancies involving lymphocytes and the lymphatic system. There are two main types of HL: classical Hodgkin Lymphoma (cHL), that accounts for 95% of cases, and nodular lymphocyte-predominant HL (Eichenauer et al. 2014; Küppers and Hansmann 2005). cHL is characterized by the presence of Reed-Sternberg cells, which express CD30, a cell membrane receptor. Additionally, anaplastic large-cell lymphoma (ALCL), a type of nHL which is characterized by the presence of large pleomorphic cells that express CD30 and T-cell markers, is composed of four different clinical variations. Two types present as systemic disease and are considered aggressive lymphomas, while the other two types present as localized disease. ~5% of patients in early stages of disease and 30–40% in advanced stages relapse after front-line therapy (Wahl et al. 2002; Francisco et al. 2003). The standard of care for patients with relapsed or refractory disease is salvage therapy with second-line high-dose chemotherapy (HDCT) followed by autologous stem cell transplant (ASCT) in those eligible for transplant (Wahl et al. 2002; Francisco et al. 2003). However, only 50% of patients are cured with standard salvage therapies (Younes

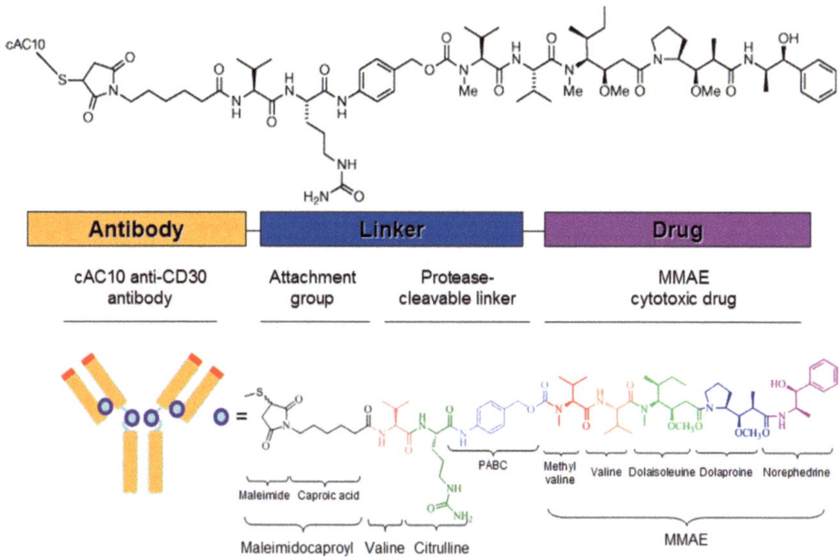

Fig. 2 Adcetris Structure

et al. 2010) and patients who relapse after salvage HDCT/ASCT generally have poor prognosis.

The first ADC to receive marketing for this indication was brentuximab vedotin (Adcetris®) (ADCETRIS 2019). Adcetris was developed by the biotechnology company Seattle Genetics Inc. and approved by the FDA in 2011 for the treatment of some hematologic malignancies such as cHL and other T-cell lymphomas. Adcetris targets the CD30 membrane receptor, a tumor necrosis factor receptor superfamily member. CD30 is an ideal target for ADC-based therapy due its high levels of expression on specific tumor cells (including cHL and systemic ALCL) and restricted expression to a small population of cells, predominantly activated B cells and T cells. In addition, CD30 is consistently expressed on cHL cells and systemic ALCL cells, irrespective of disease stage, line of therapy, and/or transplant status.

In August 2011, the FDA granted accelerated approval for the use of Adcetris in the treatment of relapsed HL and ALCL. In August 2015, FDA granted approval for the treatment of classical HL patients at high risk of relapse or progression after autologous stem cell transplant (ASCT). The FDA's approval was based on positive results from the Phase III clinical trial Aethera. In November 2017, the FDA approved Adcetris as a treatment for patients with cutaneous T-cell lymphoma (CTCL) who have received prior systemic therapy. This approval is for patients with primary cutaneous anaplastic large cell lymphoma (pcALCL) and CD30-expressing mycosis fungoides. In March 2018, the FDA granted Adcetris priority review. FDA's priority review program grants an expedited six-month review of drugs that could provide safe and effective therapy when there are no satisfactory alternative therapies on the market or when the new drug offers significant improvement over marketed products. Adcetris also received the breakthrough designation, which is also designed to expedite development and review of new drugs intended to treat serious, life-threatening diseases. This designation was granted for expansion of use to include first-line treatment of stage III and IV cHL in combination with chemotherapy (adriamycin, vinblastine, and dacarbazine), based on the ECHELON-1 clinical trial. Later in November of the same year the FDA approved Adcetris in combination with chemotherapy for previously untreated sALCL or other CD30-expressing peripheral T-cell lymphomas (PTCL) in adults, based on the ECHELON-2 clinical trial (Han et al. 2012). This was the first drug approved for previously untreated/newly diagnosed PTCL and was approved under the Real-Time Oncology Review (RTOR) program which was designed to speed the review of drugs by allowing the FDA to access key data prior to official submission.

Brentuximab is a chimeric IgG monoclonal antibody specific for CD30 antigens that is derived from the fusion of the variable heavy and light region of the murine anti-CD30 antibody with the constant gamma 1-heavy and kappa-light region of the human immunoglobin. Early phase clinical trials of the anti-CD30 antibody showed no responses in HL and minimal response in ALCL, however, conjugating the antibody to valine-citrulline mono-methyl auristatin E (vcMMAE) created a highly active compound. The drug, MMAE, is a potent microtubule disruptive cytotoxic agent that inhibits microtubule assembly, tubulin-dependent GTP hydrolysis, and polymerization, leading to apoptosis and cell death. The protease cleavable linker that covalently attaches MMAE to brentuximab consists of a thiolreactive maleimidocaproyl spacer, a dipeptide valine-citrulline linker, and a self-immolative p-amino benzyloxycarbonyl, designed to be stable in the blood stream but to be cleaved intracellularly. Similar to other targeted therapies, the ADC binds to a receptor on the surface of cells, in this case antigen CD30 on the surface of CD30-expressing cells. The resultant complex is endocytosed into the cell, matured into a lysosome where MMAE is then released via proteolytic cleavage. The free MMAE binds to tubulin, disrupting the microtubule activity and resulting in G2/M cell cycle arrest and apoptosis.

The pharmacokinetic profile of Adcetris has been characterized by measuring the serum concentrations of the total antibody (sum of ADC and brentuximab), the ADC alone, and the plasma concentrations of released MMAE (ADCETRIS 2019; Pro et al. 2012). The AUCs of the total antibody, antibody-drug conjugate, and free MMAE are dose proportional in the therapeutic single-dose range of 1.2–2.7 mg/kg and show minimal systemic accumulation when administered every 3 weeks. Peak concentrations are obtained at the end of intravenous infusion for the ADC and total antibody and approximately 2 to 3 days post-dose for MMAE. Steady-state was achieved for both the ADC and MMAE by 21 days, and the half-life (t1/2) was 4 to 6 days and 3 to 4 days, respectively. The elimination of brentuximab vedotin and the unconjugated brentuximab occurs through proteolytic degradation into amino acids and recycling into other proteins. The elimination of the unconjugated small molecule MMAE occurs through the liver, primarily via oxidation by CYP3A4 and

CYP3A5, and kidneys. As shown by a clinical excretion study, the primary excretion route of MMAE is via feces (72%, over one week). The rest of the MMAE was found in urine. Unchanged MMAE (approximately 23.5%) was the primary species excreted in both feces and urine. In patients with hepatic impairment systemic exposure of MMAE is increased 2.3-fold compared with normal individuals and in patients with severe renal impairment (Clcr <30 mL/minute), systemic exposure of MMAE is increased 1.9-fold compared with individuals with normal renal function. Gender, race, and age do not have meaningful effects on pharmacokinetics of Adcetris. Brentuximab vedotin showed substantial efficacy and an acceptable safety profile when administered at 1.8 mg kg^{-1} every 3 weeks (Pro et al. 2012) and the recommended dose in combination with chemotherapy for previously untreated Stage III or IV cHL is 1.2 mg/kg up to a maximum of 120 mg every 2 weeks for a maximum of 12 doses.

Clinical trials have reported the development of some adverse reactions after IV delivery of Adcetris (ADCETRIS 2019). The most common adverse reactions (≥20%) with monotherapy were peripheral neuropathy, fatigue, nausea, diarrhea, neutropenia, upper respiratory tract infection, and pyrexia. The most common adverse reactions (≥20%) in combination with AVD were peripheral neuropathy, neutropenia, nausea, constipation, vomiting, fatigue, diarrhea, pyrexia, alopecia, decreased weight, abdominal pain, anemia, and stomatitis. The most common adverse reactions (≥20%) in combination with CHP were anemia, neutropenia, peripheral neuropathy, lymphopenia, nausea, diarrhea, fatigue or asthenia, mucositis, constipation, alopecia, pyrexia, and vomiting. Also, as with many other IV administered drugs, Adcetris can cause infusion reactions.

Adcetris was granted FDA approval in 2011 for relapsed cHL based on a Study 1 (NCT00848926) clinical trial (Younes et al. 2012). The efficacy of ADCETRIS in patients with cHL who relapsed after autologous hematopoietic stem cell transplantation was evaluated in one open-label, single-arm, multicenter trial. 102 patients were treated with 1.8 mg/kg of Adcetris intravenously over 30 minutes every 3 weeks. Study 2 (NT00866047) clinical trial, a open-label, single-arm, multicenter trial, was conducted to assess the efficacy of Adcetris in patients with relapsed sALCL after prior therapy. 58 patients were treated with 1.8 mg/kg of Adcetris administered intravenously over 30 minutes every 3 weeks. Patients had received a median of 2 prior therapies; 26% of patients had received prior autologous hematopoietic stem cell transplantation. 50% of patients were relapsed, and 50% of patients were refractory to their most recent prior therapy. An independent review facility (IRF) performed efficacy evaluations which included overall response rate (ORR = complete response [CR] + partial response [PR]) and duration of response. Study 1 demonstrated an ORR of 73% with a median duration of response of 6.7 months. Study 2 demonstrated an ORR of 86% with a median duration of response of 12.6 months.

The efficacy of Adcetris in patients with cHL at high risk of relapse or disease progression post-auto-HSCT was studied in AETHERA clinical trial, a randomized, double-blind, placebo-controlled study. 329 patients were randomized 1:1 to receive placebo or Adcetris beginning 30–45 days post-auto-HSCT. Patients in the placebo arm with progressive disease could receive Adcetris as part of a separate trial. The primary endpoint was progression-free survival (PFS) (Moskowitz et al. 2015). A total of 329 patients were enrolled and randomized (165 Adcetris, 164 placebo). Patients had received a median of 2 prior systemic therapies excluding autologous hematopoietic stem cell transplantation. The median PFS follow-up time from randomization was 22 months (range, 0–49). AETHERA demonstrated a statistically significant improvement in PFS in the Adcetris arm compared with the placebo arm.

The efficacy of Adcetris in patients with primary cutaneous anaplastic large cell lymphoma (pcALCL) or mycosis fungoides (MF) requiring systemic therapy was studied in ALCANZA, a randomized, open-label, multicenter clinical trial. 131 patients were randomized 1:1 to receive Adcetris or physician's choice of methotrexate or bexarotene. The randomization was stratified by baseline disease diagnosis (MF or pcALCL) (Prince et al. 2017). The efficacy results were based on the proportion of patients achieving an overall response (CR + PR) that lasts at least 4 months (ORR4). The percentage of patients that achieved ORR4 increased by ~42% and the percentage of patients that achieved CR increased by ~15%. Additionally, the percentage of patients that achieved PFS increased by 13%.

The efficacy of ADCETRIS in combination with chemotherapy for the treatment of patients with previously untreated Stage III or IV cHL was evaluated in a randomized, open-label, 2-arm, multicenter trial, ECHELON-1 (Connors et al. 2018; Ramchandren et al. 2019). Of the 1334 total patients, 664 patients were randomized to the Adcetris + doxorubicin [A], vinblastine [V] and dacarbazine [D] (ADCETRIS + AVD) arm and 670 patients were randomized to the doxorubicin [A] + bleomycin [B] + vinblastine [V] + dacarbazine [D] (ABVD) arm. Efficacy was established based on modified PFS and patients in the Adcetris + AVD arm had 4.9% decrease in probability of modified PFS. Additionally, Adcetris + AVD reduced the need for chemotherapy and transplant by 33% compared to the ABVD group.

The efficacy of Adcetris in combination with chemotherapy for the treatment of adult patients with previously untreated, CD30-expressing PTCL was evaluated in ECHELON-2 trial, a multicenter, randomized, double-blind, actively controlled trial (Horwitz et al. 2019; Jacobsen et al. 2015). Of the 452 total patients, 226 patients were randomized to the Adcetris + cyclophosphamide (C) + doxorubicin (H) + prednisone (P) (Adcetris + CHP) and 226 patients were randomized to the CHOP arm (cyclophosphamide, doxorubicin, vincristine, and prednisone). The primary endpoint for the study was PFS per independent review facility (IRF). Patients treated with Adcetris + CHP experienced a 29% reduction in progression and a 34% reduction in the risk of death.

Several phase III trials are in progress to confirm efficacy and safety of Adcetris in randomized studies, both as a single agent and in combination with approved agents (NCT03138499, NCT03907488, NCT02166463) according to clinicaltrial.gov.

KADCYLA (Fig. 3)

Breast cancer is the second most common cancer diagnosis for women in the US. Human epidermal growth factor receptor 2 (HER2) belongs to the epidermal growth factor receptor (EGFR) family and is overexpressed in 20–30% of all breast cancers (Penault-Llorca et al. 2005). HER2+ cancers have been shown to be far more aggressive than HER2 cancers leading to a worse prognosis (3 year median survival compared to 6–7 years for HER2 cancer) (Penault-Llorca et al. 2005). Targeted therapy (such as Herceptin, Perjeta, Tykerb, etc.) has been shown to significantly improve the prognosis of such types of cancer. However, despite the remarkable specificity of HER2 targeted therapies in its ability to target the HER2 receptor, most patients who are initially responsive to treatment eventually experience disease progression, highlighting the need for alternative therapies (Verma et al. 2012). Kadcyla (ado-trastuzumab emtansine), an antibody drug conjugate, was developed by Genentech (a member of the Roche group) to meet this challenge (Chowdhury and Ellis 2014; Kadcyla 2019). The ADC incorporates the chemotherapeutic agent mertansine (DM1), with Herceptin to create a drug-enhanced antibody that can improve the tumor selectivity and reduce the systemic toxicity of DM1 and can confer higher potency to Herceptin (Verma et al. 2012).

Kadcyla was evaluated under the FDA's priority review program and first granted approval in 2013 by the FDA for the treatment of late stage, HER2+ metastatic breast cancer after prior treatment with Herceptin (trastuzumab) and a taxane based on the Phase III EMILIA (NCT00829166) clinical trial (Amiri-Kordestani et al. 2014; Ballantyne and Dhillon 2013). Later, in 2019, under FDA's Real-Time Oncology Review (RTOR) program, Kadcyla was granted approval for adjuvant treatment of patients with HER2+ early breast cancer (EBC) who have residual

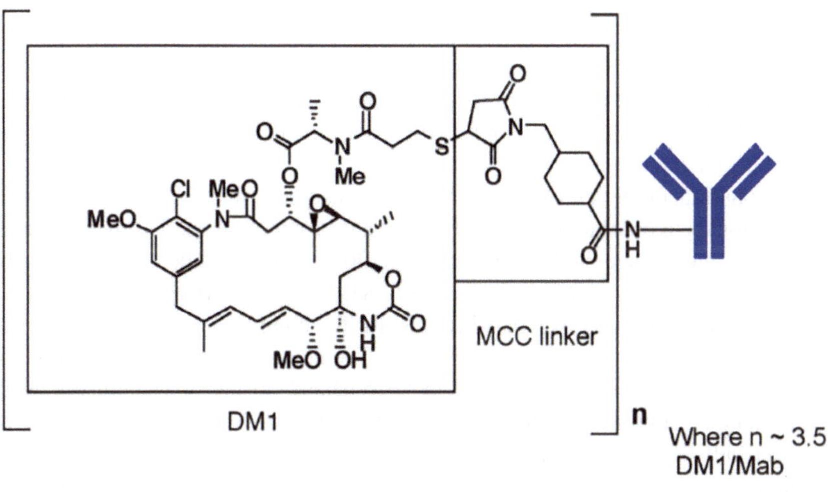

Fig. 3 Kadcyla structure

invasive disease after neoadjuvant taxane and Herceptin (trastuzumab)-based treatment based on the Phase III KATHERINE (NCT01772472) clinical trial. For this indication, Kadcyla was also granted the Breakthrough Therapy Designation. Under the RTOR program and due to the Breakthrough Therapy Designation, the drug was approved just 12 weeks after completion of submission (Pazdur 2019).

Similar to other antibody-drug conjugates, Kadcyla is composed of a monoclonal antibody, a humanized anti-HER2 IgG1, a drug, mertansine (DM1), and a linker joining the two (Fig 3) (Chowdhury and Ellis 2014). The antibody in the complex is Herceptin (chemical name trastuzumab), produced by Chinese hamster ovary (CHO) cells, which is sold as a separate drug by Genentech for the treatment of HER2+ cancers including early and metastatic breast cancer and stomach or gastric cancer. It has been approved for combination therapy with a chemotherapeutic (stomach and breast cancer) or as single treatment after the patient has received other chemotherapeutic drugs (early and metastatic breast cancer). DM1, the drug in the conjugate, is an anti-mitotic agent and is a derivative of mytansine. Originally derived from the bark of the African shrub *Maytenus ovatus,* DM1 functions by inhibiting tubulin polymerization which interferes with mitosis and promotes apoptosis in cells leading to cell death (Kupchan 1972). DM1 is known to be significantly more potent than other drugs in its class, such as paclitaxel, but Phase I and II trials continue to show an insufficient therapeutic index (Chowdhury and Ellis 2014; Issell and Crooke 1978). The linker used in the conjugate, (-succinimidyl-4-(N-maleimidomethyl) cyclohexane-1-carboxylate) (SMCC), is a non-reducible thioether linker. This linker is used to ensure high systemic exposure of the bioconjugate but not of DM1 alone, which is highly toxic at high doses (Chowdhury and Ellis 2014). Within the conjugate, the estimated drug to antibody ratio (DAR) is 3.5.

The mechanism of action of Kadcyla is dual as both the antibody and the appended drug play a role in cell death (Barok et al. 2014; Poon et al. 2013). The antibody, trastuzumab, binds to HER2 receptors on the cell membrane and is thought to block intracellular signaling pathways which then attenuates downstream signal transduction and promotes apoptosis and cell proliferation inhibition. After binding to HER2, the complex then undergoes receptor-mediated endocytosis and lysosomal degradation of the linker resulting in release of DM1 from the ADC. This allows DM1 to bind to tubulin and disrupt the microtubule networks, resulting in apoptosis and cell cycle arrest.

The pharmacokinetics of Kadcyla were measured using ELISA and LC-MS methodologies in order to measure the conjugated complex as well as the free drugs and metabolites (Kadcyla 2019; Girish et al. 2012; Stinchcombe et al. 2017). The C_{max} for the complex was found to be 72.6–83.4 μg/mL (variations due to differences in clinical studies). The half-life ($t_{1/2}$) for the complex was found to be about four days and the clearance was ~.68 L/day. The central volume of distribution of the complex was found to be 3.13 L. After repeated dosing of IV infusion every 3 weeks, no accumulation of the complex was observed. Total trastuzumab showed a longer half-life and slower clearance, which can be attributed to a slow rate of deconjugation of the complex. Plasma concentrations of DM1 were low, with an average of 5 ng/mL in patients dosed with 3.6 mg/kg Kadcyla, corresponding to only .3% of the total conjugated DM1 delivered, indicating good chemical stability of the linker *in vivo.* DM1 was shown to be metabolized by CYP3A/5 *in vitro,* but did not inhibit or induce major CYP450 enzymes. Body weight was found to be the only factor that had a meaningful clinical effect on exposure of the ADC and neither mild to moderate renal impairment nor age or race affected the pharmacokinetics of Kadcyla.

Similar to other HER2 targeted therapies such as Herceptin and Perjeta, Kadcyla is known to cause left ventricular dysfunction (Kadcyla 2019; Girish et al. 2012; Stinchcombe et al. 2017). Additionally, Kadcyla can cause embryo-fetal toxicity and thus should not be administered to pregnant women. Clinical trials with the ADC have also shown hepatotoxicity, pulmonary toxicity, neurotoxicity and thrombocytopenia. As with other drugs administered via IV infusion, delivery of Kadcyla can also cause infusion-related and hypersensitivity reactions. Phase I clinical trials have shown the maximum tolerated dose to be 3.6 mg/kg.

As mentioned previously, Kadcyla was granted FDA approval for metastatic HER2+ breast cancer based on the Phase III EMILIA clinical trial (Guerin et al. 2015). 991 participants were enrolled in the randomized, multicenter, open-label trial designed to compare the safety and efficacy of ado-trastuzumab emtansine with the combination of capetitabine + lapatinib. Patients were randomized 1:1 to receive either 3.6 mg/kg ado-trastuzumab emtansine via IV infusion in 21-day cycles or 1250 mg/day laptinib (oral, daily) plus 1000 mg/m^2 capecitabine (oral, twice-daily, days 1–14) until disease progression or unacceptable toxicity. The primary endpoints of the trial were progression-free survival (PFS), overall survival, and safety. EMILIA demonstrated a 50% improvement in median PFS, a 5.8 months improvement in overall survival, and a decrease in dose modifications and discontinuations for patients treated with Kadcyla compared to lapatinib + capecitabine (Welslau et al. 2014).

As mentioned previously, Kadcyla was granted FDA approval for HER2+ early breast cancer based on the Phase III KATHERINE clinical trial (Verma et al. 2012). 1486 patients were enrolled in the randomized, multicenter, open-label trial. Patients were randomized (1:1) to receive either 3.6 mg/kg ado-trastuzumab emtansine or 6 mg/kg trastuzumab once every 21 days for 294 days. The primary endpoint for the trial was invasive disease-free survival and after a median follow-up of 40 months, the patients treated with Kadcyla demonstrated 50% reduction in the risk of recurrence compared to patients treated with Herceptin.

Currently, Kadcyla is in clinical trials for other HER2+ cancers including metastatic colorectal cancer, lung cancer, bladder cancer, urinary tract cancers, non-small cell lung carcinoma, and gastric cancer. Additionally, it is also in clinical trial for combination with other chemotherapeutics such as palbocilib (NCT03530696), copanlisib (NCT04042051), and afatinib (NCT04158947). As of December 2019, KADCYLA is being investigated in >60 clinical trials according to clinicaltrials.gov.

BESPONSA (Fig. 4)

Acute lymphoblastic leukemia (ALL) is an aggressive type of blood cancer, where the bone marrow manufactures an excess of immature white blood cells (B-cell lymphocytes), and has a poor prognosis following relapse, indicating a high unmet need (Pui et al. 2004; Pui and Evans 2006). Cluster of differentiation-22 (CD22) is a cell-surface regulatory antigen belonging to the lectin family and is found on the surface of mature and some immature B-cells. This sugar-binding transmembrane protein serves as an inhibitory receptor for B-cell receptor signaling which then stops overactivation of the immune system and prevents the development of autoimmune disease. CD22 is present in >90% of patients with B-cell ALL, thus directing therapy to CD22-expressing malignant B-cells can significantly improve anticancer efficacy and overall prognosis in patients with relapsed or refractory B-cell ALL (Dörner and Goldenberg 2007).

Inotuzumab ozogamicin (Besponsa, CMC-544), manufactured by Pfizer, is composed of an anti-CD22 humanized monoclonal antibody, a selectively stable linker, and a chemotherapeutic drug (BESPONSA 2018). In 2015, Besponsa was granted the FDA breakthrough therapy designation for ALL based on promising results from the Phase III INO-VATE trial and due to the fact that five-year survival for adults with relapsed or refractory ALL is less than 10% (Lamb 2017). Besponsa received the orphan drug designation which aimed to encourage the development of drugs for rare diseases. It was then approved by the FDA on August 17, 2017 for adults with relapsed or refractory B-cell precursor acute lymphoblastic leukemia (Fala n.d.).

The antibody component of Besponsa, inotuzumab, is a CD22-targeting immunoglobulin G4 humanized monoclonal antibody (Lamb 2017; Fala n.d.). Similar to Mylotarg, the other ADC manufactured by Pfizer (see above), the antibody is

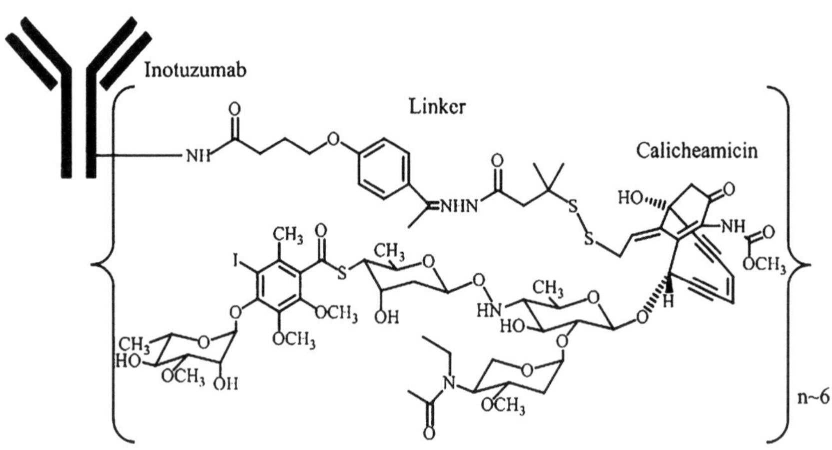

Fig. 4 Besponsa structure

connected to the chemotherapeutic calicheamicin via an acid-labile linker (4-(4′-acetylphenoxy butanoic acid), acetyl butyrate). While stable at physiological pH, the linker hydrolyzes in low pH environments, resulting in release of the drug. The drug, calicheamicin, is a potent DNA-binding cytotoxic agent and a semi-synthetic derivative of N-acetyl-γ-calichaemicin dimethyl hydrazine dichloride (Kunz et al. 2012). Free calicheamicin is known to have a narrow therapeutic window and delayed toxicity *in vivo,* thus conjugation to an antibody to control and target delivery can significantly improve the therapeutic effect by directing toxicity and reducing side effects.

Internalization of Besponsa into cells begins with the binding of the monoclonal antibody to the CD22 antigen expressed on the surface of cells (Lamb 2017; Fala n.d.; Ricart 2011). Similar to Mylotarg, the antibody only plays a role in targeting and cell-uptaking, not in toxicity due to the fact that it poorly fixes complement. Thus, the cytotoxicity and resulting therapeutic effect of the ADC is due solely to the appended drug, calicheamicin. Following the binding to CD22 antigens expressed on the surface of leukemic blasts, the complex is internalized into cells. The acid-labile linker is cleaved in lysosomes at acidic pH, releasing calicheamicin that binds to the minor groove in DNA and produces site-specific double-stranded breaks by forming p-benzene diradicals. Scission of the double stranded DNA leads to inhibition of DNA synthesis and eventual apoptosis.

The pharmacokinetics of Besponsa was characterized using a 2-compartment model with linear and time-dependent clearance compartments (BESPONSA 2018; Advani et al. 2010). The half-life ($t_{1/2}$) for the complex was found to be about 12.3 days and the clearance at steady state was ~0.03 L/hr. The total volume of distribution in humans was found to be 12 L, and the model predicted a 5.3 times accumulation after 4 cycles. The C_{max} for Besponsa was found to be 308 ng/mL. Calichaemicin is metabolized via a nonenzymatic reduction and serum levels in clinical trials were too low to quantify. Although no intrinsic factor (age, sex, and/or race) had any clinically significant effect on Besponsa's pharmacokinetics, body surface area (BSA) did play a role in Besponsa's disposition, thus the drug is dosed based on BSA.

As the conjugated drug is the same in both ADCs, Besponsa is known to be associated with adverse events similar to those seen with Mylotarg (BESPONSA 2018; Advani et al. 2010). In the INO-VATE ALL trial, the major safety concern was hepatotoxicity including hepatic veno-occlusive disease (VOD) where toxicity was observed in 14% of patients in the Besponsa arm (Fala n.d.). Additionally, a combination of hepatotoxicity and infection led to an increased risk of post-transplant non-relapse mortality. The risk of VOD was greater in patients who underwent human stem cell transplant (HSCT) before Besponsa and in patients with ongoing or prior liver disease. Other adverse events included complications due to myelosuppression such as thrombocytopenia, neutropenia, infection and hemorrhagic events. Besponsa also caused an increase in QT interval where patients with a history of QT prolongation or who were on medicinal products known to prolong the QT interval were at greater risk. Finally, similar to all IV infusion therapies, Besponsa can cause infusion-related reactions. In pregnant woman, Besponsa can also cause embryo-fetal toxicity as shown by in vivo studies which caused embryo-fetal harm at 0.4 times the exposure of the maximum recommended dose. The maximum tolerated dose was found to be 1.8 mg/m^2 based on dose finding trials.

The Phase 3 INO-VATE ALL clinical trial sought to compare Besponsa vs. standard of care chemotherapy in adult patients with relapsed or refractory CD22+ B-cell ALL (Lamb 2017). 326 patients were enrolled in the global, open-label trial. Patients were randomized 1:1 to receive either IV infusion of Besponsa in fractionated doses or investigators choice of chemotherapy (FLAG -Fludarabine, cytarabine and G-CSF-, HiDAC -high dose cytarbine-, or cytarabine and mitoxantrone). The primary endpoints for the trial were percentage of participants with complete remission or complete remission with incomplete hematologic recovery (CR/CRi) and overall survival (OS). Besponsa had a 51.3% improvement in rate of CR/CRi, but only showed a 1.5 month improvement in OS and did not meet the prespecified boundary for statistical significance.

Due to the CD22 antibody fragment of Besponsa, the drug is only effective against B-cell related cancers as they most abundantly express the CD22 antigen on their surface. However, the drug is still in clinical trials for a variety of related cancers and other diseases. As of January 2020, Besponsa is being investigated in >20 clinical trials, including Phase IV trials, according to clinicaltrials.gov.

POLIVY (Fig. 5)

Non-Hodgkin's Lymphoma (NHL), which encompasses over 60 different type of cancers, is the most common blood cancer among adults in the US. Diffuse large B-cell lymphoma (DLBCL) is the most common type of NHL in the US (1 out of 3 NHL cases are DLBCL). Prognosis of DLBCL is poor as 40% of patients are unresponsive/have recurrence after treatment with the front-line of care (Miyazaki 2016). The next treatment option is hematopoietic stem cell transplant (HSCT), (Bhatt and Vose 2014) but due to the age of patients many are ineligible to receive HSCT, and there is no other standard of care. For patients treated with stem cell therapy, 50% result in relapse, after which the prognosis is further worsened, and treatment options are limited. Polivy in combination with bendamustine plus a rituximab product is the first and only FDA-approved treatment regimen for refractory or relapsed DLBCL (Palanca-Wessels et al. 2015). The identified antigen target is CD79b, a B-cell antigen receptor complex associated protein beta chain expressed in the majority of malignant lymphomas (Polson et al. 2007; Chu and Arber 2001; Roschewski et al. 2014).

Polivy (polatuzumab vedotin or polatuzumab vedotin-piiq) was developed by Genentech using proprietary technology from Seattle Genetics, for the treatment of hematological malignancies (POLIVY 2019). It is composed of a monoclonal antibody against the CD79b receptor conjugated to the microtubule-disrupting anti-mitotic agent monomethyl auristatin (MMAE) through a cleavable linker (Dornan et al. 2009). In 2016 Polivy received the orphan drug designation, assigned to drugs

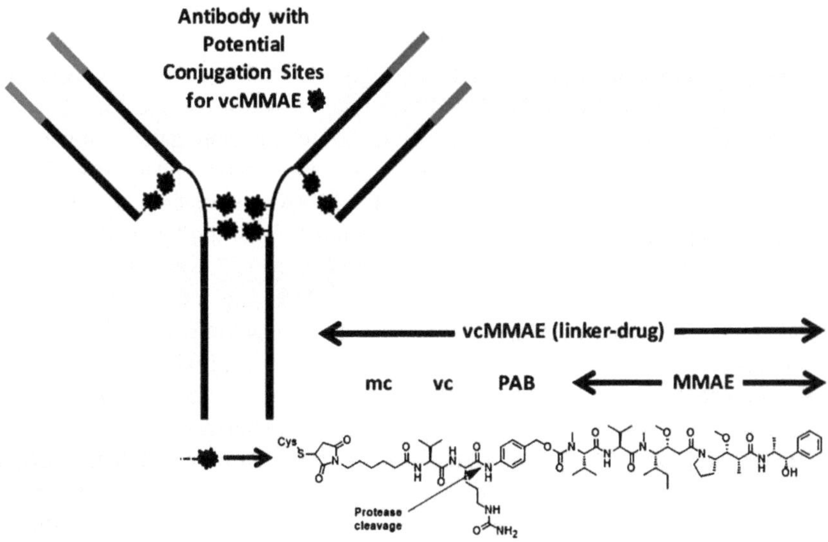

Fig. 5 Polivy Structure, an anti-CD79b (*note both Polivy® (polatuzumab vedotin-piiq) and Padcev® (enfortumab vedotin-ejfv) possess the same mc-vc-PABC-MMAE linker-drug construct as Adcetris®)

for rare diseases, for treatment of DLBCL. In September 2017 it was granted the Breakthrough Therapy Designation by the FDA and in June 2019 received the first FDA accelerated global approval for the treatment of adults with relapsed/refractory DLBCL, in combination with bendamustine plus rituximab. Accelerated approval was granted for this indication based on complete response rate from GO29365 clinical trial. Continued approval for this indication may be contingent upon verification and description of clinical benefit in confirmatory trials. The use in combination with bendamustine plus rituximab is also under regulatory review for relapsed/refractory DLBCL in the EU and is in ongoing phase 1b/2 development in this setting or relapsed/refractory follicular lymphoma (FL) in several countries.

Polivy is a CD79b-directed antibody-drug conjugate (ADC) consisting of three components: (1) the humanized immunoglobulin G1 (IgG1) monoclonal antibody specific for human CD79b; (2) the small molecule anti-mitotic agent MMAE; and (3) a protease-cleavable linker maleimidocaproyl-valine-citrulline-p-aminobenzyloxycarbonyl (mc-vc-PAB) that covalently attaches MMAE to the polatuzumab antibody. Polivy has an approximate molecular weight of 150 kDa and is produced by chemical conjugation of the antibody and small molecule components. The antibody is produced by mammalian (Chinese hamster ovary) cells, and the small molecule components are produced by chemical synthesis.

Polivy targets CD79b receptor, a B-cell specific surface protein, which is a component of the B-cell receptor in dividing B cells. The monoclonal antibody binds to CD79b, is then internalized, and the linker is cleaved by lysosomal proteases to enable intracellular delivery of MMAE. MMAE binds to microtubules and kills dividing cells by inhibiting cell division and inducing apoptosis through the inhibition of mitosis and tubulin polymerization.

The pharmacokinetic profile of Polivy was assessed in order to measure the antibody-conjugated MMAE (acMMAE) complex as well as the free drug (Palanca-Wessels et al. 2015; POLIVY 2019). The Cmax for the conjugated complex was found to be 803 ± 233 ng/mL, whereas for the unconjugated MMAE was 6.82 ± 4.73 ng/mL. The plasma exposure of acMMAE and unconjugated MMAE increased proportionally over a dose range from 0.1 to 2.4 mg/kg. acMMAE AUC were predicted to increase by approximately 30% at cycle 3 over cycle 1 AUC and achieved more than 90% of the cycle 6 AUC. The AUC after a first dose of 1.8 mg/kg (maximum tolerated dose) was 1860 ± 966 for the acMMAE and 52.3 ± 18.0 for the unconjugated antibody. The acMMAE half-life is approximately 12 days at Cycle 6, with predicted clearance of 0.9 L/day. The unconjugated MMAE half-life is approximately 4 days after the first dose. The central volume of distribution of the complex was found to be 3.15 L. Catabolism has not been studied in humans, however, Polivy is expected to be catabolized into small peptides, amino acids, unconjugated MMAE, and unconjugated MMAE-related catabolites. No dedicated clinical drug–drug interaction studies with Polivy in humans have been conducted. Age, sex, and race/ethnicity, as well as renal impairments, do not alter the pharmacokinetic profile of the ADC. In mild hepatic impairment, there was a 40% increase in MMAE exposure, which was not clinically significant. Body weight was identified as a variable for volume of distribution and clearance of acMMAE in a population PK analysis. This suggests that a body-weight based dosing will lessen the pharmacokinetic variability than a fixed-dose schedule. MMAE is a CYP3A4 substrate; thus, its exposure may increase or decrease if co-administered with drugs that are inhibitors or inducers of CYP3A4.

Intravenous Polivy, as monotherapy or in combination with immunotherapy, had an acceptable tolerability profile in patients with NHL in phase 1 or 1b/2 trials (Palanca-Wessels et al. 2015; POLIVY 2019). As other drugs administered via IV infusion, the administration can cause infusion-related reactions. When administered as monotherapy in patients with NHL (phase 1 study at 2.4 mg/kg dose, trial NCT01290549) the most frequent adverse events were neutropenia (40% of patients), anemia (11%) and peripheral sensory neuropathy (9%). Peripheral sensory neuropathy most often resulted in therapy being interrupted whereas neutropenia was the most common reason for its dosing being delayed (24%). Other serious adverse events occurred in 38% of patients, such as diarrhea, lung infection, lung disorder and disease progression.

The efficacy of Polivy was evaluated in an open-label, multicenter clinical trial (GO29365, NCT02257567) that included a cohort of 80 patients with relapsed or refractory DLBCL after one prior regimen (POLIVY 2019). Patients were randomized 1:1 to receive either Polivy in combination with bendamustine and a rituximab product (BR) or BR alone. The primary endpoint of the study was the complete response rate at the end of the treatment. The results from the study demonstrated that 40% of patients treated with Polivy plus BR achieved a complete response (CR), compared to 18% of the patients treated with BR alone. Approximately 45%

of patients that were treated with Polivy plus BR achieved objective response at the end of treatment, compared to only 18% of patients treated with BR alone. The response lasted for at least six months in 64% of the Polivy plus BR-treated patients that achieved either CR or partial response, compared to 30% of those treated with BR alone. The patients treated with Polivy plus BR achieved an overall survival of 12.4 months compared to 4.7 months in patients treated with BR alone.

A new clinical trial (NCT04231877) to study the side effects of Polivy when given with combination chemotherapy for the treatment of patients with untreated large B-cell lymphoma that grows and spreads quickly has been submitted in January 2020 and is the only one currently in progress (clinicaltrial.gov).

PADCEV (Fig. 6)

Bladder cancer is the sixth most common cancer in the U.S. with urothelial cancer accounting for more than 90% of bladder cancers. Platinum-containing chemotherapy, programmed death receptor-1 (PD-1) and programmed death ligand 1 (PD-L1) inhibitors are standard treatments for patients with bladder cancer but the initial therapy fails during the latter stages of cancer (Sridhar 2017). PADCEV (enfortumab vedotin) represents a new type of therapy for adult patients with locally advanced or metastatic urothelial cancer who have previously received the standard treatments. PADCEV is the first drug that targets Nectin-4, a cell surface adhesion protein expressed on bladder cancer cells only (Challita-Eid et al. 2016).

PADCEV was co-developed by Astellas and Seattle Genetics and it was granted Accelerated Approval by the FDA, which approved the drug to fill an unmet medical need based on the promising results of the phase II EV201 clinical trial, that predicted the clinical benefits to patients (Rosenberg et al. 2019; PACDEV 2019). Further phase III clinical trials will be required to verify and describe PADCEV's clinical benefits. FDA granted this application Priority Review and the Breakthrough Therapy designation, which expedited the development and review of the drug. PADCEV received approval in December 2019, approximately three months before the goal date.

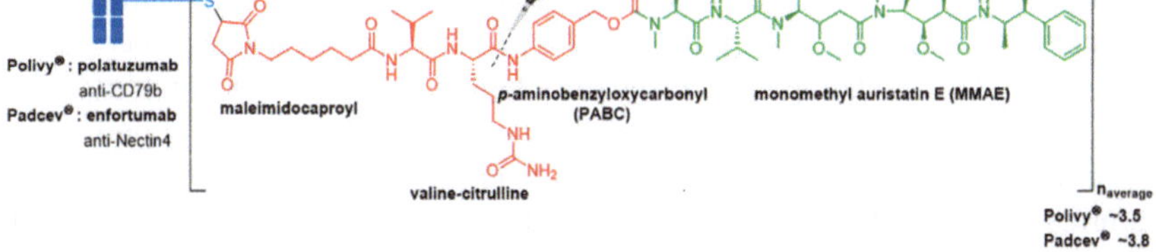

Fig. 6 Padcev Structure, an anti-Nectin4 (*note both Polivy® (polatuzumab vedotin-piiq) and Padcev® (enfortumab vedotin-ejfv) possess the same mc-vc-PABC-MMAE linker-drug construct as Adcetris®)

PADCEV is produced by chemical conjugation of a fully human anti-Nectin-4 IgG1 kappa monoclonal antibody and the small molecule monomethyl auristatin E (MMAE), a microtubule disrupting agent, through maleimidocaproyl valine-citrulline (vc), a protease-cleavable linker. The product has a drug-to-antibody ratio of approximately 3.8:1, which means that 4 molecules of MMAE are attached to each antibody molecule. The molecular weight is approximately 150 kDa. The antibody is produced by mammalian (Chinese hamster ovary) cells and the small molecule components are produced by chemical synthesis. Nonclinical data suggest that the anticancer activity of PADCEV is due to the binding of the ADC to Nectin-4-expressing cells, (Challita-Eid et al. 2016) followed by internalization of the ADC-Nectin-4 complex, and the release of MMAE via proteolytic cleavage. Release of MMAE disrupts the microtubule network by binding to tubulin within the cell, subsequently inducing cell cycle arrest and apoptotic cell death.

Pharmacokinetics was assessed by three Phase 1 studies and one Phase 2 study. PADCEV's pharmacokinetic profile was characterized after single and multiple doses in patients with locally advanced or metastatic urothelial carcinoma and other solid tumors (PACDEV 2019). The ADC Cmax was 28 μg/mL while MMAE Cmax was 4.8 ng/mL. Peak ADC concentrations were observed near the end of intravenous infusion while peak MMAE concentrations were observed approximately 2 days after PADCEV dosing. Minimal accumulation of the ADC and MMAE was observed following repeat administration of PADCEV in patients. Steady-state concentrations of the ADC and MMAE were reached after 1 treatment cycle. The steady-state volume of distribution of ADC is 11 liters following administration. ADC and MMAE exhibited multi-exponential declines with an elimination half-life of 3.4 days and 2.4 days, respectively. The mean clearance (CL) of PADCEV and free MMAE in patients was 0.10 L/h and 2.7 L/h, respectively. PADCEV catabolism has not been studied in humans; however, it is expected to undergo catabolism to small peptides, amino acids, unconjugated MMAE, and unconjugated MMAE-related catabolites. PADCEV releases MMAE via proteolytic cleavage, and MMAE is primarily metabolized by CYP3A4 in vitro. The excretion of PADCEV is not fully characterized. Following a single-dose of another ADC that contains MMAE, 17% of the total MMAE administered was recovered in feces and 6% in urine over a 1 week period, primarily as unchanged drug. A similar excretion profile of MMAE is expected after PADCEV administration.

Serious adverse reactions occurred in 46% of patients and the most common serious adverse reactions (≥3%) were urinary tract infection (6%), cellulitis (5%), febrile neutropenia (4%), diarrhea (4%), sepsis (3%), acute kidney injury (3%), dyspnea (3%), and rash (3%) (PACDEV 2019). Fatal adverse reactions occurred in 3.2% of patients, including acute respiratory failure, aspiration pneumonia, cardiac disorder, and sepsis. The most common adverse reaction leading to discontinuation was peripheral neuropathy (6%). Adverse reactions leading to dose interruption occurred in 64% of patients with the most common adverse reactions being peripheral neuropathy (18%), rash (9%) and fatigue (6%). The most common adverse reactions (≥20%) were fatigue, peripheral neuropathy, decreased appetite, rash, alopecia, nausea, dysgeusia, diarrhea, dry eye, pruritus and dry skin. The toxicity incidence and severity may increase after co-administration with a strong CYP3A4 inhibitor as this can increase free MMAE exposure.

The efficacy of PADCEV was evaluated in EV-201 (NCT03219333), single-arm, multicenter trial that enrolled 125 patients with locally advanced or metastatic urothelial cancer who received prior treatment with a PD-1 or PD-L1 inhibitor and platinum-based chemotherapy (PACDEV 2019). The median duration of exposure to PADCEV was 4.6 months. The major efficacy outcome measures in EV201 clinical trial were confirmed objective response rate (ORR) and duration of response (DOR) which were 44% and 7.6 months respectively.

Two clinical trials (NCT04223856 and NCT03288545) are currently studying a combination of PADCEV and pembrolizumab alone or with platinum chemotherapy to treat patients with urothelial cancer. The study will compare these drugs to the standard of care. The studies will also look at the side effects of PADCEV alone and with other anticancer therapies.

Enhertu (Fig. 7)

As discussed previously, breast cancer is the leading type of cancer in women and accounts for the second highest cause of cancer related death in women (de Melo Gagliato et al. 2016; Tandon et al. 1989). Human epidermal receptor 2 (HER2) is a tyrosine kinase receptor growth-promoting oncogene that is known to be overexpressed on the surface of invasive cancer cells and plays a role in tumor cell proliferation, adhesion, migration, and differentiation. HER2 overexpression accounts is found in 15–20% of breast cancers and 20% of advanced gastric cancers making it an ideal target (Janjigian et al. 2012; Slamon et al. 1987). Although targeted therapy (such as Herceptin (trastuzumab), Perjeta (pertuzumab), etc.) has been shown to improve the prognosis, resistance to therapies resulting in disease progression highlights the need for alternate therapies (Barok et al. 2014; de Melo Gagliato et al. 2016).

Fam-trastuzumab derutecan-nxki (Enhertu, DS-8201), manufactured by Daiichi Sankyo and distributed by AstraZeneca and Daiichi Sankyo, is composed of an anti-HER2 humanized monoclonal antibody, a tetrapeptide-based linker, and a

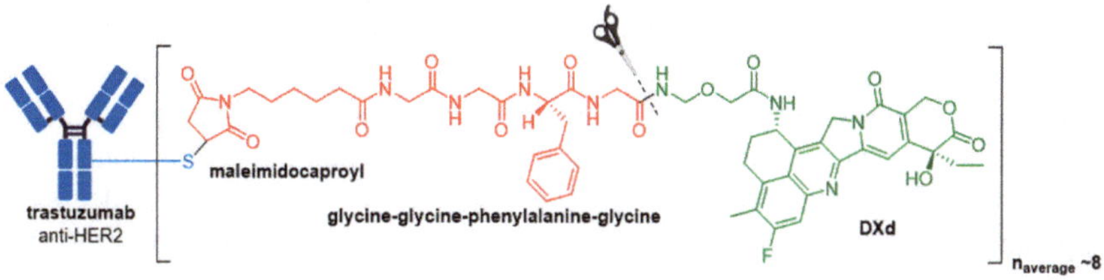

Fig. 7 Enhertu structure, anti-HER2 fam-trastuzumab. a protease cleavable tetrapeptide-based linker, and DXd as the drug payload

topoisomerase I inhibitor (Hacker 2020; ENHERTU 2019). The ADC was approved by the FDA in 2019 based on the DESTINY-Breast01 phase II clinical trial for the treatment of unresectable or metastatic HER2-positive breast cancer in adult patients who have received two or more prior anti-HER2 regimens. The ADC received fast track designation in December 2016 and breakthrough therapy designation in August 2017 (Keam 2020). Although the trial received accelerated approval, continued approval is dependent on substantiation of clinical benefit in a subsequent phase III confirmatory trial.

Similar to Kadcyla, the antibody component of Enhertu has a similar amino acid sequence as trastuzumab and is a humanized anti-HER2 IgG1 produced by Chinese hamster ovary (CHO) cells. However the linker and the drug payload differ between the two ADCs. Enhertu harbors a camptothecin analog (a topoisomerase I inhibitor), DXd, as the toxic payload, which has a different mechanism of action than T-DM1 in Kadcyla. Enhertu is the first ADC to reach clinical trials that uses a topoisomerase I inhibitor. Additionally, the linker binding the antibody to the drug is a novel self-immolative cleavable linker. The goal of using this linker is to attenuate the dose-limiting toxicity associated with Kadcyla (Doi et al. 2017). Enhertu has a drug to antibody ratio (DAR) of 8.

As the antibody component of Enhertu is similar to trastuzumab, Enhertu binds to HER2 on cancer cells. After cell internalization, the enzymatically cleavable peptide-based linker is selectively cleaved by lysosomal proteases in tumor cells (Nakada et al. 2019). After release into the cell, the membrane permeable drug payload, DXd, binds and stabilizes the complex formed between DNA and topoisomerase I (Liu et al. 2000), preventing DNA replication and leading to cell cycle arrest and apoptosis. Unlike Kadcyla, the anticancer activity of Enhertu is dependent on HER2 expression rather than HER2 amplification (Keam 2020).

The pharmacokinetics of Enhertu based on a phase I dose-escalation clinical trial revealed a half-life ($t_{1/2}$) of 6 days for the complex, 10.8 days for the drug alone, and 6.78 days for trastuzumab, at the recommended dose (5.4 mg/kg, established form phase II trials) (ENHERTU 2019; Doi et al. 2017). The clearance for the complex was 10.1 mL/day/kg. The volume of distribution at steady state was found to be 75.2 mL/kg. The C_{max} was found to be 127 μg/mL for Enhertu, 116 μg/m for total antibody, and 10.8 μg/m for the drug. Low concentrations of free drug were observed and the concentration of trastuzumab deruxtecan and total antibody were similar at all time points assessed. DXd is a substrate for CYP3A enzymes and is metabolized by CYP3A4, as demonstrated by *in vitro* studies.

Similar to KADCYLA and other HER2 targeted therapies, Enhertu is known to cause left ventricular dysfunction, which can lead to congestive heart failure (Modi et al. 2020). Other serious adverse events seen during the phase 1 and phase 2 clinical trials include interstitial lung disease (ILD)/pneumonitis and neutropenia including febrile neutropenia. Nausea, fatigue, vomiting, alopecia, constipation, decreased appetite, anemia, neutropenia, diarrhea, leukopenia, cough, and thrombocytopenia were the most common adverse reactions seen in >20% of the patients. Additionally, as with all therapeutic proteins, Enhertu poses a risk of immunogenicity and is known to cause embryo-fetal toxicity.

The phase 2 DESTINY-Breast01 clinical trial sought to evaluate the efficacy of trastuzumab deruxtecan in adult patients with unresectable or metastatic breast cancer who had previous treatment with trastuzumab emtansine (KADCYLA) (Modi et al. 2020). The open-label, single-group trial enrolled a total of 253 patients and was composed of two parts: (1) establishing pharmacokinetics followed by a dose finding stage and (2) evaluating the safety and efficacy of the recommended dose from the dose finding stage. In the pharmacokinetic stage of part 1, patients were randomized 1:1:1 to receive 5.4 mg/kg, 6.4 mg/kg, or 7.4 mg/kg trastuzumab deruxtecan. After pharmacokinetic analysis, 5.4 mg/kg and 6.4 mg/kg were used in the dose-finding stage. A benefit-risk profile was then established and used to choose the 5.4 mg/kg dose in part 2, where patients were split into two cohorts: disease progression after administration of KADCYLA or discontinuation of KADCYLA for other reasons. The primary end point of the study was complete response plus partial response in patients who received the recommended dose. Secondary endpoints included response duration, progression-free survival, and overall survival, among others. Enhertu monotherapy resulted in a 60.3% overall response rate and a median duration of response of 14.8 months. The median duration of treatment was 7 months and the majority of adverse reactions were Grade 1 or 2.

In addition to the phase 2 DESTINY-Breast01 clinical trial, a phase 2 clinical trial was concurrently active to determine the effect of Enhertu in gastric cancer (DESTINY-Gastric01). Additionally, Enhertu is currently in several other clinical trials to evaluate efficacy in NSCLC, gastroesophageal junction cancer, esophageal cancer, biliary tract cancer, and colorectal cancer. Additionally three phase 3 trials are underway to confirm the safety and efficacy of the drug in breast cancer (Keam 2020).

Conclusion

As of now, 7 ADCs have been approved by the FDA and more than new 60 ADCs are at different clinical stages. Many more clinical trials are studying the effect of the approved ADCs for new indications as well as use in combination therapies.

Combination therapies can potentially reduce drug resistance, improve drug efficacy, shrink tumor metastasis and growth, and increase the cancer survival rate. For instance, the combination of trastuzumab emtansine with pembrolizumab (a PD-1 inhibitor) and Brentuximab vedotin with PD-L1 or PD-1 inhibitors are being studied in metastatic breast cancer subjects and patients with refractory or relapsed Hodgkin lymphoma, respectively. The clinical data emerging from these studies provide important insights into the mechanistic basis of ADC design, and the opportunity to better understand the impact of changes in ADC properties on therapeutic activity, efficacy, and safety, opening a new window for ADC design research.

A systemic evaluation of the ADC composition and 20 years of clinical experience are applied in next generation ADC development, including the incorporation of fully humanized monoclonal antibodies to overcome immunogenicity, utilization of potently toxic payloads with diverse mechanisms of action, optimization of linkers, and conjugation methods to increase plasma stability and maximize drug delivery directly to tumors. Due to the use of monoclonal antibodies, most ADCs under development or in clinical trials are for oncological and hematological indications. However, some drug developers are trying to expand the application of ADCs to other important disease areas using the same approach in order to achieve a targeted therapy. Thus ADCs provide a new approach to chemotherapy as well as other indications.

References

ADCERTIS [Package Insert]: Seattle Genetics, Inc., Bothell (2019)

Advani, A., et al.: Safety, pharmacokinetics, and preliminary clinical activity of inotuzumab ozogamicin, a novel immunoconjugate for the treatment of B-cell non-Hodgkin's lymphoma: results of a phase I study. J. Clin. Oncol. **28**(12), 2085–2093 (2010)

Alley, S.C., Okeley, N.M., Senter, P.D.: Antibody–drug conjugates: targeted drug delivery for cancer. Curr. Opin. Chem. Biol. **14**(4), 529–537 (2010)

Amadori, S., et al.: Gemtuzumab ozogamicin versus best supportive care in older patients with newly diagnosed acute myeloid leukemia unsuitable for intensive chemotherapy: results of the randomized phase III EORTC-GIMEMA AML-19 trial. J. Clin. Oncol. **34**, 972–979 (2016)

Amiri-Kordestani, L., et al.: FDA approval: ado-trastuzumab emtansine for the treatment of patients with HER2-positive metastatic breast cancer. Clin. Cancer Res. **20**(17), 4436–4441 (2014)

Ballantyne, A., Dhillon, S.: Trastuzumab emtansine: first global approval. Drugs. **73**(7), 755–765 (2013)

Barok, M., Joensuu, H., Isola, J.: Trastuzumab emtansine: mechanisms of action and drug resistance. Breast Cancer Res. **16**(2), 209 (2014)

BESPONSA (Inotuzumab Ozogamicin) [Package Insert]: Wyeth Pharmaceuticals LLC a subsidiary of Pfizer Inc., Philadelphia (2018)

Bhatt, V.R., Vose, J.M.: Hematopoietic stem cell transplantation for non-Hodgkin lymphoma. Hematology/Oncology Clinics. **28**(6), 1073–1095 (2014)

Birrer, M.J., et al.: Antibody-drug conjugate-based therapeutics: state of the science. JNCI J. Natl. Cancer Inst. **111**(6), 538–549 (2019)

Breccia, M., Lo-Coco, F.: Gemtuzumab ozogamicin for the treatment of acute promyelocytic leukemia: mechanisms of action and resistance, safety and efficacy. Expert. Opin. Biol. Ther. **11**(2), 225–234 (2011)

Bross, P.F., et al.: Approval summary: gemtuzumab ozogamicin in relapsed acute myeloid leukemia. Clin. Cancer Res. **7**(6), 1490–1496 (2001)

Challita-Eid, P.M., Satpayev, D., Yang, P., An, Z., Morrison, K., Shostak, Y., Raitano, A., Nadell, R., Liu, W., Lortie, D.R., Capo, L.: Enfortumab vedotin antibody–drug conjugate targeting nectin-4 is a highly potent therapeutic agent in multiple preclinical cancer models. Cancer Res. **76**(10), 3003–3013 (2016)

Chowdhury, R., Ellis, P.: Trastuzumab (Herceptin®) and Ado-Trastuzumab Emtansine (Kadcyla®): treatments for HER2-positive breast cancer. In: Handbook of Therapeutic Antibodies, pp. 2041–2068. Wiley (2014)

Chu, P.G., Arber, D.A.: CD79: a review. Appl. Immunohistochem. Mol. Morphol. **9**(2), 97–106 (2001)

Connors, J.M., et al.: Brentuximab vedotin with chemotherapy for stage III or IV Hodgkin's lymphoma. N. Engl. J. Med. **378**(4), 331–344 (2018)

De Kouchkovsky, I., Abdul-Hay, M.: Acute myeloid leukemia: a comprehensive review and 2016 update. Blood Cancer J. **6**(7), e441–e441 (2016)
de Melo Gagliato, D., et al.: Mechanisms of resistance and sensitivity to anti-HER2 therapies in HER2+ breast cancer. Oncotarget. **7**(39), 64431–64446 (2016)
Doi, T., et al.: Safety, pharmacokinetics, and antitumour activity of trastuzumab deruxtecan (DS-8201), a HER2-targeting antibody–drug conjugate, in patients with advanced breast and gastric or gastro-oesophageal tumours: a phase 1 dose-escalation study. Lancet Oncol. **18**(11), 1512–1522 (2017)
Dornan, D., et al.: Therapeutic potential of an anti-CD79b antibody–drug conjugate, anti–CD79b-vc-MMAE, for the treatment of non-Hodgkin lymphoma. Blood. **114**(13), 2721–2729 (2009)
Dörner, T., Goldenberg, D.M.: Targeting CD22 as a strategy for treating systemic autoimmune diseases. Ther. Clin. Risk Manag. **3**(5), 953–959 (2007)
Eichenauer, D.A., et al.: Hodgkin's lymphoma: ESMO clinical practice guidelines for diagnosis, treatment and follow-up. Ann. Oncol. **25**(suppl_3), iii70–iii75 (2014)
ENHERTU (fam-trastuzumab deruxtecan-nxki) [package insert]: Daiichi Sankyo Inc., Basking Ridge (2019)
Fala, L..: Besponsa (Inotuzumab Ozogamicin) Approved for Adults with Relapsed or Refractory B-Cell Precursor Acute Lymphoblastic Leukemia
Francisco, J.A., et al.: cAC10-vcMMAE, an anti-CD30–monomethyl auristatin E conjugate with potent and selective antitumor activity. Blood. **102**(4), 1458–1465 (2003)
Giles, F., Estey, E., O'Brien, S.: Gemtuzumab ozogamicin in the treatment of acute myeloid leukemia. Cancer. **98**(10), 2095–2104 (2003)
Girish, S., et al.: Clinical pharmacology of trastuzumab emtansine (T-DM1): an antibody–drug conjugate in development for the treatment of HER2-positive cancer. Cancer Chemother. Pharmacol. **69**(5), 1229–1240 (2012)
Guerin, M., Sabatier, R., Goncalves, A.: Trastuzumab emtansine (Kadcyla (®)) approval in HER2-positive metastatic breast cancers. Bull. Cancer. **102**(4), 390–397 (2015)
Hacker, C.: Fam-Trastuzumab Deruxtecan-Nxki (Enhertu®). LWW (2020)
Han, H., et al.: The pharmacokinetics of BRENTUXIMAB VEDOTIN (SGN-35), an antibody-drug conjugate (ADC). T.: PII-1. Clin. Pharmacol. Ther. **91** (2012)
Hauswirth, A., et al.: Expression of the target receptor CD33 in CD34+/CD38−/CD123+ AML stem cells. Eur. J. Clin. Investig. **37**(1), 73–82 (2007)
Horwitz, S., et al.: Brentuximab vedotin with chemotherapy for CD30-positive peripheral T-cell lymphoma (ECHELON-2): a global, double-blind, randomised, phase 3 trial. Lancet. **393**(10168), 229–240 (2019)
Ikemoto, N., et al.: Calicheamicin-DNA complexes: warhead alignment and saccharide recognition of the minor groove. Proc. Natl. Acad. Sci. **92**(23), 10506–10510 (1995)
Issell, B.F., Crooke, S.T.: Maytansine. Cancer Treat. Rev. **5**(4), 199–207 (1978)
Jacobsen, E.D., et al.: Brentuximab vedotin demonstrates objective responses in a phase 2 study of relapsed/refractory DLBCL with variable CD30 expression. Blood. **125**(9), 1394–1402 (2015)
Janjigian, Y., et al.: Prognosis of metastatic gastric and gastroesophageal junction cancer by HER2 status: a European and USA international collaborative analysis. Ann. Oncol. **23**(10), 2656–2662 (2012)
Jen, E.Y., et al.: FDA approval: gemtuzumab ozogamicin for the treatment of adults with newly diagnosed CD33-positive acute myeloid leukemia. Clin. Cancer Res. **24**(14), 3242–3246 (2018)
Kadcyla (ado-trastuzumab emtansine) [package insert]: Genentech, Inc, South San Francisco (2019)
Keam, S.J.: Trastuzumab Deruxtecan: First Approval. Drugs. **80**, 1–8 (2020)
Khongorzul, P., et al.: Antibody–drug conjugates: a comprehensive review. Mol. Cancer Res. **18**(1), 3–19 (2020)
Kunz, A., et al.: Calicheamicin derivative-carrier conjugates. Google Patents (2012)
Kupchan, S.: Maytansine, a novel antileukemic ansa macrolide from Maytenus ovatus. J. Am. Chem. Soc. **94**, 1354–1356 (1972)

Küppers, R., Hansmann, M.-L.: The Hodgkin and reed/Sternberg cell. Int. J. Biochem. Cell Biol. **37**(3), 511–517 (2005)
Lamb, Y.N.: Inotuzumab ozogamicin: first global approval. Drugs. **77**(14), 1603–1610 (2017)
Liu, L.F., et al.: Mechanism of action of camptothecin. Ann. N. Y. Acad. Sci. **922**(1), 1–10 (2000)
Miyazaki, K.: Treatment of diffuse large B-cell lymphoma. J. Clin. Exp. Hematop. **56**(2), 79–88 (2016)
Modi, S., et al.: Trastuzumab deruxtecan in previously treated HER2-positive breast cancer. N. Engl. J. Med. **382**(7), 610–621 (2020)
Moskowitz, C.H., et al.: Brentuximab vedotin as consolidation therapy after autologous stem-cell transplantation in patients with Hodgkin's lymphoma at risk of relapse or progression (AETHERA): a randomised, double-blind, placebo-controlled, phase 3 trial. Lancet. **385**(9980), 1853–1862 (2015)
Mylotarg (Gemtuzumab Ozogamicin) [Package Insert]: Wyeth Pharmaceuticals LLC, a subsidiary of Pfizer Inc., Philadelphia (2020)
Nakada, T., et al.: The latest research and development into the antibody–drug conjugate,[fam-] trastuzumab deruxtecan (ds-8201a), for HER2 cancer therapy. Chem. Pharm. Bull. **67**(3), 173–185 (2019)
PACDEV (enfortumab vedotin-ejfv) [Package Insert]: Agensys, Inc and Seattle Genetics, Inc, Bothell (2019)
Palanca-Wessels, M.C.A., et al.: Safety and activity of the anti-CD79B antibody–drug conjugate polatuzumab vedotin in relapsed or refractory B-cell non-Hodgkin lymphoma and chronic lymphocytic leukaemia: a phase 1 study. Lancet Oncol. **16**(6), 704–715 (2015)
Pazdur, R.: FDA Oncology Update (2019)
Penault-Llorca, F., et al.: Incidence and implications of HER2 and hormonal receptor overexpression in newly diagnosed metastatic breast cancer (MBC). J. Clin. Oncol. **23**(16_suppl), 764–764 (2005)
Petersdorf, S.H., et al.: A phase 3 study of gemtuzumab ozogamicin during induction and post-consolidation therapy in younger patients with acute myeloid leukemia. Blood. **121**(24), 4854–4860 (2013)
POLIVY [polatuzumab vedotin-piiq]: Genentech, Inc., South San Francisco (2019)
Polson, A.G., et al.: Antibody-drug conjugates targeted to CD79 for the treatment of non-Hodgkin lymphoma. Blood. **110**(2), 616–623 (2007)
Poon, K.A., et al.: Preclinical safety profile of trastuzumab emtansine (T-DM1): mechanism of action of its cytotoxic component retained with improved tolerability. Toxicol. Appl. Pharmacol. **273**(2), 298–313 (2013)
Prince, H.M., et al.: Brentuximab vedotin or physician's choice in CD30-positive cutaneous T-cell lymphoma (ALCANZA): an international, open-label, randomised, phase 3, multicentre trial. Lancet. **390**(10094), 555–566 (2017)
Pro, B., et al.: Brentuximab vedotin (SGN-35) in patients with relapsed or refractory systemic anaplastic large-cell lymphoma: results of a phase II study. J. Clin. Oncol. **30**(18), 2190–2196 (2012)
Pui, C.-H., Evans, W.E.: Treatment of acute lymphoblastic leukemia. N. Engl. J. Med. **354**(2), 166–178 (2006)
Pui, C.-H., Relling, M.V., Downing, J.R.: Acute lymphoblastic leukemia. N. Engl. J. Med. **350**(15), 1535–1548 (2004)
Ramchandren, R., et al.: Brentuximab vedotin plus chemotherapy in North American subjects with newly diagnosed stage III or IV Hodgkin lymphoma. Clin. Cancer Res. **25**(6), 1718–1726 (2019)
Ricart, A.D.: Antibody-drug conjugates of calicheamicin derivative: gemtuzumab ozogamicin and inotuzumab ozogamicin. Clin. Cancer Res. **17**(20), 6417–6427 (2011)
Roschewski, M., Staudt, L.M., Wilson, W.H.: Diffuse large B-cell lymphoma—treatment approaches in the molecular era. Nat. Rev. Clin. Oncol. **11**(1), 12–23 (2014)
Rosenberg, J.E., O'Donnell, P.H., Balar, A.V., McGregor, B.A., Heath, E.I., Yu, E.Y., Galsky, M.D., Hahn, N.M., Gartner, E.M., Pinelli, J.M., Liang, S.Y.: Pivotal trial of enfortumab vedotin

in urothelial carcinoma after platinum and anti-programmed death 1/programmed death ligand 1 therapy. J. Clin. Oncol. **37**(29), 2592–2600 (2019)

Scheinberg, D.A., et al.: A phase I trial of monoclonal antibody M195 in acute myelogenous leukemia: specific bone marrow targeting and internalization of radionuclide. J. Clin. Oncol. **9**(3), 478–490 (1991)

Selby, C., Yacko, L.R., Glode, A.E.: Gemtuzumab ozogamicin: back again. J. Adv. Pract. Oncol. **10**(1), 68–82 (2019)

Sievers, E.L., Linenberger, M.: Mylotarg: antibody-targeted chemotherapy comes of age. Curr. Opin. Oncol. **13**(6), 522–527 (2001)

Sievers, E.L., Senter, P.D.: Antibody-drug conjugates in cancer therapy. Annu. Rev. Med. **64**, 15–29 (2013)

Sievers, E.L., et al.: Efficacy and safety of gemtuzumab ozogamicin in patients with CD33-positive acute myeloid leukemia in first relapse. J. Clin. Oncol. **19**(13), 3244–3254 (2001)

Slamon, D.J., et al.: Human breast cancer: correlation of relapse and survival with amplification of the HER-2/neu oncogene. Science. **235**(4785), 177–182 (1987)

Sridhar, S.S.: Evolving treatment of advanced urothelial cancer. J. Oncol. Pract. **13**(5), 309–315 (2017)

Stasi, R.: Gemtuzumab ozogamicin: an anti-CD33 immunoconjugate for the treatment of acute myeloid leukaemia. Expert. Opin. Biol. Ther. **8**(4), 527–540 (2008)

Stinchcombe, T., et al.: Efficacy, Safety, and Biomarker Results of Trastuzumab Emtansine (T-DM1) in Patients (pts) with Previously Treated HER2-Overexpressing Locally Advanced or Metastatic Non-small Cell Lung Cancer (mNSCLC). American Society of Clinical Oncology (2017)

Strebhardt, K., Ullrich, A.: Paul Ehrlich's magic bullet concept: 100 years of progress. Nat. Rev. Cancer. **8**(6), 473–480 (2008)

Taksin, A., et al.: High efficacy and safety profile of fractionated doses of Mylotarg as induction therapy in patients with relapsed acute myeloblastic leukemia: a prospective study of the alfa group. Leukemia. **21**(1), 66–71 (2007)

Tandon, A.K., et al.: HER-2/neu oncogene protein and prognosis in breast cancer. J. Clin. Oncol. **7**(8), 1120–1128 (1989)

Verma, S., et al.: Trastuzumab emtansine for HER2-positive advanced breast cancer. N. Engl. J. Med. **367**(19), 1783–1791 (2012)

Wahl, A.F., et al.: The anti-CD30 monoclonal antibody SGN-30 promotes growth arrest and DNA fragmentation in vitro and affects antitumor activity in models of Hodgkin's disease. Cancer Res. **62**(13), 3736–3742 (2002)

Welslau, M., et al.: Patient-reported outcomes from EMILIA, a randomized phase 3 study of trastuzumab emtansine (T-DM1) versus capecitabine and lapatinib in human epidermal growth factor receptor 2–positive locally advanced or metastatic breast cancer. Cancer. **120**(5), 642–651 (2014)

Younes, A., et al.: Brentuximab vedotin (SGN-35) for relapsed CD30-positive lymphomas. N. Engl. J. Med. **363**(19), 1812–1821 (2010)

Younes, A., et al.: Results of a pivotal phase II study of brentuximab vedotin for patients with relapsed or refractory Hodgkin's lymphoma. J. Clin. Oncol. **30**(18), 2183–2189 (2012)

Application of ADCs to Therapeutic Areas Outside of Oncology

Anna M. Papinska, Mitalee Tamhane, and Mihir Shah

Abstract Antibody drug conjugates (ADCs) have been successful in oncology, mainly due to enhanced specificity and reduced side effects that burden highly cytotoxic cancer drugs. The interest to apply the ADC approach in non-oncological indications has been growing in recent years as other properties of molecules, such as efficacy and pharmacokinetics, can also be improved through conjugation to an antibody. While the most advanced ADC therapies in other therapeutic areas depend heavily on the progress made in oncology, new ideas and applications are also emerging. In this review, we summarize the current knowledge and development status of ADC candidates in non-oncological areas. We focus primarily on immunology and infectious disease but other therapeutic areas, such as atherosclerosis and ophthalmology, are also discussed.

Keywords Antibody-drug conjugates (ADCs) · Immunology · Infectious disease · Graft-versus-host disease (GVHD) · HIV/AIDS · Rheumatoid arthritis (RA) · Atherosclerosis · Age-related macular degeneration (AMD) · Posterior capsule opacification (PCO) · Proliferative vitreoretinopathy (PVR)

Introduction

Development of small molecule drugs is often limited by the lack of specificity, high toxicity, and suboptimal pharmacokinetic (PK) properties. Targeted delivery with a highly specific antibody has a potential to improve efficacy and safety by reducing off-target and increasing on-target effects. With the recent success of antibody-drug conjugates (ADCs) in oncology, the interest in drug conjugates is

A. M. Papinska (✉) · M. Tamhane · M. Shah
Nonclinical and Translational Sciences, Allergan, an AbbVie Inc. Company, Irvine, CA, USA
e-mail: annapapinska@gmail.com; mitalee.tamhane@allergan.com; mihir.shah@allergan.com

H. J. Gukasyan, A. Kulkarni (eds.), *Antibody Drug Conjugates*, AAPS Advances in the Pharmaceutical Sciences Series 52, https://doi.org/10.1007/978-3-031-89319-3_2

also growing in other therapeutic areas, especially in immunology, where pathological mechanisms are often similar to oncology, as well as infectious disease, and others.

Cytotoxic compounds used in oncology can often be repurposed for other indications that require targeted cell-killing, like autoimmune diseases. For example, brentuximab vedotin (Adcetris®, Seattle Genetics), one of the few FDA-approved ADCs in oncology, is being explored for treatment of graft versus host disease (GVHD) and rheumatoid arthritis (RA). However, outside oncology the balance between establishing optimal drug to antibody ratio (DAR) for efficacy versus safety is critical, so even the repurposed compounds may need some modifications. Conjugates used in oncology typically have a high cytotoxic payload to effectively kill targeted cancer cells. In non-oncology indications, even though the ADC carries the same cytotoxic small molecule, the payload can be much lower, just enough to generate efficacy without inducing any toxicity.

Similar to cytotoxic drugs, ADC approach can also be used to reduce off-target effects of a variety of non-cytotoxic drugs. For example, targeted delivery with an antibody is expected to reduce side effects of steroids and enable long-term therapy. But antibody conjugation offers other benefits as well. The pharmacokinetic (PK) properties of a clinical stage antibiotic-antibody conjugate (AAC), were significantly improved compared to unconjugated drug. The potency of a novel HIV treatment candidate was enhanced after conjugation to an antibody, likely due to the interplay of two different mechanisms of action: binding of the virus by the antibody and effect of the small molecule inhibitor on the host cell.

While ADCs in oncology focus mainly on improving safety of the highly cytotoxic drugs, research in non-oncological indications provides many insights into the effects of antibody conjugation on other properties of the molecules, such as PK and potency. Even though development of ADCs in oncology provides learnings that we can build on, especially with respect to manufacturing and safety profile, ADCs in non-oncology indications have a higher bar in terms of safety profile and risk tolerance. In this chapter, we discuss ongoing efforts in the field of ADCs in treatment of immune diseases, bacterial and HIV-related infections, and potential to treat other diseases such as atherosclerosis, choroidal neovascularization, proliferative vitreoretinopathy and posterior capsule opacification (Table 1).

Immunology

Immune diseases and cancer have a lot in common and are often treated using similar approaches with an overall goal to attenuate the excessive cell response or proliferation. Antibody therapies are commonly used to treat both autoimmune diseases and cancer. More than two dozen antibody therapies have been approved for treatment of immune diseases ranging from transplant rejection to RA (Yasunaga et al. 2017). However, in many cases, antibody therapies alone are insufficient to treat immune disorders. On the other hand, strong small molecule immunosuppressants

Table 1 ADC compounds in development for non-oncology indications

Therapeutic area	ADC Name	Company	Indication	Stage of development	References
Immunology	ABBV-3373 (anti-TNF—glucocorticoid receptor modulator steroid)	Abbvie	Rheumatoid arthritis	Phase 2	NCT03823391
	Brentuximab vedotin (anti-CD30—MMEA)	Seattle genetics	Systemic sclerosis	Phase 1/2	NCT03222492 NCT03198689
	Brentuximab vedotin (anti-CD30—MMEA)	Seattle genetics	Graft vs host disease	Phase 1	NCT01596218
	Brentuximab vedotin (anti-CD30—MMEA)	Seattle genetics	Graft vs host disease	Phase 1	NCT01940796 DeFilipp et al. (2018)
	Brentuximab vedotin (anti-CD30—MMEA)	Seattle genetics	Graft vs host disease	Phase 1	NCT01700751
	Anti-CD117—saporin	–	Pre-transplant conditioning	Preclinical	Czechowicz et al. (2019)
	Anti-CD45—amanitin	Magenta therapeutics	Pre-transplant conditioning	Preclinical	Palchaudhuri et al. (2019)
	Anti-CD45—saporin	Magenta therapeutics	Pre-transplant conditioning	Preclinical	Palchaudhuri et al. (2016)
	A7R—SN-38 A7R—MMAE	–	Rheumatoid arthritis	Preclinical	Yasunaga et al. (2017)
	Tocilizumab-alendronate	–	Rheumatoid arthritis	Preclinical	Lee et al. (2017)
	L19—IL-10	–	Rheumatoid arthritis	Preclinical	Trachsel et al. (2007)
	Anti-CD163 -dexamethasone	–	NA	Preclinical	Graversen et al. (2012)
	Fc—CGS-21680	–	NA	Preclinical	Chiang et al. (2014)
	Anti-CXCR4—dasatinib	–	NA	Preclinical	Wang et al. (2015)

(continued)

Table 1 (continued)

Therapeutic area	ADC Name	Company	Indication	Stage of development	References
Infectious disease	DSTA4637S (IgG1- dmDNA31)	Genentech	Anti-S. Aureus (bacteremia)	Phase 1	Peck et al. (2019) NCT02596399 NCT03162250
	P4/D10—doxorubicin	–	HIV	Preclinical	Johansson et al. (2006)
	bNAb—aplaviroc	–	HIV	Preclinical	Gavrilyuk et al. (2013)
Cardiovascular disease	Anti-CD11a—LXR agonist	–	Atherosclerosis	Preclinical	Lim et al. (2015)
Ophthalmology	Anti- αvβ3 integrin—Mitomycin C	–	Choroidal neovascularization	Preclinical	Kamizuru et al. (2001)
	Anti-transferrin receptor—ricin	–	Proliferative vitreoretinopathy	Preclinical	Jaffe et al. (1990)
	Anti-transferrin receptor—ricin	–	Posterior capsule opacification	Preclinical	Goins et al. (1994)

such as steroids have good efficacy but are often associated with severe side effects. To minimize off-target effects but retain high efficacy, more attention has been recently focused on targeted delivery of potent small molecules through conjugation to antibodies. Several ADCs for treatment of immune diseases are under development with brentuximab vedotin (Adcetris®, Seattle Genetics) currently leading the race. Below, we summarize the development of the most advanced ADCs intended for treatment of immune diseases.

Graft Versus Host Disease

Acute and chronic GVHD are common complications associated with allogeneic hematopoietic stem cell transplant and are primarily treated with strong immunosuppressants such as steroids. Around 50% of patients don't respond to this therapy and require further treatment with second-line immunosuppressants (Flowers and Martin 2015). The second-line therapy is currently the focus of researchers because of the lack of an established standard of care (Jacobsohn and Vogelsang 2007). New therapeutic approaches target specific subsets of T-cells that are known or suspected to fuel the disease. ADCs offer an advantage to efficiently deliver drugs to these selected T-cell populations.

Brentuximab vedotin (Adcetris®, Seattle Genetics) is one of the few ADCs that have gained regulatory approval and are available on the market. Currently, brentuximab vedotin is indicated for treatment of difficult cases of Hodgkin lymphoma and systemic anaplastic large cell lymphoma (Brentuximab vedotin Package Insert 2014). Due to its high specificity and efficacy, several indications outside of oncology have also been of interest, including chronic GVHD, systemic sclerosis, and lupus erythematosus.

Brentuximab vedotin comprises of an anti-CD30 antibody conjugated to highly cytotoxic monomethyl auristatin E (MMAE)—a microtubule disrupting agent—that is too toxic to be administered by itself. This inherent toxicity is reduced by targeted delivery to the CD30 positive cells by the antibody portion of the conjugate. Following binding to CD30, the whole complex—ADC and CD30—is internalized, MMAE is released from the conjugate, and the cell undergoes apoptosis (Younes et al. 2012). Thus, brentuximab vedotin has the potential to be effective in any disease with increased activity of immune cells expressing CD30.

Seattle Genetics has tested brentuximab vedotin for treatment of steroid refractory GVHD in two clinical trials (NCT01596218—acute and NCT01940796—chronic GVHD), since this disease is associated with increased levels of CD30 on the alloreactive T-cells (Blazar et al. 2004; Chen et al. 2012).

The aim of the first trial was to determine the maximum tolerated dose in patients with steroid-refractory acute GVHD (Chen et al. 2017). The initial design—weekly dosing for 3 weeks followed by dosing every 3 weeks, a dosing regimen that was well-tolerated in patients with lymphoma—was revised after 2 out of 6 patients in the first two dose groups (0.6 mg/kg [N = 3] and 0.9 mg/kg [N = 3]) experienced serious neutropenia, likely due to MMAE accumulation. The trial continued with dosing every 2 weeks for a total of 4 doses at 0.6 kg/kg (N = 10) and 0.8 mg/kg (N = 18). Dose limiting toxicity was identified at 0.8 mg/kg after one patient experienced sepsis and multiorgan failure. Overall, the treatment improved patient outcomes (the response rate was 38.2% on day 28) and survival (41% of patients survived until month 6 and 38% survived until month 12) but the response did not correlate with the expression of CD30 on CD8-positive, CD4-positive, or regulatory T-cells. Even though results of this phase 1 trial are promising, more investigation is needed to determine potential of brentuximab vedotin for treatment of refractory GVHD.

The results from the second clinical study of brentuximab vedotin for treatment of refractory GVHD—this time for the chronic disease—were quite similar (DeFilipp et al. 2018). The patients were dosed every 3 weeks for a total of up to 16 doses, at dose levels ranging from 0.6 to 1.8 mg/kg. However, only 2 out of 17 patients received all 16 doses. All other patients discontinued the therapy early, primarily due to toxicities. The investigators observed two dose limiting toxicities: posterior reversible encephalopathy syndrome and sepsis. However, the authors state that the maximum tolerated dose could not be established because the trial was terminated early due to toxicities. The treatment somewhat improved the patient outcomes, with 47% of participants who partially responded to the therapy, with no complete responders. Nonetheless, treatment reduced the dose of steroids for 65% of patients. Similar to the previous results, the investigators found no correlation between CD30 levels and response to treatment; thus, these preliminary assessments showed that CD30 level may be a poor prognostic or efficacy marker for brentuximab vedotin treatment of GVHD under these conditions.

Seattle Genetics also initiated a pilot clinical trial of brentuximab vedotin for prevention of GVHD after unrelated allogeneic stem cell transplantation that focused on establishing safety and maximum tolerated dose when given with tacrolimus and methotrexate in patients with acute myeloid leukemia, acute lymphoblastic leukemia, or myelodysplastic syndromes (NCT01700751). The study was completed in 2016, but the results have not been published.

Overall, even though brentuximab vedotin improved the outcomes of patients with both refractory acute and chronic GVHD, it showed several toxicities, which may limit its use in these patient populations.

Pre-transplant Conditioning

Bone marrow transplant can be a potential cure for both malignant and non-malignant blood and immune diseases. However, many patients refuse this radical treatment as it requires patient preparation (conditioning) that can lead to many severe side effects such as organ toxicities, cancer, and infertility. Typical conditioning treatment includes chemotherapy, monoclonal antibody therapy, and radiation aimed to prevent transplant rejection and to prepare marrow for new cells (Gyurkocza and Sandmaier 2014).

A CD117-conjugated ADC carrying saporin—a cytotoxic ribosome-inactivating agent—was shown to lead to almost complete depletion of hematopoietic stem cells and successful engraftment of donor cells in mice, without causing lymphopenia or severe immunosuppression, likely due to the absence of CD117 on the lymphocytes (Czechowicz et al. 2019). In addition, a human CD117-saporin ADC inhibited human hematopoietic stem cells in vitro, suggesting that this compound has potential for further evaluation in the clinic.

An ADC driven by anti-CD45 antibody that carries amanitin showed potency in depleting human hematopoietic stem cells and immune cells in vitro and in vivo (Palchaudhuri et al. 2019). Administration of this compound to 4 mouse models of leukemia improved animal survival. The variant of the ADC specifically engineered to shorten its half-life, to avoid harming the incoming graft, also showed potent depletion of cells in the marrow in non-human primates. Treatment with another CD45 conjugate carrying saporin also enabled engraftment of donor cells in a mouse model of sickle-cell anemia (Palchaudhuri et al. 2016). This treatment had minimal toxicity and enabled rapid recovery of T and B cells but, as with all CD45 directed treatments, it was associated with severe depletion of lymphocytes that may leave the patients susceptible to infections.

Another CD45-targeting ADC is currently under development for conditioning to enable autologous stem cell transplant in diseases such as multiple sclerosis. A single dose of this ADC was well tolerated and efficacious in 3 mouse models of autoimmune disease (Gillard et al. 2019). A humanized version of the ADC was effective in depleting hematopoietic stem cells and immune cells in humanized NSG mice (immunodeficient mouse model). Single doses of this ADC were well tolerated and effective in cell killing in non-human primates.

Rheumatoid Disease

RA is an autoimmune disease with chronic inflammation that damages joints and the spine. It is often regarded as a rare disease even though statistics show that it occurs in approximately 1% of the worldwide population (Kvien 2004). Current RA therapies are burdened with low specificity, toxicities, high cost, and lack of efficacy. In contrast to traditional immunosuppressants such as methotrexate and cyclosporin A, emerging biologics target specific molecules (such as TNF) or processes (such as activation of immune response) and offer better specificity. But they often carry a risk of infections due to aggressive immunosuppression and are more expensive due to high development and manufacturing costs (Atzeni et al. 2017). Tolerability, efficacy, and bioavailability can potentially be improved with ADCs via targeted delivery to the site of inflammation.

Steroids are still commonly used to treat RA but they are only appropriate for short-term treatment and can be associated with numerous adverse effects due to targeting many different cell types. Frequent use of steroids is also often associated with steroid resistance. Targeted delivery of steroids can potentially alleviate the side effects and offer a viable long-term therapy.

AbbVie is heading the clinical development of two ADCs for targeted delivery of steroids for treatment of RA. Phase 2 study evaluating safety, tolerability, PK, and efficacy of ABBV-3373 in patients with RA is currently recruiting and is expected to be completed in Q3 2020 (NCT03823391). ABBV-3373 is an intravenously administered ADC comprised of an antibody targeting TNF and a glucocorticoid receptor modulator steroid. TNF is a well-established target that is abundant in the synovial fluids and synovial membrane of RA patients (Blüml et al. 2012). Thus, antibodies targeting TNF should theoretically deliver the drug load directly to the site of action, reducing any side effects. The phase 2 study is designed to compare ABBV-3373 against blockbuster Humira® (adalimumab, AbbVie). Another TNF targeted steroid conjugate—ABBV-154—is also featured in AbbVie's pipeline for the treatment of RA but no details about this product are available other than that it is currently being tested in phase 1 trials.

Brentuximab vedotin, in addition to treatment of GVHD, is also currently undergoing clinical investigation for systemic sclerosis (scleroderma)—a form of an uncommon rheumatic disease that is driven by autoimmune mechanisms, affects connective tissues, and results in fibrosis of the skin, internal organs, and vasculature (NCT03222492 and NCT03198689). Treatment options for this orphan disease are still very limited (Denton and Khanna 2017). Pathogenesis of systemic sclerosis shares many similarities with chronic GVHD. For example, high levels of CD30 were found in the skin of both patient populations (Amedei et al. 2013; D'Elios et al. 1997). Serum levels of soluble CD30 also correlate with the clinical features of systemic sclerosis (Giacomelli et al. 1997). Both clinical studies lead by Seattle Genetics are still recruiting.

Recently, Yasunaga and colleagues developed ADCs that target IL-7R with the aim to overcome steroid resistance and treat RA. IL-7R signaling is associated with steroid resistance in both cancer and autoimmune diseases (Yasunaga et al. 2017). Since monoclonal antibodies alone proved insufficient to fight RA induced inflammation in mice (Hartgring et al. 2010), Yasunaga's group developed two ADCs to target and eliminate IL-7R-positive steroid-resistant immune cells, using two well-known cytotoxic compounds: SN-38—an active metabolite of the topoisomerase I inhibitor Irinotecan (marketed product used for treatment of colon and rectum cancer); and MMAE—a potent antimitotic agent also used in the approved brentuximab vedotin. As expected, ADC carrying more cytotoxic MMAE showed better efficacy in the mouse model of RA than the ADC with SN-38. This research showed that both steroid-sensitive and steroid-resistant cells can be targeted by IL-7R-directed ADCs.

Lee and colleagues have taken an even more aggressive approach to targeting RA. Both the antibody and the drug in their ADC are designed to reduce inflammation and improve symptoms of RA. In this innovative approach, tocilizumab—a humanized antibody that binds to IL-6 receptor and inhibits IL-6 signaling and a subsequent increase in inflammatory response—was conjugated to alendronate, an anti-inflammatory drug that binds to the bone surface and targets macrophages (Lee et al. 2017). The molecule was designed to deliver and release alendronate at the target site to reduce its impact on healthy cells. The authors tested the ADC in vitro and in vivo in a mouse model of RA (CIA mouse model). The results showed that the ADC had a greater effect on ameliorating RA than either of the treatments alone.

Arthritis-specific antigens, such as those expressed due to active angiogenesis and remodeling in affected tissue, also offer an attractive target for drug delivery. Some components of the remodeled extracellular matrix are abundant in the inflamed tissue and virtually absent from normal tissues (Schultz 2019). Borsi and colleagues developed L19 antibody specific to alternatively spliced extra domain B (EDB) of fibronectin, to target new vessels formed in tumors (Borsi et al. 2002). The EDB is identical in mice and human so this antibody can be used in both nonclinical and clinical development. Trachsel and colleagues explored other uses of L19 and showed that L19-IL-10 conjugate performed superior to IL-10 conjugated to an antibody of irrelevant specificity (negative control) in a mouse model of arthritis (CIA model) (Trachsel et al. 2007). Clinical development of recombinant IL-10 for treatment of arthritis was discontinued due to lack of efficacy but targeted delivery to the site of inflammation may improve the therapeutic index and is worth investigating further (Dela Cruz et al. 2004).

Other

In this section, we summarize data for novel immunomodulatory ADCs that are in pre-clinical stages of development and for which specific indication had not yet been selected.

CD163-Dexamethasone

Dexamethasone is one of the most widely used steroid treatment. Graversen and colleagues reported a new dexamethasone conjugate that targets CD163—a hemoglobin scavenger receptor in macrophages to potentially reduce the side effects associated with the use of dexamethasone (Graversen et al. 2012). CD163 is a macrophage-specific surface receptor that is highly expressed in macrophages residing in tissues (such as liver, spleen, and bone marrow) and infiltrating sites of inflammation such as joints in RA and atherosclerotic lesions (Komohara et al. 2006; Li et al. 2004). The CD163-dexamethasone ADC was approximately 50 times more potent in lowering serum TNF-α and IL-1 levels than free dexamethasone when administered to lipopolysaccharide-induced rat model of acute inflammation. The ADC also had a much lower effect on the non-target organs compared to free dexamethasone. This molecule has a potential to be effective in TNF-α driven inflammatory diseases such as RA and acute inflammatory diseases such as sepsis and acute hepatitis where steroids are typically not recommended due to severe immunosuppression. More uses for this CD163 delivery platform are summarized in a review by Graversen and Moestrup (Graversen and Moestrup 2015). The molecule was presumably under development for treatment of non-alcoholic steatohepatitis by a Danish company Cytoguide (Graversen's affiliation), however as of 2016, no further reports of development were identified, and the program was most likely discontinued.

Fc-CGS-21680

Small molecule adenosine 2a receptor (A2AR) agonists, such as CGS-21680 (CGS), are burdened with short half-lives and potential for off-target toxicities (de Lera et al. 2014). To solve these issues, Chiang and colleagues conjugated CGS to Fc, a constant region of immunoglobulin (Chiang et al. 2014), a strategy that was previously successfully used to prolong in vivo stability of etanercept, a fusion protein TNF-a inhibitor. Mice with pulmonary inflammation survived longer and had reduced inflammation in the lungs when treated with FC-CGS conjugate compared with CGS or Fc alone. This effect was likely, at least in part, due to longer residence time compared to the unconjugated CGS. The off-target disposition of CGS, particularly in the brain, was also reduced with the conjugate. The authors hypothesize that Fc facilitates distribution of the ADC primarily to the site of inflammation since lung drug levels were higher in the model of pulmonary inflammation than in healthy mice. With supporting results, Fc may prove to be useful for targeting to sites of inflammation.

HCLX-Dasatinib

Dasatinib is a kinase inhibitor indicated for treatment of BCR-ABL-dependent chronic myelogenous leukemia. Like most kinase inhibitors, the use of dasatinib is limited to treatment of cancer due to its off-target toxicities resulting from the lack of selectivity to particular kinase family members. To remedy this, an ADC tagged with anti-CXCR4 antibody for targeted delivery of dasatinib to T-cells was developed (Wang et al. 2015). Even though CXCR4 is expressed on other cell types such as hematopoietic stem cells, B-cells, and monocytes, antagonism of these cells with dasatinib is not expected to cause any severe clinical effects (Kohler et al. 2008; Signoret et al. 1997). An antibody—HCLX—was developed based on the trastuzumab scaffold that selectively binds to human CXCR4. Because the target of dasatinib is intracellular, it was also confirmed that the antibody is internalized by human T-cells. The resulting HCLX-dasatinib conjugate successfully reduced activation of human T-cells by inhibiting Lck (tyrosine kinase) in vitro. These results suggest that HCLX may be useful in targeted delivery of dasatinib to T-cells for treatment of immune disorders, however animal studies are needed to ensure that the off-target toxicity profile is acceptable. Because CXCR4 expression differs significantly between human and rodents, these studies will need to be conducted in higher species.

Infectious Disease

ADC development in immunology has been successful primarily due to the potential to repurpose highly cytotoxic compounds, too toxic to be administered alone, for targeted elimination of immune cell subtypes. Infectious disease is another therapeutic area that can benefit from a similar highly specific pathogen-killing approach. In this section, we discuss anti-infective molecules that have been evaluated in preclinical models and in the clinic, mainly in treatment of bacterial and HIV infections.

Bacterial Infections

Among numerous pathogens that invade human body, Staphylococcus aureus (S. aureus) is responsible for most bacterial infections in hospitals and communities (Mariathasan and Tan 2017). The clinical manifestations most commonly seen with S. aureus are bacteremia, infective endocarditis, as well as skin and soft tissue, pulmonary and device-related infections. Treatment of S. aureus infections has become increasingly difficult due to rapid evolution of methicillin-resistant S. aureus (MRSA) strains, development of resistance to all known B-lactam antibiotics, and dose limiting toxicities of widely used antibiotics like vancomycin (Mariathasan and Tan 2017; Lehar et al. 2015). Spread of antibiotic resistance is caused by the occurrence of "persisters"—antibiotic tolerant cells that are ubiquitous among bacteria such as S. aureus (Fisher et al. 2017).

A novel approach to fight antibiotic resistance and improve safety is through development of antibody-antibiotic conjugates (AAC), which allow for pathogen-specificity to broad-spectrum antibiotics. Genentech is currently developing THIOMAB™ (an antibody technology platform) based AAC, DSTA4637S for treatment of serious methicillin-sensitive S. aureus (MSSA) or MRSA infections (Peck et al. 2019). The AAC consists of a human IgG1 monoclonal antibody, a protease cleavable valine-citrulline linker and a rifamycin class antibiotic (4-dimethyla minopiperidino-hydroxybenzoxazino rifamycin [dmDNA31]) with an average drug-to-antibody ratio (DAR) of approximately 2. The antibody binds specifically to an antigen that is abundant and highly expressed on S. aureus (β-O-linked N-acetylglucosamine [GlcNAc] residues on ribitol phosphate units of wall teichoic acids of S. aureus) and is not expressed on human tissues, indicating the potential for safe and targeted delivery of DSTA4637S (Mariathasan and Tan 2017); (Wang-Lin et al. 2018).

In a mouse model of bacteremia, DSTA4637A (liquid formulation of DSTA4637S) was more potent than vancomycin (Lehar et al. 2015). The disposition of DSTA4637A in rats and monkeys was similar to that of the antibody, with slower clearance and longer half-life (Deng et al. 2019). Even though factors such as drug load and increase in hydrophobicity can have an impact on the PK properties of the conjugates (Zhou et al. 2016), low drug load (DAR=2) of DSTA4637A and hydrophilic nature of dmDNA31, may have preserved the PK properties of the antibody.

In a first-in-human, single-ascending-dose phase 1 study (NCT02596399), DSTA4637S was administered intravenously to thirty healthy male and female volunteers, randomized into five dose groups of 5, 15, 50, 100, and 150 mg/kg of DSTA4637S or placebo (4 active:2 placebo). The AAC was well-tolerated with no serious adverse events, no anti-drug antibodies, and no deaths. Overall, DSTA4637S has the potential to treat infections, where current therapies have failed, by maximizing efficacy and minimizing off-target toxicity. However, impact of the disease status on safety, deconjugation process, and PK has yet to be explored, and the ongoing phase 1b study in patients with S. aureus bacteremia (NCT03162250) should help provide more insights.

HIV/AIDS

Acquired immunodeficiency syndrome (AIDS) caused by HIV is a disease that impacts the body's own immune system, resulting in opportunistic infections and development of a range of cancers. Majority of people develop HIV-related infections within 5–10 years if left untreated (World Health Organization 2020). Antiretroviral drugs which suppress HIV-1 replication have been the cornerstone of treatment for HIV-related infections. Highly active antiretroviral therapy (HAART) which consists of a combination of three or more antiretroviral drug classes such as reverse transcriptase inhibitors, protease inhibitors and fusion inhibitors, reduced global mortality rate related to HIV-infections (Pereira and Paridaen 2004), but is associated with high treatment cost, complications related to drug-drug interactions, lack of adherence to treatment, and numerous side effects (Awi and Teow 2018; Gavrilyuk et al. 2013). In addition, persistence of virus-infected cells is a major hurdle (Berger 2011).

Even though the HAART is highly efficacious in suppressing HIV replication, it does not kill cells that have already been infected, resulting in a rapid rebound of infection upon stopping the treatment. Antibody therapies may help to circumvent this hurdle because they not only interfere with viral replication, but also target the virus-infected cells, and accelerate their clearance in-vivo ((Awi and Teow 2018), (Lu et al. 2016)).

A combination of ADCs carrying small molecule drugs, such as HAART or other antiviral cytotoxic molecules, could drastically improve efficacy. For example, doxorubicin (a small molecule anticancer anthracycline) conjugated to anti-gp 120 antibody P4/D10 (a neutralizing monoclonal antibody against the HIV-1 outer envelope) provided amplified protection in an HIV-1/MuLV murine challenge model, at almost 8 times lower dose than that observed with the antibody alone (Johansson et al. 2006).

In another study, broadly neutralizing antibodies (bNAbs) conjugated to a small molecule inhibitor of HIV-1 coreceptor CCR5—aplaviroc—neutralized the activities of all 117 different viral strains that were evaluated (Gavrilyuk et al. 2013). The ADC was more potent than the parent bNab alone, with IC_{50}s lower than 50 μg/mL against all the evaluated viral strains. The enhanced potency is likely due to the interplay of two different mechanisms of action: binding of the virus by the antibody and blocking of the CCR5 receptor on the host cell by the small molecule inhibitor.

Therefore, ADCs could be beneficial in treatment of HIV/AIDs by providing means to target virus-infected cells and treat persistent infections, in addition to suppressing replication of the virus. However, antibody-based HIV drugs also face many downsides such as low throughput and time-consuming manufacturing, high cost, and, most importantly, rapid development of viral resistance (Awi and Teow 2018).

Other Therapeutic Areas

Outside cancer, immunology, and infectious disease, the development of ADCs has been limited. A few programs in early discovery as well as preclinical and clinical stages of development in areas such as atherosclerosis and ophthalmology are described here.

Atherosclerosis

Atherosclerosis is a chronic disease that's responsible for nearly 35% of deaths in the United States (Roger et al. 2011). It is primarily managed with lipid-lowering statins and lifestyle modifications; but with diabetes and obesity on the rise, there's a need for pharmacological interventions targeting alternative pathways (Libby and Everett 2019). Atherosclerosis is also treated with anti-inflammatory agents that often cause side effects and lack efficacy, because of the complexity of inflammatory pathways (Chistiakov et al. 2018). ADCs have a potential for specific targeting of relevant inflammatory molecules with little to no risk of adverse effects.

In the last few decades, liver x receptors (LXR), were identified as suitable targets for treatment of atherosclerosis, due to their role in inflammation. When activated in macrophages and monocytes, LXRs suppress the inflammatory response; however, when activated in the liver, LXRs increase hepatic triglyceride levels, one of the risk factors for atherosclerosis (Zhu and Li 2009). So, selective targeting of LXRs in macrophages and monocytes is key to the successful development of LXR agonists. Lim and colleagues generated a targeted anti-CD11a IgG-LXR agonist ADC that causes selective activation of LXR in monocyte/ macrophage cells but not hepatocytes due to absence of CD11 expression in the liver. CD11a overexpression in monocytes has been shown to correlate with atherosclerotic coronary stenosis. The ADC showed selective targeting, internalization, and activation of LXR in human monocyte/macrophage THP-1 cells and no internalization or LXR activation in human liver (HepG2) cells (Lim et al. 2015). These in vitro data support potential site-specific activation of LXR agonist in macrophages in vivo while avoiding the side effects in liver cells.

Choroidal Neovascularization

Choroidal neovascularization (CNV) is commonly observed in age-related macular degeneration (AMD), a chronic, progressive, degenerative ocular disease, and is a result of the emergence and penetration of new blood vessels from the choroid into the Bruch's membrane leading to hemorrhage, vascular leakage, and eventually sudden vision loss (Shah et al. 2019). While the introduction of intravitreally delivered anti-vascular endothelia growth factor (VEGF) monotherapies has resulted in significant reduction in disease burden, these therapies are limited by high treatment burden (Apte 2020), treatment resistance (Holz et al. 2015), and poor long-term outcomes (Maguire et al. 2016).

ADC approach offers a potential to improve efficacy by double-targeting, for instance, with an anti-VEGF antibody conjugated to verteporfin—a photosensitizing agent used in treatment of AMD that closes up the leaking blood vessels when exposed to light. However, when tested in cultured VEGF-expressing murine endothelial cells, the ADC had similar cellular effects to free verteporfin (Mayo et al. 2003). Another conjugate that targets a membrane-bound CD105—an antigen that is overexpressed in CNV membranes from AMD patients—and carries mitomycin C (a potent cytotoxic DNA crosslinker), inhibited proliferation of human umbilical vein endothelial cells (HUVECs) (Yasukawa et al. 2000). A similar mitomycin C conjugate targeting $\alpha v\beta 3$ integrin that is minimally expressed on normal blood vessels but highly expressed in CNV tissue, also showed inhibitory effect in HUVECs and in vivo in a rodent laser CNV model (Kamizuru et al. 2001).

Despite these somewhat positive data, little progress has been made in the development of ADCs for CNV in the past 15 years, possibly due to availability of effective and safe anti-VEGF therapies. More careful consideration will have to be given to the use of relevant preclinical models to not only understand the targets better but also carefully gauge the risk-benefit profile.

Proliferative Vitreoretinopathy

Proliferative vitreoretinopathy (PVR) is characterized by growth of proliferative membranes within the vitreous cavity as well as on the retinal surface in the posterior segment of the eye which in turn can lead to blindness (Idrees et al. 2019). It is primarily treated through vitrectomy but the visual outcomes post-surgery remain poor, warranting a need for pharmacological interventions.

The proliferative membrane formation in PVR is supported by many growth factors and/or pathogenic immune factors which represent good potential targets for active drug targeting. Using vitreous and subretinal fluid samples from PVR patients, transferrin receptor, associated with cell proliferation, was identified as a promising target for drug delivery. An immunotoxin (ricin) bound to a monoclonal antibody against transferrin receptor in human retinal pigmented epithelial cells (RPE) showed reduced DNA synthesis along with a decrease in proliferating RPE cells (Jaffe et al. 1990). Despite the potential of targeting this membrane antigen in PVR, there has been no further progress made in the past three decades.

Posterior Capsule Opacification

Posterior capsule opacification (PCO), sometimes referred to as secondary cataract, is a complication that arises after cataract surgeries and occurs in almost every child and roughly half of the adults within a few years of surgery. Currently, PCO is managed using neodymium:YAG (Nd:YAG) laser capsulotomy and, in rare instances, surgical capsulotomy. As these therapies are associated with potential complications and high costs, there's a need for new interventions that inhibit or delay PCO formation. The cell proliferation of the remaining lens epithelium on the lens capsule surface that causes PCO can be specifically targeted via ADCs without causing side effects on other ocular cells.

Multiple efforts were made to develop ADCs for PCO, starting with Goins and colleagues developing an anti-transferrin receptor antibody conjugated to ricin, an immunotoxin, which showed decreased cell proliferation in human lens epithelial cells in vitro (Goins et al. 1994). Similarly, Kelleher and colleagues developed an ADC consisting of a murine monoclonal antibody (4197X) with the cytotoxic portion of ricin, which specifically targets a glycoprotein epitope present on the surface of human epithelial cells. This immunotoxin inhibited protein synthesis and cell proliferation in remnant human lens epithelial cells collected during cataract surgery. 4197X-Ricin A showed 200-fold more potency relative to pure ricin.

This success led to the initiation of a phase 1/2 clinical trial of 2 doses of the immunotoxin (now called MDX-RA). The lower dose (50 units) was well tolerated with a relatively safe profile and efficacy in the inhibition of PCO up to 24 months after surgery. The higher dose (175 units) was associated with postoperative intraocular inflammation even though it was transient in nature (Clark et al. 1998). Based on these results, a phase 3 trial of MDX-RA was initiated in 680 patients undergoing primary cataract surgery. Initial results from 8 patients in the low dose group (50 units) and 9 patients in the high dose group (100 units) showed efficacy in reducing PCO after cataract surgery; however, the low dose group had more postoperative inflammation relative to the high dose group which led to discontinuation of the phase 3 trials (Castillo 1998). In conclusion, for locally delivered ADCs in therapeutic areas such as ophthalmology, safety and tolerability studies need to be designed more critically as the risk tolerance for adverse effects may be lower compared to systemic administration.

Conclusion

Development of tumor-specific ADCs in oncology has successfully delivered 5 approved therapeutics. This success has turned the attention to the potential use of ADCs in non-oncological indications, especially immune diseases which share similar disease-driving mechanisms with cancer. Repurposed ADCs approved for cancer treatment, such as brentuximab vedotin, lead the race to become the first non-oncology ADC, but several research groups are working on novel disease-specific conjugates as well. Aside from immune disease, ADCs for infectious, cardiovascular and ophthalmologic diseases have also gained some momentum, but they are even farther away from approval.

Similar to oncology, ADCs offer a path to enhance specificity and reduce toxicity of a variety of drugs in non-oncological applications. Traditionally, conjugation to an antibody was used to improve properties of potent cell-killing drugs that are too toxic to be administered by themselves. Even though this premise is still used in areas such as immunology to, e.g., improve safety profile of steroids, more research is now focused on boosting the efficacy of molecules by delivering them specifically to the target site. In the infectious disease area, this approach was taken one step further—by combining 2 separate mechanisms of action: one delivered by the antibody, and the other—by the drug. ADCs also offer a second life to compounds that failed due to poor PK; residence time of molecules with short half-life can be improved by piggybacking on the antibody's properties, even though conjugation process can have a significant impact on the PK of the antibody.

Despite the challenges described in this chapter, the future of ADCs for non-oncological indications seems promising, with many preclinical and clinical studies conducted on new and existing ADCs that continue to tackle issues such as unacceptable toxicity, lack of efficacy, and suboptimal PK.

References

Amedei, A., Pimpinelli, N., Grassi, A., Bella, C.D., Niccolai, E., Brancati, S., et al.: Skin CD30(+) T cells and circulating levels of soluble CD30 are increased in patients with graft versus host disease. Autoimmun. Highlights. **5**(1), 21–26 (2013)

Apte, R.S.: Reducing treatment Burden in AMD. Cell. **180**(6), 1033 (2020)

Atzeni, F., Masala, I.F., di Franco, M., Sarzi-Puttini, P.: Infections in rheumatoid arthritis. Curr. Opin. Rheumatol. **29**(4), 323–330 (2017)

Awi, N.J., Teow, S.Y.: Antibody-mediated therapy against HIV/AIDS: where are we standing now? J Pathog. **2018**, 8724549 (2018)

Berger, E.A.: Targeted cytotoxic therapy: adapting a rapidly progressing anticancer paradigm for depletion of persistent HIV-infected cell reservoirs. Curr. Opin. HIV AIDS. **6**(1), 80–85 (2011)

Blazar, B.R., Levy, R.B., Mak, T.W., Panoskaltsis-Mortari, A., Muta, H., Jones, M., et al.: CD30/ CD30 ligand (CD153) interaction regulates $CD4^+$ T cell-mediated graft-versus-host disease. J. Immunol. **173**(5), 2933 (2004)

Blüml, S., Scheinecker, C., Smolen, J.S., Redlich, K.: Targeting TNF receptors in rheumatoid arthritis. Int. Immunol. **24**(5), 275–281 (2012)

Borsi, L., Balza, E., Bestagno, M., Castellani, P., Carnemolla, B., Biro, A., et al.: Selective targeting of tumoral vasculature: comparison of different formats of an antibody (L19) to the ED-B domain of fibronectin. Int. J. Cancer. **102**(1), 75–85 (2002)

Brentuximab vedotin Package Insert (2014).

Castillo, E.: MDX-RA Medarex Inc. IDrugs. **1**(4), 476–479 (1998)

Chen, Y.B., McDonough, S., Hasserjian, R., Chen, H., Coughlin, E., Illiano, C., et al.: Expression of CD30 in patients with acute graft-versus-host disease. Blood. **120**(3), 691–696 (2012)

Chen, Y.-B., Perales, M.-A., Li, S., Kempner, M., Reynolds, C., Brown, J., et al.: Phase 1 multicenter trial of brentuximab vedotin for steroid-refractory acute graft-versus-host disease. Blood. **129**(24), 3256–3261 (2017)

Chiang, M.-J., Holbert, M.A., Kalin, J.H., Ahn, Y.-H., Giddens, J., Amin, M.N., et al.: An Fc domain protein-small molecule conjugate as an enhanced immunomodulator. J. Am. Chem. Soc. **136**(9), 3370–3373 (2014)

Chistiakov, D.A., Melnichenko, A.A., Grechko, A.V., Myasoedova, V.A., Orekhov, A.N.: Potential of anti-inflammatory agents for treatment of atherosclerosis. Exp. Mol. Pathol. **104**(2), 114–124 (2018)

Clark, D.S., Emery, J.M., Munsell, M.F.: Inhibition of posterior capsule opacification with an immunotoxin specific for lens epithelial cells: 24 month clinical results. J. Cataract Refract. Surg. **24**(12), 1614–1620 (1998)

Czechowicz, A., Palchaudhuri, R., Scheck, A., Hu, Y., Hoggatt, J., Saez, B., et al.: Selective hematopoietic stem cell ablation using CD117-antibody-drug-conjugates enables safe and effective transplantation with immunity preservation. Nat. Commun. **10**(1), 617 (2019)

D'Elios, M.M., Romagnani, P., Scaletti, C., Annunziato, F., Manghetti, M., Mavilia, C., et al.: In vivo CD30 expression in human diseases with predominant activation of Th2-like T cells. J. Leukoc. Biol. **61**(5), 539–544 (1997)

de Lera, R.M., Lim, Y.-H., Zheng, J.: Adenosine A2A receptor as a drug discovery target. J. Med. Chem. **57**(9), 3623–3650 (2014)

DeFilipp, Z., Li, S., Kempner, M.E., Brown, J., Del Rio, C., Valles, B., et al.: Phase I trial of brentuximab vedotin for steroid-refractory chronic graft-versus-host disease after allogeneic hematopoietic cell transplantation. Biol Blood Marrow Transplant. **24**(9), 1836–1840 (2018)

Dela Cruz, J.S., Huang, T.H., Penichet, M.L., Morrison, S.L.: Antibody-cytokine fusion proteins: innovative weapons in the war against cancer. Clin. Exp. Med. **4**(2), 57–64 (2004)

Deng, R., Zhou, C., Li, D., Cai, H., Sukumaran, S., Carrasco-Triguero, M., et al.: Preclinical and translational pharmacokinetics of a novel THIOMAB antibody-antibiotic conjugate against Staphylococcus aureus. MAbs. **11**(6), 1162–1174 (2019)

Denton, C.P., Khanna, D.: Systemic sclerosis. Lancet. **390**(10103), 1685–1699 (2017)

Fisher, R.A., Gollan, B., Helaine, S.: Persistent bacterial infections and persister cells. Nat. Rev. Microbiol. **15**(8), 453–464 (2017)

Flowers, M.E.D., Martin, P.J.: How we treat chronic graft-versus-host disease. Blood. **125**(4), 606–615 (2015)

Gavrilyuk, J., Ban, H., Uehara, H., Sirk, S.J., Saye-Francisco, K., Cuevas, A., et al.: Antibody conjugation approach enhances breadth and potency of neutralization of anti-HIV-1 antibodies and CD4-IgG. J. Virol. **87**(9), 4985–4993 (2013)

Giacomelli, R., Cipriani, P., Lattanzio, R., Di Franco, M., Locanto, M., Parzanese, I., et al.: Circulating levels of soluble CD30 are increased in patients with systemic sclerosis (SSc) and correlate with serological and clinical features of the disease. Clin. Exp. Immunol. **108**(1), 42–46 (1997)

Gillard, G.O., Proctor, J.L., Brooks, M.L., Lamothe, T.L., Hyzy, S.L., Mikse, O., et al.: A novel targeted approach to achieve immune system reset: CD45-targeted antibody drug conjugates enable autologous HSCT, ameliorate disease in autoimmune models, potently kill human immune cells from normal donors and ms patients, and achieve immune depletion in Non-Human Primates (NHP). Blood. **134**(Supplement_1), 3208 (2019)

Goins, K.M., Ortiz, J.R., Fulcher, S.F., Handa, J.T., Jaffe, G.J., Foulks, G.N., et al.: Inhibition of proliferating lens epithelium with antitransferrin receptor immunotoxin. J. Cataract Refract. Surg. **20**(5), 513–516 (1994)

Graversen, J.H., Moestrup, S.K.: Drug trafficking into macrophages via the endocytotic receptor CD163. Membranes. **5**(2), 228–252 (2015)

Graversen, J.H., Svendsen, P., Dagnæs-Hansen, F., Dal, J., Anton, G., Etzerodt, A., et al.: Targeting the hemoglobin scavenger receptor CD163 in macrophages highly increases the anti-inflammatory potency of dexamethasone. Mol. Ther. **20**(8), 1550–1558 (2012)

Gyurkocza, B., Sandmaier, B.M.: Conditioning regimens for hematopoietic cell transplantation: one size does not fit all. Blood. **124**(3), 344–353 (2014)

Hartgring, S.A., Willis, C.R., Alcorn, D., Nelson, L.J., Bijlsma, J.W., Lafeber, F.P., et al.: Blockade of the interleukin-7 receptor inhibits collagen-induced arthritis and is associated with reduction of T cell activity and proinflammatory mediators. Arthritis Rheum. **62**(9), 2716–2725 (2010)

Holz, F.G., Tadayoni, R., Beatty, S., Berger, A., Cereda, M.G., Cortez, R., et al.: Multi-country real-life experience of anti-vascular endothelial growth factor therapy for wet age-related macular degeneration. Br. J. Ophthalmol. **99**(2), 220 (2015)

Idrees, S., Sridhar, J., Kuriyan, A.E.: Proliferative vitreoretinopathy: a review. Int. Ophthalmol. Clin. **59**(1), 221–240 (2019)

Jacobsohn, D.A., Vogelsang, G.B.: Acute graft versus host disease. Orphanet J. Rare Dis. **2**, 35 (2007)

Jaffe, G.J., Earnest, K., Fulcher, S., Lui, G.M., Houston, L.L.: Antitransferrin receptor immunotoxin inhibits proliferating human retinal pigment epithelial cells. Arch. Ophthalmol. (Chicago, Ill: 1960). **108**(8), 1163–1168 (1990)

Johansson, S., Goldenberg, D.M., Griffiths, G.L., Wahren, B., Hinkula, J.: Elimination of HIV-1 infection by treatment with a doxorubicin-conjugated anti-envelope antibody. AIDS (London, England). **20**(15), 1911–1915 (2006)

Kamizuru, H., Kimura, H., Yasukawa, T., Tabata, Y., Honda, Y., Ogura, Y.: Monoclonal antibody-mediated drug targeting to choroidal neovascularization in the rat. Invest. Ophthalmol. Vis. Sci. **42**(11), 2664–2672 (2001)

Kohler, R.E., Comerford, I., Townley, S., Haylock-Jacobs, S., Clark-Lewis, I., McColl, S.R.: Antagonism of the chemokine receptors CXCR3 and CXCR4 reduces the pathology of experimental autoimmune encephalomyelitis. Brain Pathol. **18**(4), 504–516 (2008)

Komohara, Y., Hirahara, J., Horikawa, T., Kawamura, K., Kiyota, E., Sakashita, N., et al.: AM-3 K, an anti-macrophage antibody, recognizes CD163, a molecule associated with an anti-inflammatory macrophage phenotype. J. Histochem. Cytochem. **54**(7), 763–771 (2006)

Kvien, T.K.: Epidemiology and burden of illness of rheumatoid arthritis. PharmacoEconomics. **22**(2 Suppl 1), 1–12 (2004)

Lee, H., Bhang, S.H., Lee, J.H., Kim, H., Hahn, S.K.: Tocilizumab–Alendronate conjugate for treatment of rheumatoid arthritis. Bioconjug. Chem. **28**(4), 1084–1092 (2017)
Lehar, S.M., Pillow, T., Xu, M., Staben, L., Kajihara, K.K., Vandlen, R., et al.: Novel antibody-antibiotic conjugate eliminates intracellular S. aureus. Nature. **527**(7578), 323–328 (2015)
Li, W., Xu, L.H., Yuan, X.M.: Macrophage hemoglobin scavenger receptor and ferritin accumulation in human atherosclerotic lesions. Ann. N. Y. Acad. Sci. **1030**, 196–201 (2004)
Libby, P., Everett, B.M.: Novel Antiatherosclerotic Therapies. Arterioscler. Thromb. Vasc. Biol. **39**(4), 538–545 (2019)
Lim, R.K., Yu, S., Cheng, B., Li, S., Kim, N.J., Cao, Y., et al.: Targeted delivery of LXR agonist using a site-specific antibody-drug conjugate. Bioconjug. Chem. **26**(11), 2216–2222 (2015)
Lu, C.L., Murakowski, D.K., Bournazos, S., Schoofs, T., Sarkar, D., Halper-Stromberg, A., et al.: Enhanced clearance of HIV-1-infected cells by broadly neutralizing antibodies against HIV-1 in vivo. Science. **352**(6288), 1001–1004 (2016)
Maguire, M.G., Martin, D.F., Ying, G.S., Jaffe, G.J., Daniel, E., Grunwald, J.E., et al.: Five-year outcomes with anti-vascular endothelial growth factor treatment of neovascular age-related macular degeneration: the comparison of age-related macular degeneration treatments trials. Ophthalmology. **123**(8), 1751–1761 (2016)
Mariathasan, S., Tan, M.W.: Antibody-antibiotic conjugates: a novel therapeutic platform against bacterial infections. Trends Mol. Med. **23**(2), 135–149 (2017)
Mayo, G.L., Melendez, R.F., Kumar, N., McKinnon, S.J., Glickman, R.D.: Antibody-targeted photodynamic therapy. Am. J. Ophthalmol. **136**(6), 1151–1152 (2003)
Palchaudhuri, R., Saez, B., Hoggatt, J., Schajnovitz, A., Sykes, D.B., Tate, T.A., et al.: Non-genotoxic conditioning for hematopoietic stem cell transplantation using a hematopoietic-cell-specific internalizing immunotoxin. Nat. Biotechnol. **34**(7), 738–745 (2016)
Palchaudhuri, R., Pearse, B.R., Proctor, J.L., Hyzy, S.L., Aslanian, S., Bhat, A., et al.: Non-genotoxic conditioning using amanitin antibody-drug conjugates targeting CD45 effectively deplete human and non-human primate hematopoietic stem cells and immune cells. Biol. Blood Marrow Transplant. **25**(3), S32 (2019)
Peck, M., Rothenberg, M.E., Deng, R., Lewin-Koh, N., She, G., Kamath, A.V., et al.: A phase 1, randomized, single-ascending-dose study to investigate the safety, tolerability, and pharmacokinetics of DSTA4637S, an anti-staphylococcus aureus thiomab antibody-antibiotic conjugate, in healthy volunteers. Antimicrob. Agents Chemother. **63**(6), e02588-18 (2019)
Pereira, C.F., Paridaen, J.T.: Anti-HIV drug development—an overview. Curr. Pharm. Des. **10**(32), 4005–4037 (2004)
Roger, V.L., Go, A.S., Lloyd-Jones, D.M., Adams, R.J., Berry, J.D., Brown, T.M., et al.: Heart disease and stroke statistics—2011 update: a report from the American Heart Association. Circulation. **123**(4), e18–e209 (2011)
Schultz, C.: Targeting the extracellular matrix for delivery of bioactive molecules to sites of arthritis. Br. J. Pharmacol. **176**(1), 26–37 (2019)
Shah, M., Cabrera-Ghayouri, S., Christic, L.-A., Held, K.S., Viswanath, V.: Translational preclinical pharmacologic disease models for ophthalmic drug development. Pharm. Res. **36**(4), 58 (2019)
Signoret, N., Oldridge, J., Pelchen-Matthews, A., Klasse, P.J., Tran, T., Brass, L.F., et al.: Phorbol esters and SDF-1 induce rapid endocytosis and down modulation of the chemokine receptor CXCR4. J. Cell Biol. **139**(3), 651–664 (1997)
Trachsel, E., Bootz, F., Silacci, M., Kaspar, M., Kosmehl, H., Neri, D.: Antibody-mediated delivery of IL-10 inhibits the progression of established collagen-induced arthritis. Arthritis Res Ther. **9**(1), R9 (2007)
Wang, R.E., Liu, T., Wang, Y., Cao, Y., Du, J., Luo, X., et al.: An immunosuppressive antibody–drug conjugate. J. Am. Chem. Soc. **137**(9), 3229–3232 (2015)
Wang-Lin, S.X., Zhou, C., Kamath, A.V., Hong, K., Koppada, N., Saad, O.M., et al.: Minimal physiologically-based pharmacokinetic modeling of DSTA4637A, a novel THIOMAB antibody antibiotic conjugate against Staphylococcus aureus, in a mouse model. MAbs. **10**(7), 1131–1143 (2018)

World Health Organization, HIV/AIDS; https://www.who.int/features/qa/71/en/. last Accessed 30 Jan 2020
Yasukawa, T., Kimura, H., Tabata, Y., Miyamoto, H., Honda, Y., Ikada, Y., et al.: Active drug targeting with immunoconjugates to choroidal neovascularization. Curr. Eye Res. **21**(6), 952–961 (2000)
Yasunaga, M., Manabe, S., Matsumura, Y.: Immunoregulation by IL-7R-targeting antibody-drug conjugates: overcoming steroid-resistance in cancer and autoimmune disease. Sci. Rep. **7**(1), 10735 (2017)
Younes, A., Yasothan, U., Kirkpatrick, P.: Brentuximab vedotin. Nat. Rev. Drug Discov. **11**(1), 19–20 (2012)
Zhou, C., Lehar, S., Gutierrez, J., Rosenberger, C.M., Ljumanovic, N., Dinoso, J., et al.: Pharmacokinetics and pharmacodynamics of DSTA4637A: a novel THIOMAB™ antibody antibiotic conjugate against Staphylococcus aureus in mice. MAbs. **8**(8), 1612–1619 (2016)
Zhu, Y., Li, Y.: Liver X receptors as potential therapeutic targets in atherosclerosis. Clin. Invest. Med. **32**(5), E383–E394 (2009)

Regulatory Framework for the Manufacturing and Marketing of ADCs

Hriday Gondalia

Abstract Developing effective anticancer drugs is crucial to address the severity of malignant tumors and the limited efficacy of current treatments. However, the planning and execution of nonclinical studies to support the development of anticancer pharmaceuticals lack globally recognized, calibrated guidelines. This gap arises due to the unique challenges presented by advanced therapies like Antibody Drug Conjugates (ADCs), which often require different considerations than other pharmaceuticals. Nonclinical evaluations of anticancer drugs serve to identify pharmacologic properties, establish safe initial dosing, and understand toxicological profiles. Meanwhile, clinical studies involve patients with fatal and progressive conditions, leading to close proximity of dose levels to adverse effects. Consequently, regulatory bodies face challenges in establishing suitable and harmonized regulations for ADCs.

This chapter explores regulations for ADCs and their impact on approval processes. For exemplification purposes examining how Kadcyla, an ADC used to treat HER2-positive breast cancer, aligns with current ADC regulations will be staged in a structured, stepwise manner. The discourse emphasizes the necessity for harmonization initiatives among various regulatory bodies on national, subregional, regional, and international levels. Such initiatives aim to promote informed, transparent, and consistent decision-making processes, enhancing global public health through effective medicine access. However, the proliferation of separate projects can hinder harmonization efforts, highlighting the importance of coordinated collaboration to avoid duplicative work and ensure meaningful standardization. As these initiatives evolve, interdependence among nations and organizations becomes more pronounced, simplifying international collaboration and trade.

H. Gondalia (✉)
University of Southern California Mann School of Pharmacy and Pharmaceutical Sciences, Biopharmaceutical Science BS, Business Finance & Entrepreneurship Minor, Los Angeles, CA, USA

Managing Director at True Drug Pharma Pvt. Ltd, Surat, Gujarat, India

H. J. Gukasyan, A. Kulkarni (eds.), *Antibody Drug Conjugates*, AAPS Advances in the Pharmaceutical Sciences Series 52, https://doi.org/10.1007/978-3-031-89319-3_3

Keywords Anticancer drugs · Antibody drug conjugates (ADCs) · Regulatory bodies · Harmonization initiatives · Nonclinical studies · Clinical studies · Kadcyla · Public health · Global collaboration

Introduction

Background

It is desirable to provide new, effective anticancer drugs to patients more quickly due to the seriousness of malignant tumors, the high mortality rate associated with these diseases, and the limited efficacy of current anti-cancer therapies. Regarding the planning and execution of nonclinical studies to support the development of anti-cancer pharmaceuticals like ADCs in clinical trials for the treatment of patients with advanced disease and few therapeutic options, there are no globally recognized goals or recommendations. Non-clinical evaluations for anticancer drugs are conducted for 3 reasons. Firstly, it is to identify the pharmacologic properties of a pharmaceutical. Secondly, to establish a safe initial dose level for the first human exposure, and finally to understand the toxicological profile of a pharmaceutical (e.g., identification of target organs, exposure-response relationships, and reversibility). On the other hand, clinical studies for anticancer drugs involve cancer patients whose condition is fatal and progressive. Furthermore, the dose levels in these clinical studies often are close to or at the adverse effect dose levels. Due to this, the type, timing, and flexibility called for in the design of nonclinical studies of anticancer pharmaceuticals can differ from those elements in nonclinical studies for other pharmaceuticals. Hence creating new regulations for advanced therapies such as ADCs can be tough for global regulatory bodies to navigate. For example, the US FDA's 2025 shift away from mandatory animal testing for biologics, which inherently includes monoclonal antibodies (mAbs), could also significantly accelerate and reshape ADC development in the context of preclinical safety evaluation. ADCs, which combine mAbs with cytotoxic payloads would traditionally require extensive animal testing to establish safety boundaries and off-target effects. mAbs often show species-specific pharmacodynamics and toxicities, making translation from animal models uncertain. Moving forward, human based systems (e.g., organoids, humanized cell lines) or progressive computational, modeling and simulation, or machine learning approaches which offer more predictive New Approach Methodologies (NAMs) for safety profile understanding, especially for payload-related toxicities, should be considered for development of ADCs. While mAb components of ADCs may benefit from NAMs, cytotoxic payloads and linker stability still pose hurdles. ADC sponsors must validate NAMs for their specific technology and constructs, selected payloads, and anticipate case-by-case evaluations, especially for first-in-class ADCs in this context, as cytotoxic payloads and linker technologies still pose an uncertainty.

Scope

This chapter talks about regulations for antibody drug conjugates and the impact of current discussions of regulatory bodies on the regulations of approval for ADCs. This chapter also looks at how Kadcyla aligns with the current regulations on ADCs. Kadcyla is an ADC used to treat HER2-positive breast cancer.

The Regulatory Hierarchy

Current Conversation Between the Different Regulatory Bodies

Over the last few decades, numerous harmonization initiatives have been established at the national, subregional, regional, and international levels. The common goal shared by each of them is to promote public health in the most resource-effective way possible through a better informed, transparent, and consistent decision-making process, though each has a unique objective, structure, and process (Olson 2014). Patients will have easier access to new safe and effective medicines as a result of this process, which is founded on sound scientific expertise and judgment.

While it is encouraging that health issues are becoming more prominent in world affairs and that there is a growing interest on the part of nations to collaborate and harmonize technical standards, the proliferation of these unrelated projects can occasionally be frustrating and counterproductive. If none of these initiatives are properly coordinated, the problem of work duplication and the introduction of new standards will persist, and the goal of harmonization will not be reached.

All of these bilateral, regional, and international projects have evolved and grown over time. They are now interdependent, with the same players involved. Countries engage in bilateral cooperation (for example, the EU/US bilateral inspection cooperation has been extended to other countries), regional initiatives, and global cooperation (e.g., EMA and US FDA collaboration with WHO to support developing countries), characterized in Diagram 1. The WHO and the ICH have made a number of commendable efforts to improve harmonization initiative coordination. As a result, every country follows the same set of rules, making international collaboration and trading easier (Olson 2014).

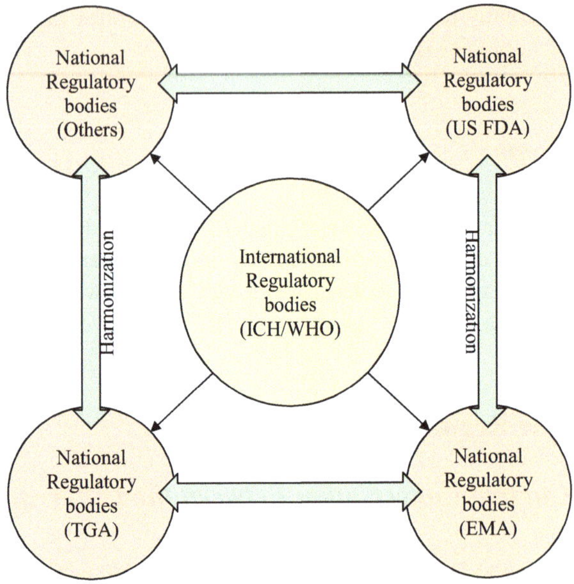

Diagram 1 Design of Global harmonization attempts

Studies to Support Non-clinical Evaluation

Pharmacology (Nonclinical Evaluation for Anticancer Pharmaceuticals (S9) 2013)

Preliminary characterization of the drug's mechanism(s) of action, schedule dependencies, and anti-tumor activity should have been done prior to Phase I studies.

Although the drug need not be studied using the same tumor types that are intended for clinical evaluation, appropriate models should be chosen based on the target and mechanism of action.

These studies can serve as nonclinical proof of concept, schedule and dose escalation schemes, test species selection information, start dose selection assistance, selection of investigational biomarkers assistance, and, if pertinent, justification for pharmaceutical combinations.

The assessment of a drug's safety for humans may be aided by knowledge of its secondary pharmacodynamic properties, which may then be appropriately investigated.

Safety Pharmacology (Nonclinical Evaluation for Anticancer Pharmaceuticals (S9) 2013)

Prior to the start of clinical studies, an evaluation of the drug's impact on essential organ functions (such as the cardiovascular, respiratory, and central nervous systems) should be available; such parameters could be included in general toxicology studies. In general, comprehensive clinical observations after dosing and suitable electrocardiographic measurements in non-rodents are regarded as sufficient. It is not necessary to conduct independent safety pharmacology studies to support research on patients with advanced cancer. The appropriate safety pharmacology studies outlined in ICH S7A and/or S7B should be taken into consideration when specific concerns have been identified that could expose patients to appreciable additional risks during clinical trials. Such studies won't be required to support clinical trials or for marketing in the absence of a specific risk.

ICH S7A (Safety Pharmacology Studies for Human Pharmaceuticals S7A 2001)

This document recommends the use of safety pharmacology core batteries. The safety pharmacology core battery's purpose is to investigate the effects of the test substance on vital functions. The cardiovascular, respiratory, and central nervous systems are commonly regarded as vital organ systems that should be studied in the core battery. Based on scientific evidence, the core battery should be supplemented or not implemented in some cases. Certain tests or explorations of specific organs, systems, or functions should be scientifically justified. ICH S7A mentions some key tests for safety pharmacology core batteries.

Firstly, The effects of the test substance on the central nervous system should be properly evaluated. It is necessary to assess motor activity, behavioral changes, coordination, sensory/motor reflex responses, and body temperature. A functional observation battery (FOB), modified Irwin's, or another appropriate test can be used.

Secondly, The effects of the test substance on the cardiovascular system should be carefully evaluated. The electrocardiogram, blood pressure, and heart rate should all be checked. Methods for repolarization and conductance abnormalities, as well as in vivo, in vitro, and/or ex vivo evaluations, should be considered.

Finally, The effects of the test substance on the respiratory system should be evaluated properly. Respiratory rate and other respiratory function measures (e.g., tidal volume or hemoglobin oxygen saturation) should be assessed. In general, clinical observation of animals is insufficient to assess respiratory function; thus, these parameters should be quantified using appropriate methodologies.

The ICH Guideline on Safety Pharmacology Studies for Human Pharmaceuticals (ICH S7A) is extended and supplemented by the S7B guideline. When appropriate (e.g., when adverse clinical events, a new patient population, or a new route of administration raise concerns that have not previously been addressed), this guideline applies to new chemical entities for human use and marketed pharmaceuticals. ICH S7A describes the conditions under which studies are not required. Hence to get more detailed information on the Studies highlighted in S7A look at S7B.

Non-clinical Studies for Kadcyla (Leighton 2013)

1. Effects of N^{2}'-deacetyl-N^{2}'-(3-mercapto-1-oxopropyl) maytansine (DM1) on Cloned hERG Potassium Channels Expressed in Human Embryonic Kidney Cells
2. A Single Dose Intravenous Cardiovascular Safety Pharmacology Study of Trastuzumab-MCC-DM1 (PRO132365) Administered to Female Cynomolgus Monkeys with a 3 Week Recovery Period

3. Cross-Reactivity of Trastuzumab-MCC-DM1 (PRO132365) with Human and Cynomolgus Monkey Tissues In Vitro
4. Hemolytic Potential and Blood Compatibility of Trastuzumab-MCC-DM1 (PRO132365)

A Clinical Study Done for Kadcyla (A Study of Trastuzumab Emtansine (T-DM1) 2014) A Study of Trastuzumab Emtansine (T-DM1) Sequentially With Anthracycline-based Chemotherapy, as Adjuvant or Neoadjuvant Therapy for Patients With Early Stage Herceptin (HER)2-positive Breast Cancer.

Pharmacokinetics (Nonclinical Evaluation for Anticancer Pharmaceuticals (S9) 2013)

The selection of doses, scheduling, and dose escalation during Phase I studies can be made easier by evaluating a small number of pharmacokinetic parameters in the animal species used for nonclinical studies, such as peak plasma/serum levels, area under the curve (AUC), and half-life. Normally, research on the pharmaceutical's animal absorption, distribution, metabolism, and excretion should be conducted concurrently with clinical development.

Pharmacokinetics Tests conducted for Kadcyla (Leighton 2013)

1. In Vitro Plasma Protein Binding Study of DM1 (a Maytansinoid) in Sprague-Dawley Rat, Cynomolgus Monkey, and Human Plasma

Metabolism Tests Included (Leighton 2013)

1. Evaluation of P450 induction potential by DM1 using Human Cryopreserved Hepatocytes
2. Inhibition of catalytic activities of cytochrome P450 in Human Liver Microsomes by DM1.
3. Metabolite radioprofiling and Identification in plasma, bile, and urine for rats following IV administration of Tmab-MCC-[^{3}H]DM1

General Toxicology (Nonclinical Evaluation for Anticancer Pharmaceuticals (S9) 2013)

Assessing the drug's safety is the main goal of Phase I clinical trials in patients with advanced cancer. Dosing to a maximum tolerated dose (MTD) and dose-limiting toxicity are examples of Phase I assessments (DLT). To support the clinical use of an anticancer pharmaceutical, toxicology studies to establish a no observed adverse effect level (NOAEL) or no effect level (NOEL) are not deemed necessary. An approximate clinical schedule for the drug should be examined in toxicology studies because it can have a significant impact on how toxic the drug is. Additionally, this is covered in sections "Duration and schedule of toxicology studies to support initial clinical trials" and "Duration of toxicity studies to support continued clinical development and marketing" of the ICH S9. To determine whether severe adverse effects are reversible or irreversible, an assessment of the potential for recovery from toxicity should be given.

If there is severe toxicity at approximately clinical exposure and recovery cannot be predicted by scientific assessment, a study with a terminal non-dosing period is required. This scientific evaluation may take into account the size and severity of the pathologic lesion as well as the organ system's capacity for regeneration. If a recovery study is required, it must be made available to support clinical development. It is not regarded as necessary to show complete recovery. Rodents and non-rodents are typically used in general toxicology testing for small molecules.

Alternative strategies may be appropriate in some situations, as determined on a case-by-case basis (e.g., for genotoxic drugs targeting rapidly dividing cells, a repeat-dose toxicity study in one rodent species might be considered sufficient, provided the rodent is a relevant species). For information on the number of species to be studied for biopharmaceuticals, see ICH S6 page 13. As necessary, a toxicokinetic analysis should be carried out.

Kadcyla: Carried Out Toxicity Tests for Single and Multiple Doses, as Well as, Toxicokinetic Tests

Single Dose Tests (Leighton 2013)

1. Single Dose Intravenous Toxicity Study of Trastuzumab-MCC-DM1 (PRO132365) in Cynomolgus Monkeys with a 3-Week Recovery Period
2. Single Dose Intravenous Toxicity Study of Trastuzumab-MCC-DM1 (PRO132365) with 5–7% Unconjugated Maytansinoid in Cynomolgus Monkeys with a 3-Week Recovery Period

Repeat Dose Tests (Leighton 2013)

1. Trastuzumab-SMCC-DM1: Dose Range Finding Study in Cynomolgus Monkeys
2. Multiple Dose Intravenous Toxicity Study of Trastuzumab-MCC-DM1 (PRO132365) Administered to Cynomolgus Monkeys Once Every 3 Weeks for 4 Doses, with a 3- or 6-week Recovery Period
3. An Intravenous Chronic Toxicity Study of Trastuzumab-MCC-DM1 (PRO132365) Administered to Cynomolgus Monkeys Once Every 3 Weeks for 8 Doses, with a 6-Week Recovery Phase

Toxicokinetics Tests (Leighton 2013)

1. Reaction Phenotyping of DM1 Using Human Liver Microsomes and Recombinant Cytochrome P450 (rCYP) and Metabolite Profiling in Human
2. *In Vitro* Interaction Studies of DM1 (Maytansine) with Human MDR1 (ABCB1/P-gp) ABC (Efflux) Transporter
3. Pharmacokinetic Comparability of Trastuzumab-MCC-DM1 (undisclosed word) in Female Cynomolgus Monkeys

4. Evaluation of the Pharmacokinetics of PR0132365 (Trastuzumab-SMCC-DM1) Following a Single Intravenous Bolus Dose in Cynomolgus Monkeys with Preliminary Tolerability (Non-GLP)

Reproduction Toxicology (Nonclinical Evaluation for Anticancer Pharmaceuticals (S9) 2013)

To inform patients who are or may become pregnant about potential risks to the developing embryo or fetus, an embryofetal toxicology assessment is carried out. Although these studies are expected to be available when the marketing application for an anticancer drug is submitted, they are not regarded as necessary to support clinical trials meant to treat patients with advanced cancer. Additionally, these studies are not regarded as necessary for the marketing of pharmaceuticals that have been shown to be developmental toxins or to be genotoxic and to target rapidly dividing cells (such as bone marrow or crypt cells) in general toxicity studies.

According to ICH S5 (Detection of Toxicity to Reproduction for Medicinal Products and Toxicity to Male Fertility S5(R2) 1994), studies on embryofetal toxicity for small molecules are typically carried out in two species (R2). A confirmatory study in a different species is typically not necessary when an embryofetal developmental toxicity study reveals embryofetal lethality or teratogenicity.

An evaluation in one pharmacologically relevant species should typically be sufficient for biopharmaceuticals. This evaluation could be carried out using ICH S6-compliant study designs or by assessing the toxicity during the organogenesis phase. If supported by science, alternative approaches might be deemed appropriate. Alternative strategies could involve evaluating the literature, placental transfer, the biopharmaceutical direct or indirect effects, or other elements.

It is not necessary to conduct research on fertility and early embryonic development to support clinical trials or the marketing of drugs for the treatment of advanced cancer patients. The assessment of fertility impairment should be based on data from general toxicology studies on the impact of the pharmaceutical on reproductive organs. To support clinical trials or for the marketing of pharmaceuticals for the treatment of patients with advanced cancer, a pre- and postnatal toxicology study is typically not required.

ICH S6 (Preclinical Safety Evaluation of Biotechnology-Derived Pharmaceuticals S6(R1) 2011)

The product, clinical indication, and intended patient population all influence the need for reproductive/developmental toxicity studies. The study design and dosing schedule may be changed depending on species specificity, immunogenicity, biological activity, and/or a long elimination half-life. Concerns about potential developmental immunotoxicity, which may apply particularly to certain monoclonal antibodies with long-lasting immunological effects, could be addressed in a study design modified to assess neonatal immune function.

Tests Performed for Kadcyla (Leighton 2013):

1. Study #95–039-1450. GN1450. Intravenous embryo-fetal development study in the cynomolgus monkey.
2. Study #95–238-1450. Late gestational toxicity, placental transfer, and secretion in milk study with GN1450 in the cynomolgus monkey
3. Study #95–038-1450. GN1450. Intravenous menstrual cycle study (fertility evaluation) in the female cynomolgus monkey

Genotoxicity (Nonclinical Evaluation for Anticancer Pharmaceuticals (S9) 2013)

In order to support clinical trials for therapeutics intended to treat patients with advanced cancer, genotoxicity studies are not regarded as being necessary. To support marketing, genotoxicity studies should be carried out (see ICH S2). For biopharmaceuticals, the guidelines in ICH S6 should be followed. An in vivo assay might not be necessary if the in vitro tests are positive.

ICH S6 (Preclinical Safety Evaluation of Biotechnology-Derived Pharmaceuticals S6(R1) 2011)

The range and type of genotoxicity studies that are routinely performed for pharmaceuticals are not applicable to biotechnology-derived pharmaceuticals and are thus unnecessary. Furthermore, administering large amounts of peptides/proteins may produce unexpected results. These substances are not expected to interact directly with DNA or other chromosomal material. In cases where there is cause for concern about the product, studies in available and relevant systems, including newly developed systems, should be conducted (e.g., because of the presence of an organic linker molecule in a conjugated protein product). Standard genotoxicity studies are not considered appropriate for assessing the genotoxic potential of process contaminants. However, if performed for this purpose, the rationale should be provided.

Tests Performed for Kadcyla (Leighton 2013):

1. $N^{2'}$- deacetyl-$N^{2'}$-(3-mercapto-1-oxopropyl) maytansine (DM1): Salmonella—Escherichia coli/Mammalian-Microsome Reverse Mutation Assay with a Confirmatory Assay
2. $N^{2'}$- deacetyl-$N^{2'}$-(3-mercapto-1-oxopropyl) maytansine (DM1): In Vivo Rat Bone Marrow Micronucleus Assay

Carcinogenicity (Nonclinical Evaluation for Anticancer Pharmaceuticals (S9) 2013)

ICH S1A describes the appropriateness of a carcinogenicity assessment for anticancer pharmaceuticals. Carcinogenicity studies are not required to support the marketing of therapeutics intended to treat advanced cancer patients (Preclinical safety evaluation of biotechnology-derived pharmaceuticals S6(R1) 2011).

ICH S1A Guideline Says (Guideline on the Need for Carcinogenicity Studies of Pharmaceuticals S1A 1995):
Unequivocally genotoxic compounds, in the absence of other data, are presumed to be trans-species carcinogens, implying a hazard to humans. Such compounds need not be subjected to long-term carcinogenicity studies. However, if such a drug is intended to be administered chronically to humans a chronic toxicity study (up to 1 year) may be necessary to detect early tumorigenic effects. Assessment of the genotoxic potential of a compound should take into account the totality of the findings and acknowledge the intrinsic value and limitations of both in vitro and in vivo tests. The test battery approach of in vitro and in vivo tests is designed to reduce the risk of false negative results for compounds with genotoxic potential. A single positive result in any assay for genotoxicity does not necessarily mean that the test compound poses a genotoxic hazard to humans (ICH Guidance on Specific Aspects of Regulatory Genotoxicity Tests).

Tests Performed for Kadcyla: Not required and not submitted.

Immunotoxicity (Nonclinical Evaluation for Anticancer Pharmaceuticals (S9) 2013)

The design components of general toxicology studies are considered sufficient to evaluate immunotoxin potential and support marketing for the majority of anticancer pharmaceuticals. Additional endpoints (such as immunophenotyping by flow cytometry) may be included in the study design for immunomodulatory pharmaceuticals (Nonclinical evaluation for anticancer pharmaceuticals (S9) 2013).

Kadcyla did not mention any specific immunotoxicity studies they did.

Photosafety Testing (Nonclinical Evaluation for Anticancer Pharmaceuticals (S9) 2013)

An initial assessment of phototoxic potential should be conducted prior to Phase I, based on photochemical properties of the drug and information on other members in the class. If assessment of these data indicates a potential risk, appropriate protective measures should be taken during outpatient trials. If the photosafety risk cannot be adequately evaluated based on nonclinical data or clinical experience, a photosafety assessment consistent with the principles described in ICH M3 should be provided prior to marketing (Nonclinical evaluation for anticancer pharmaceuticals (S9) 2013).

ICH M3 (Guidance on Nonclinical Safety Studies for the Conduct of Human Clinical Trials and Marketing Authorization for Pharmaceuticals M3(R2) 2009):
The appropriateness of or timing of photosafety testing in relation to human exposure should be influenced by:

1. the photochemical properties (photoabsorption and photostability) of the molecule
2. information on the phototoxic potential of chemically-related compounds
3. tissue distribution
4. clinical or nonclinical findings indicative of phototoxicity.

An initial assessment of phototoxic potential should be performed based on a drug's physical/chemical properties for photoreactivity, spectral absorption properties, and pharmacologic class /SAR. If the assessment reveals a potential phototoxicity risk, it may be necessary to implement protective measures during subsequent clinical trials. Following that, an evaluation of the nonclinical drug distribution to skin and eye should be performed to provide additional information on the human risk. Prior to exposing a large number of subjects, an experimental evaluation of phototoxic potential should be performed if appropriate (Phase 3). Alternatively, to avoid the need for early assessment of eye/skin distribution studies and clinical protective measures, a direct assessment of phototoxic potential in nonclinical or clinical studies can be conducted. Photo carcinogenicity testing in rodents is generally not considered useful in support of pharmaceutical development and is not recommended. Specific testing is not recommended if the photo carcinogenic risk in patients can be adequately managed, and a warning statement can be included in the informed consent or labeling. If testing is deemed necessary, methods other than rodent bioassays should be considered and completed prior to marketing to assess photo carcinogenic potential.

Kadcyla report did not mention any photosafety testing.

Nonclinical Data to Support Clinical Trial Design and Marketing

Start Does for First Administration in Humans (Nonclinical Evaluation for Anticancer Pharmaceuticals (S9) 2013)

Finding a dose that is expected to have pharmacologic effects and is generally safe to use is the aim of choosing the start dose. All available nonclinical data, such as pharmacokinetics, pharmacodynamics, and toxicity, should be used to support the start dose in a scientific manner (see Note 2). Interspecies scaling of the animal doses to an equivalent human dose is typically based on normalization to body surface area for the majority of systemically administered small molecules. Interspecies scaling based on body weight, AUC, or other exposure parameters may be appropriate for both small molecules and biopharmaceuticals. When choosing the start dose for biopharmaceuticals with immune agonistic properties, a minimally anticipated biologic effect level (MABEL) should be taken into account.

Note 2 in the ICH S9 states: Setting a start dose at 1/10 the Severely Toxic Dose in 10% of Animals (STD 10) in rodents is a common strategy for many small molecules. If the non-rodent is the most suitable species, then a starting dose of 1/6 of the highest non-severely toxic dose (HNSTD) is recommended. The highest dose level that does not result in evidence of lethality, life-threatening toxicities, or irreversible findings is known as the HNSTD.

Kadcyla reported this in their non-clinical multiple dose studies:

Kadcyla conducted Study 04–0977-1459, a repeat-dose GLP toxicity study in Cynomolgus monkeys. The researchers administered four IV doses of trastuzumab emtansine every 3 weeks, followed by various recovery times (A Study of Trastuzumab Emtansine in Participants With Human Epidermal Growth Factor Receptor (HER)2 Immunohistochemistry (IHC)-Positive, Locally Advanced or Metastatic Non-Small Cell Lung Cancer (NSCLC) 2019). Evaluation doses were 0, 3, 10, and 30 mg/kg. None of these animals displayed outward toxicity. The sciatic nerve and the dorsal funiculus of the spinal cord both showed microscopic evidence of axonal degeneration at dose levels of 10 and 30 mg/kg. Both necropsy times revealed these changes in the animals from the two higher dose groups, but the effects were more pronounced in the 30 mg/kg group. The possibility of an irreversible or slowly reversible neurotoxicity is raised by this finding. Neurological examinations during life did not predict this toxicity. At any of the three necropsy time points, animals that received 3 mg/kg ADC did not exhibit any histopathological changes. Kadcyla determined that 10 mg/kg of trastuzumab emtansine administered as four injections to Cynomolgus monkeys every 3 weeks was the highest non-severely toxic dose based on the overall tolerability and reversibility of the results (A Study of Trastuzumab-MCC-DM1 2015).

Dose Escalation and Highest Dose in Clinical Trial (Nonclinical Evaluation for Anticancer Pharmaceuticals (S9) 2013)

Generally, the highest dose or exposure investigated in a clinical trial in cancer patients is not constrained by the highest dose or exposure tested in nonclinical studies. Smaller than usual dose increments (fractional increments rather than dose doubling) should be taken into account when a steep dose- or exposure-response curve for severe toxicity is seen in nonclinical toxicology studies, or when there is no preceding marker of severe toxicity available.

In the case of ***Kadcyla***:

Title of Study
A Study of Trastuzumab-MCC-DM1 Administered Intravenously to Patients With HER2-Positive Metastatic Breast Cancer Who Have Previously Received a Trastuzumab-Containing Regimen (A Study of Trastuzumab Emtansine (T-DM1) 2014).

Results (Leighton 2013)
The results of this study suggest that the maximum tolerated dose (MTD) was determined to be 3.6 mg/kg IV q3w in the phase 1 trial (Study TDM3569g).

Duration and Schedule of Toxicology Studies to Support Initial Clinical Trials (Nonclinical Evaluation for Anticancer Pharmaceuticals (S9) 2013)

A new toxicology study is not necessary to support continued treatment after the completion of the completed toxicology studies in Phase I clinical trials because treatment can be continued based on the patient's response. Nonclinical study designs should be carefully chosen to account for potential initial clinical trial dosing schedule variations. Although it is not anticipated that the precise clinical schedule will always be followed, the information from the toxicology studies should be enough to support the clinical dose and schedule and to identify any potential toxicity. The half-life in the test species and the anticipated (or known) half-life in humans, for instance, can both be taken into account. Other variables might include receptor saturation, toxicity profile, and exposure assessment. Examples of nonclinical treatment schedules for small molecules or biopharmaceuticals that are frequently used in the development of anticancer medications are shown in Table 1 (taken from ICH S9 (Nonclinical evaluation for anticancer pharmaceuticals (S9) 2013)). An additional toxicology study in a single species is typically sufficient when the available toxicology data does not support a change in clinical schedules.

Table 1 Examples of treatment schedules for anticancer pharmaceuticals to support initial clinical trials (Nonclinical evaluation for anticancer pharmaceuticals (S9) 2013)

Clinical schedule	Examples of nonclinical treatment schedule
Once every 3–4 weeks	Single dose
Daily for 5 days every 3 weeks	Daily for 5 days
Daily for 5–7 days, alternating weeks	Daily for 5–7 days, alternating weeks (2-dose cycles)
Once a week for 3 weeks, 1 week off	Once a week for 3 weeks
Two or three times a week	Two or three times a week for 4 weeks
Daily	Daily for 4 weeks
Weekly	Once a week for 4–5 doses

Duration of Toxicity Studies to Support Continued Clinical Development and marketing (Nonclinical Evaluation for Anticancer Pharmaceuticals (S9) 2013)

For patients with advanced cancer to proceed to Phase II and begin second or first line therapy, the nonclinical data supporting Phase I and the clinical Phase I data would typically be sufficient. Results from repeat dose studies lasting 3 months and conducted in accordance with the intended clinical schedule should be presented before beginning Phase III studies in order to support the continued development of an anticancer pharmaceutical for patients with advanced cancer. Three-month nonclinical studies are thought to be sufficient to support marketing for the majority of pharmaceuticals used to treat patients with advanced cancer.

An analysis of the current clinical data should be done to support any changes being considered to the clinical schedule. The factors covered in section "Duration and schedule of toxicology studies to support initial clinical trials" of the ICH S9 should be taken into account if the clinical data are insufficient on their own to justify the change in schedule.

Kadcyla did a study titled:

An Intravenous Chronic Toxicity Study of Trastuzumab-MCC-DM1 (PRO132365) Administered to Cynomolgus Monkeys Once Every 3 Weeks for 8 Doses, with a 6-Weeks Recovery Phase (A Study of Trastuzumab Emtansine in Participants With Human Epidermal Growth Factor Receptor (HER)2 Immunohistochemistry (IHC)-Positive, Locally Advanced or Metastatic Non-Small Cell Lung Cancer (NSCLC) 2019).

Postmarket requirements and commitments include studies and clinical trials that sponsor conducted after marketing approval to gather additional information about Kadcyla's safety, efficacy, or optimal use. Studies or clinical trials a sponsor has committed to conduct as a condition for approval may be required (A Study of Trastuzumab-MCC-DM1 2015).

Combination of Pharmaceuticals (Nonclinical Evaluation for Anticancer Pharmaceuticals (S9) 2013)

Drugs intended for combination use should undergo thorough individual toxicology assessments. Prior to beginning the clinical study, data to support a justification for the combination should be offered. In general, toxicology research into the efficacy of drug combinations used to treat patients with advanced cancer is not necessary. A nonclinical study assessing the combination is typically not warranted if the human toxicity profile of the pharmaceuticals has been characterized. A pharmacology study to support the justification for the combination should be provided for studies in which at least one of these compounds is in early stage development (i.e., the human toxicity profile has not been characterized). Based on a few select safety endpoints, such as mortality, clinical signs, and body weight, this study should demonstrate increased activity without a significant increase in toxicity. It should be decided, in light of the information at hand, whether or not a focused toxicology study of the combination is necessary.

A example of a study Kadcyla submitted (A Study of Trastuzumab Emtansine (Trastuzumab-MCC-DM1, T-DM1) 2013):

A Study of Trastuzumab Emtansine (Trastuzumab-MCC-DM1, T-DM1) in Combination With Pertuzumab Administered to Patients With Human Epidermal Growth Factor Receptor-2 (HER2)-Positive Locally Advanced or Metastatic Breast Cancer Who Have Previously Received Trastuzumab.

Non-clinical Studies to Support Trials in Pediatric Populations (Nonclinical Evaluation for Anticancer Pharmaceuticals (S9) 2013)

The typical approach for studying the majority of anticancer medications in young patients is to first establish a dose that is generally considered safe in adult populations before evaluating a portion of that dose in early pediatric clinical studies. This population is also covered by the nonclinical testing suggestions made elsewhere in this document. Studies on young animals are rarely done to support the inclusion of pediatric populations in cancer treatment. Only when previous animal studies and human safety data are deemed insufficient for a safety evaluation in the intended pediatric age group should the conduct of studies in juvenile animals be taken into consideration.

Kadcyla did this study (Leighton 2013):

A Study of Trastuzumab Emtansine in Participants With Human Epidermal Growth Factor Receptor (HER)2 Immunohistochemistry (IHC)-Positive, Locally Advanced or Metastatic Non-Small Cell Lung Cancer (NSCLC).

Other Considerations

Conjugated Products (Nonclinical Evaluation for Anticancer Pharmaceuticals (S9) 2013)

Pharmaceuticals that are covalently attached to carrier molecules like proteins, lipids, or sugars are known as conjugated products. The main concern is the safety of the conjugated material. A more constrained evaluation of the unconjugated material's safety, including the linker employed, is possible. It should be stated whether the conjugate is stable in both human plasma and the test species. Following administration of the conjugated material, a toxicokinetic assessment should be performed on both the conjugated and the unconjugated compound.

Tests Performed for Kadcyla: Characterization of Plasma Stability of Trastuzumab-MCC-[^{3}H]DM1 In Vitro (Postmarketing Requirements and Commitments 2016).

Evaluation of Drug Metabolites (Nonclinical Evaluation for Anticancer Pharmaceuticals (S9) 2013)

Metabolites that have been found in humans in some cases have not been verified in nonclinical studies. For patients with advanced cancer, a separate evaluation for these metabolites is typically not necessary.

Tests Performed for Kadcyla: Metabolite Radioprofiling and Identification in Plasma, Bile, and Urine from Rats Following IV Administration of Tmab-MCC-[^{3}H]DM1 (Leighton 2013).

Evaluation of impurities (Nonclinical Evaluation for Anticancer Pharmaceuticals (S9) 2013)

As discussed in ICH Q3A and Q3B, it is acknowledged that impurity standards have been based on a negligible risk. Anticancer medications may be appropriate for exceeding the established limits for the impurities listed in these ICH guidelines, and a justification should be given in the marketing application. The parent pharmaceutical's characteristics (pharmacologic properties, genotoxicity and carcinogenic potential, etc.), the duration of treatment, and the effect of impurity reduction on manufacturing could all be considered as part of the justification. Other factors to consider include the disease being treated and the patient population.

Additionally, the qualification evaluation could take into account the dose or concentration tested in nonclinical research in relation to clinical levels. Limits for genotoxic impurities have been set using a variety of strategies based on the increased lifetime risk of cancer. Such limits are inappropriate for drugs used to treat patients with advanced cancer, and higher limits should be set in light of the aforementioned justifications. In general, contaminants that are also metabolites found in animal and/or human studies are regarded as qualified.

Kadcyla: No comments on impurities and degradants of concern.

Summary of Tests Done by Kadcyla

Table of all the studies submitted (pg 10–14) **(Leighton 2013):**

Study number	Title of study
04-0962-1459	Response of BT474E1 Tumor to 3 Doses of Trastuzumab-MCC-DM1 Administered as an Every 3 Week Regimen in Beige Nude Mice
04-0962 A-1459	Dose Response of Trastuzumab-MCC-DM1 (Administered on a Weekly or Biweekly Basis) Against BT474EI Xenograft Tumors in Beige Nude Mice
05-1111	In Vitro Binding and Activity of Tmab-MCC-DM1
05 1508	Response of MMTV-HER2 Fo5 Tumor to Treatment of Trastuzumab and DM1 Approximately 30 Minutes Later in Nude Mice
07-1030 C	Response of MDA-MB-361.1 Mammary Tumor to Combination Therapy of GDC-0941 and Trastuzumab-MCC-DM1 in Taconic NCR Nude Mice
07-1629	Effect of Single Agent or Combinations of Pertuzumab and Trastuzumab-MCC-DM1 on KPL-4 Tumors Inoculated into the Mammary Fat Pad of SCID Beige Mice
10-0010	Binding of Trastuzumab-MCC-DM1 (T-DM1) to HER2 ECD by Surface Plasmon Resonance (Biacore)® Technology)
10-0046	In Vitro Biological Characterization of Trastuzumab-MCC-DM1 (T-DM1) for Fc Gamma Receptor Binding, Antibody-Dependent Cell-Mediated Cytotoxicity, and C1q Binding Activities
10-0825	In Vitro Characterization of Trastuzumab-MCC-DM1 in Breast Cancer Cell Lines
10-0826	Viability/Proliferation Assays on T-DM1 Clinical Qualification Lots
99-292 X-1459	Specificity of Trastuzumab-MCC-DM1 for HER2 Over-Expressing Mammary Tumors
99-292 ZC-1459	Response of MMTV-HER2 Fo5 Tumor to 3 Doses of Trastuzumab-MCCDM1 Administered as an Every-3-Week Regimen
99-292 ZD-1459	Dose Response of MMTV-HER2 Fo5 Tumor to Trastuzumab-MCC-DM1 Administered on a Weekly Schedule in Mice
99-292 ZF-1469	Response of MMTV-HER2 Fo5 Tumor to Control Antibody-MCC-DM1
04-0901-4747	The Influence of HERCEPTIN on MMTV-HER2 Fo5 Tumor Response to Trastuzumab-DM1 Conjugates

Study number	Title of study
05-0754-1459	Validation of a Method for the Determination of DM1 in Sprague Dawley Rat LI-Heparin Plasma Samples by LC-MS/MS
05-0755-1459	Validation of a Method for the Determination of DM1 in Cynomolgus Monkey Li-Heparin Plasma Samples by LC-MS/MS
06-0809	Development and Validation of a Method for the Determination of DM1 in Cynomolgus Monkey Plasma
08-1059	Determination of the Stability of 4 Lots of Tmab-MCC-DM1 in Human and Cynomolgus Monkey Li-Heparin Plasma by LC-MS/MS
4-herc-1-avr-3-4. HERc.1	Total Trastuzumab (PRO132365) Antigen ELISA
4-herc-1-avr-3-4.HERc.2	Trastuzumab-MCC-DM1 (PRO132365) Antigen ELISA
4-herc-1-avr-3-4.HERc.3	Trastuzumab-MCC-DM1 (PRO132365) Antibody ECLA
vr0606	Determination of the Stability of Tmab-MCC-DM1 in Human, Cynomolgus Monkey and Sprague Dawley Rat L1-Heparin Plasma by LC-MS/MS
04-0380-1459	Characterization of the Pharmacokinetics of Her-SMCC-DM1 (Low Drug Load, 2.7 Drugs/Ab), Her-SMCC-DM1 (High Drug Load, 3.6 Drugs/Ab), and Her-MC-MMAE Conjugates in Male Rats
04-0645-1605	Pharmacokinetics of Herceptin ("Her", rhu4D5), Her-MC-vc-PAB-MMAF, Her-MC-MMAF, and Her-SMCC-DM1 Following IV Administration in Athymic (nu/nu) Mice
04-0975-1459	Evaluation of the Pharmacokinetics of PR0132365 (TrastuzumabSMCC-DM1) Following a Single Intravenous Bolus Dose in Cynomolgus Monkeys with Preliminary Tolerability (Non-GLP)
05-0022-1459	Trastuzumab-SMCC-DM1 (PRO132365): Dose Range-Finding Study in Sprague-Dawley Rats
05-0278-1459	Trastuzumab-MCC-DM1 (PRO132365): Dose Ranging Pharmacokinetics Study in Non-Tumor Bearing Beige Nude Mice
05-282-1459	Trastuzumab-MCC-DM1 (PRO132365): Dose-Ranging Pharmacokinetics Study in Tumor-Bearing (BT474 E1) Beige Nude Mice
05-1104	DM1: Dose Range Finding Study in Male and Female Sprague-Dawley Rats
06-1068	A Study of the Pharmacokinetics of DM1 in Athymic Nude Mice
06-1270	Pharmacokinetics Study of MCC-DM1 in Male and Female SpragueDawley Rats
04-1022-1459	Tissue Distribution Study of Herceptin-MCC-[^{3}H] DM1 in Rats
05-0336-1459	Tissue Distribution of [^{125}I]Herceptin and [^{125}I]Herceptin-MCC-DM1 in Normal Rats
08-1136	Determination of Mass Balance and Route of Excretion Following Intravenous Administration of Free [^{3}H]-DM1 in Female Sprague-Dawley Rats
04-1099-1459	Pharmacokinetics, Metabolism, and Excretion of Herceptin-sMCC-[^{3}H]-DM1 in Rat after IV Administration

Study number	Title of study
05-1047-1459	In Vitro Plasma Protein Binding Study of DM1 (a Maytansinoid) in Sprague-Dawley Rat, Cynomolgus Monkey, and Human Plasma
05-0297-1459	Characterization of Plasma Stability of Trastuzumab-MCC-DM1, Trastuzumab-MC-vc-PAB-MMAF, and Trastuzumab-MC-MMAF In Vitro
05-0298-1459	Characterization of Plasma Stability of Trastuzumab-MCC-[^{3}H]DM1 In Vitro
08-1137	Determination of Radioactivity Excreted in the Bile Following a Single Intravenous Administration of [^{3}H]-DM1 or Unlabeled DM1 in SpragueDawley Rats
09-1060	Determination of Radioactivity in Plasma and Excreta Matrices following Intravenous Administration of Trastuzumab-MCC-[^{3}H]-DM1 or Unlabeled Trastuzumab-MCC-DM1 in Female Sprague Dawley Rats
09-1344	Inhibition of the Catalytic Activities of Cytochromes P450 in Human Liver Microsomes by DM-1
09-1929	Metabolite Radioprofiling and Identification in Plasma, Bile, and Urine from Rats Following IV Administration of Tmab-MCC-[^{3}H]DM1
09-2381	Metabolite Profiling and Identification of DM1 in Rat
09-2382	Evaluation of Cytochrome P450 Induction Potential by DM1 Using Human Cryopreserved Hepatocytes
09-2416	Reaction Phenotyping of DM1 Using Human Liver Microsomes and Recombinant Cytochrome P450 (rCYP) and Metabolite Profiling in Human
10-1207	In Vitro Interaction Studies of DM1 (Maytansine) with Human MDR1 (ABCB1/P-gp) ABC (Efflux) Transporter
10-1356	Exploratory Quantitative Determination of Lys-MCC-DM1, MCC-DM1, and DM1 in Plasma, Urine, and Bile from Rat following Intravenous Administration of Tmab-MCC-DM1
08-1136	Determination of Mass Balance ad Route of Excretion Following Intravenous Administration of Free [3 H]-DM1 in Female Sprague-Dawley Rats
08-1137	Determination of Radioactivity Excreted in the Bile Following a Single Intravenous Administration of [^{3}H]-DM1 or Unlabeled DM1 in SpragueDawley Rats
09-1060	Determination of Radioactivity in Plasma and Excreta Matrices following Intravenous Administration of Trastuzumab-MCC-[^{3}H]-DM1 or Unlabeled in Trastuzumab-MCC--DM1 Female Sprague-Dawley Rats
05-1359	Pharmacokinetic, Pharmacodynamic, Toxicokinetic, and Toxico-dynamic Modeling and Simulation Analysis of PRO32365 (Trastuzumab-MCCDM1)
07-1474	Pharmacokinetic Comparability of Trastuzumab-MCC-DM1 … in Female Cynomolgus Monkeys
07-1475	Pharmacokinetic Comparability of Trastuzumab-MCC-DM1 … in Female Sprague Dawley Rats
08-0800	Pharmacokinetic Comparability of Trastuzumab-MCC-DM1 … in Female Cynomolgus Monkeys

'…' in the table means that the words were redacted.

Study number	Title of study
03-0140-1345	Safety Evaluation of Various Herceptin-DM-1 Lots; Herceptin-SPP-DM1 and Herceptin-DM1 with an Un-Cleavable Linker (HerceptinSMCCDM1), and Free DM1 in Female Sprague-Dawley Rats
04-0976-1459	Single Dose Intravenous Toxicity Study of Trastuzumab-MCC-DM1 (PRO132365) in Cynomolgus Monkeys with a 3-Week Recovery Period
04-1214-1459	Single Dose Intravenous Toxicity Study of Trastuzumab-MCC-DM1 (PRO132365) in Sprague-Dawley Rats with a 3-Week Recovery Period
05-1157	A Dose Range-Finding Study of Intravenous DM1 in Female SpragueDawley Rats with a 3-Week Recovery Period
05-1191	Single Dose Intravenous Toxicity Study of DM1 in Sprague-Dawley Rats with a 3-Week Recovery Period
07-0389	A Single Dose 12-Day Intravenous Toxicity Study of Hu Tmab-MCCDM1, Thio Hu Tmab-MPEO-DM1, Thio Hu Tmab-MC-MMAF, and Thio Hu Tmab-MC-vc-PAB-MMAE in Female Sprague-Dawley Rats (Crl:CD(SD))

Study number	Title of study
03-0674-1459	Trastuzumab-SMCC-DM1: Dose Range Finding Study in Cynomolgus Monkeys
04-0977-1459	Multiple Dose Intravenous Toxicity Study of Trastuzumab-MCC-DM1 (PRO132365) Administered to Cynomolgus Monkeys Once Every 3 Weeks for 4 Doses, with a 3- or 6-Week Recovery Phase
07-0653	An Intravenous Chronic Toxicity Study of Trastuzumab-MCC-DM1 (PRO132365) Administered to Cynomolgus Monkeys Once Every 3 Weeks for 8 Doses, with a 6-Weeks Recovery Phase
07-0655	A 6-Week Repeat Dose Intravenous Toxicity Study of Thio mab MC MMAF, Thio Tmab vc MMAE, Thio Tmab mpeo DM1, and Tmab-MCCDM1 in Cynomolgus Monkeys
04-0669-1459	Safety Evaluation of Multiple-Dose Intravenous Herceptin-SMCC-DM1 Immuno-conjugates in Female Sprague-Dawley Rats

Study number	Title of study
09-2654	N2'-deacetyl-N2'-(3-mercapto-1-oxopropyl) maytansine (DM1): Salmonella—Escherichia coli/Mammalian-Microsome Reverse Mutation Assay with a Confirmatory Assay
09-2726	N2'-deacetyl-N2'-(3-mercapto-1-oxopropyl) maytansine (DM1): In Vivo Rat Bone Marrow Micronucleus Assay
07-0653	An Intravenous Chronic Toxicity Study of Trastuzumab-MCC-DM1 (PRO132365) Administered to Cynomolgus Monkeys Once Every 3 Weeks for 8 Doses, with a 6-Weeks Recovery Phase

Study number	Title of study
04-1031-1605	A Single Dose Intravenous Cardiovascular Safety Pharmacology Study of Trastuzumab-MCC-DM1 (PRO132365) Administered to Female Cynomolgus Monkeys with a 3 Week Recovery Period
04-1215-1605	Cross-Reactivity of Trastuzumab-MCC-DM1 (PRO132365) with Human and Cynomolgus Monkey Tissues In Vitro
04-1257-1459	Hemolytic Potential and Blood Compatibility of Trastuzumab-MCC-DM1 (PRO132365)
05-0617-1459	Safety Evaluation of Single-Dose Intravenous Herceptin-MCC-DM1 Immunoconjugates in Female Sprague-Dawley Rats
05-0775-1459	Safety Evaluation of Single-Dose Intravenous MCC-DM1 in Female Sprague-Dawley
05-0848	Single Dose Intravenous Toxicity Study of Trastuzumab-MCC-DM1 (PRO132365) with 5-7% Unconjugated Maytansinoid in Cynomolgus Monkeys with a 3-Week Recovery Period
09-0234	Effects of N2'-deacetyl-N2'-(3-mercapto-1-oxoproply) maytansine (DM1) on Cloned hERG Potassium Channels Expressed in Human Embryonic Kidney Cells

Conclusion

Chapter Summary

The ICH S9 guideline provides critical recommendations for designing nonclinical studies to support the development of anticancer pharmaceuticals, particularly for patients with advanced disease and limited treatment options. Rooted in the principles of the 3Rs (reduce, refine, and replace), the guidance aims to accelerate oncology drug development while minimizing unnecessary risks to patients and reducing the use of animals. As mentioned earlier, the recent regulatory policy shift away from traditional animal testing is going to have its own impact on ADCs and very likely biosimilar drug conjugates (BDCs) as well. Biosimilar drug conjugates are also complex biologics that combine a targeting antibody with a cytotoxic payload. While the FDA is moving toward case-by-case evaluations in the context of ADCs, some animal testing may still be required for assessing immunogenicity, establishing dose-toxicity boundaries of the conjugated payload, and confirming pharmacokinetics and biodistribution of novel biosimilars with linker-payload combinations. Sponsors should anticipate an expectation to justify the use of aforementioned NAMs and demonstrate that they provide equivalent or superior predictive value compared to animal models in the context of ADCs and BDCs.

While this framework supports many therapies, ADCs present unique challenges. Current regulations often group ADCs with either biologics or small-molecule drugs, failing to account for their dual nature. ADCs combine the targeted precision of monoclonal antibodies with the potent cytotoxicity of chemotherapy agents, making them a hybrid therapy that doesn't fit neatly into existing categories.

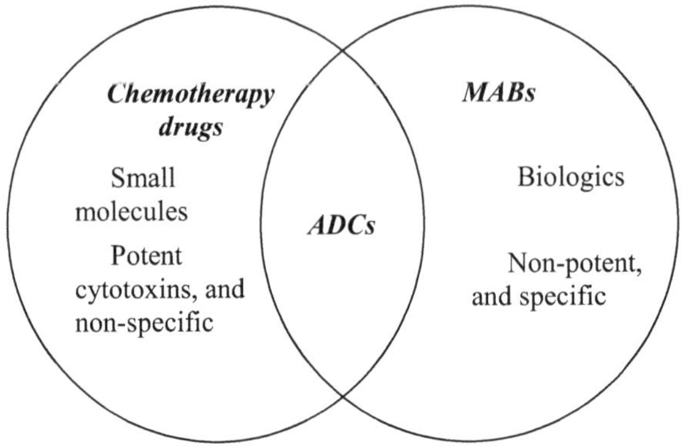

Fig. 1 The anatomy of antibody drug conjugates

As shown in Fig. 1, ADCs are composed of large biologic molecules (monoclonal antibodies) linked to synthetic small-molecule cytotoxins. This structural fusion allows ADCs to deliver highly potent chemotherapy directly to cancer cells while sparing healthy tissue. Both monoclonal antibodies and small-molecule drugs have distinct mechanisms of action and regulatory pathways, and when combined into a single therapy, ADCs inherit complexities from both. The broader anticancer drug landscape includes monoclonal antibodies in targeted therapies such as imatinib (Gleevec) and palbociclib (Ibrance), and immunotherapies like nivolumab (Opdivo) and pembrolizumab (Keytruda), which enable the immune system to identify and destroy cancer cells. Each of these therapies is governed by its regulatory framework. ADCs, however, represent a hybrid model that integrates elements from all of these modalities, resulting in both unique advantages and distinct challenges.

Due to their complexity, a dedicated regulatory framework for ADCs is essential. Such a framework would acknowledge their hybrid nature and offer clearer guidance to developers, reducing confusion and inefficiencies. Tailored regulations for ADCs will not only streamline development but also ensure better safety, efficacy, and innovation in cancer therapeutics.

References

A Study of Trastuzumab Emtansine (T-DM1) Sequentially With Anthracycline-based Chemotherapy, as Adjuvant or Neoadjuvant Therapy for Patients With Early Stage Herceptin (HER)2-positive Breast Cancer. ClinicalTrials.gov: Hoffmann-La Roche; (2014)

A Study of Trastuzumab Emtansine (Trastuzumab-MCC-DM1, T-DM1) in Combination With Pertuzumab Administered to Patients With Human Epidermal Growth Factor Receptor-2 (HER2)-Positive Locally Advanced or Metastatic Breast Cancer Who Have Previously Received Trastuzumab. In: Clinical Trials H-LR, (ed.) National Library of Medicine: Hoffmann-La Roche; (2013)

A Study of Trastuzumab Emtansine in Participants With Human Epidermal Growth Factor Receptor (HER)2 Immunohistochemistry (IHC)-Positive, Locally Advanced or Metastatic Non-Small Cell Lung Cancer (NSCLC). In: Clinical Trials H-LR, (ed.) eighth July 2019 ed. National Library of Medicine: Hoffmann-La Roche, (2019)

A Study of Trastuzumab-MCC-DM1 Administered Intravenously to Patients With HER2-Positive Metastatic Breast Cancer Who Have Previously Received a Trastuzumab-Containing Regimen. In: Clinical Trials Posting Group G, Inc., (ed.) National Library of Medicine: Genentech, Inc., (2015)

Detection of Toxicity to Reproduction for Medicinal Products & Toxicity to Male Fertility S5(R2). In: Group IEW, (ed.) Step 4 version ed. European Medicines Agency: International conference on harmonisation of technical requirements for registration of pharmaceuticals for human use (ICH), p. 20–1 (1994)

Guidance on nonclinical safety studies for the conduct of human clinical trials and marketing authorization for pharmaceuticals M3(R2). In: Group IEW, (ed.) Step 4 version ed. ICH: International conference on harmonisation of technical requirements for registration of pharmaceuticals for human use (ICH), p. 20–1 (2009)

Guideline on the need for carcinogenicity studies of pharmaceuticals S1A. In: Group IEW, (ed.) Step 4 version ed. ich.org: International conference on harmonisation of technical requirements for registration of pharmaceuticals for human use (ICH), p. 2 (1995)

Impurities in New Drug Substances Q3A (R2). In: Group IEW, (ed.) Step 4 version ed. ich.org: International conference on harmonisation of technical requirements for registration of pharmaceuticals for human use (ICH), p. 5–6 (2006)

Impurities in New Drug Substances Q3B (R2). In: Group IEW, (ed.) Step 4 version ed. ich.org: International conference on harmonisation of technical requirements for registration of pharmaceuticals for human use (ICH), p. 4–6 (2006)

Leighton JK. Pharmacology review(S). In: Services HaH, (ed.) February 1, 2013 ed. accessdata.fda.gov: U.S. Food and Drug Administration, p. 10–4,143 (2013)

Nonclinical evaluation for anticancer pharmaceuticals (S9). In: Group IEW, (ed.) Current Step 5 version ed. European Medicines Agency: International Council for Harmonisation of Technical Requirements for Pharmaceuticals for Human Use, p. 7–9 (2013)

Olson, M.K.: Regulation of safety, efficacy, and quality. In: Culyer, A.J. (ed.) Encyclopedia of Health Economics, 1st edn, pp. 240–248. Elsevier (2014)

Postmarketing Requirements and Commitments: Introduction. 12th January 2016 ed. fda.gov: U.S. Food and Drug Administration; (2016)

Preclinical safety evaluation of biotechnology-derived pharmaceuticals S6(R1). In: CHMP, editor. Step 5 ed. European Medicines Agency: International Council for Harmonisation of Technical Requirements for Pharmaceuticals for Human Use, p. 10–13 (2011)

Safety Pharmacology Studies for Human Pharmaceuticals S7A. In: Group IEW, (ed.) Step 5 version ed. European Medicine Agency: International conference on harmonisation of technical requirements for registration of pharmaceuticals for human use (ICH), p. 5–6 (2001)

Conjugation Chemistry and Chemical Design for the Discovery and Development of Antibody Drug Conjugates

Amir Assadieskandar and Golnaz Kamalinia

Abstract Antibody drug conjugation is a promising approach to improve the clinical outcomes of cancer therapy. Antibody Drug Conjugates (ADCs) allow for targeted delivery of the cytotoxic agents to the tumor cells and can diminish dose limiting toxicity that is mostly associated with unwanted effects of cancer therapeutics on normal cells. Conjugation of monoclonal antibodies (mAbs) with small cytotoxic molecules can improve the pharmacokinetic profile of the small molecule and minimize its off-target toxicity, and as a result improve therapeutic outcomes. Antibody, payload and the linker form the three major components of an ADC, and each have been extensively studied to improve the ADCs properties. Conjugation chemistry strategy is a major aspect when designing ADCs, and can have a profound impact on the potency and safety of the developed agent. Linker technology and choice of payload are other important critical attributes. The current book chapter covers an extensive evaluation of chemistry adapted for conjugation practice and various linkers that are commonly used for ADC development, in addition to a brief discussion about the payloads that have been subjected to antibody drug conjugation.

Keywords Antibody drug conjugation · Bioconjugation · Conjugation chemistry · Linker · Payload

A. Assadieskandar
School of Pharmacy, University of Southern California, 1985 Zonal Avenue, Los Angeles, California, USA

G. Kamalinia (✉)
Department of Chemistry, University of Southern California, University Park, Los Angeles, California, USA

H. J. Gukasyan, A. Kulkarni (eds.), *Antibody Drug Conjugates*, AAPS Advances in the Pharmaceutical Sciences Series 52, https://doi.org/10.1007/978-3-031-89319-3_4

Abbreviations

ADC	Antibody drug conjugate
DAIP	Dispase autolysis-inducing protein
DAR	Drug to Antibody Ratio
DBCO	Aza-dibenzocyclooctyne
FGE	Formylglycine-Generating Enzyme
IgG	Immunoglobulin G
mAbs	Monoclonal antibodies
mc	Maleimidocaproyl
mcc	Maleimidomethyl cyclohexane carboxylate
MMAE	Monomethyl auristatin E
MMAF	Monomethyl auristatin F
NHS	N-hydroxy succinimide
PAB	p-aminobenzyl alcohol
PABC	p-aminobenzyloxycarbonyl
PBDs	Pyrrolobenzodiazepines
PNA	Peptide nucleic acids
PPTase	Phosphopantetheinyl Transferase
PTAD	Phenyltriazolinediones
ReACT	Redox-activated chemical tagging technology
SMCC	Succinimidyl-4-(N-maleimidomethyl) cyclohexane-1-carboxylate
SMPP	N-succinimidyl 4-methyl-4-(2-pyridyldithio)pentanoate
SPDB	N-succinimidyl 4-(2-pyridyldithio)butanoate
SPDP	N-succinimidyl 3-(2-pyridyldithio)propanoate
SPI_P	Streptomyces papain inhibitor
SPP	N-succinimidyl 4-(2-pyridyldithio)pentanoate

Introduction

Antibody Drug Conjugates (ADCs) combine the attractive properties of monoclonal antibodies (mAbs) such as their selective specificity, targetability and favorable pharmacokinetic profile together with the extraordinary therapeutic potency of unique small molecule agents to introduce a new class of pharmaceuticals having desirable safety and efficacy profile. Conjugation of mAbs with highly potent cytotoxic molecules can improve the pharmacokinetic profile of the small molecule and minimize its off-target toxicity, and as a result improve overall therapeutic outcomes. Antibody, payload, and linkers as the three major components of an ADC have been extensively studied, refined and progressed in recent years (Fig. 1) (McCombs and Owen 2015).

Initial efforts using conventional anticancer agents, such as doxorubicin, and their conjugation with mAbs proved to be unsuccessful in clinical trials for two

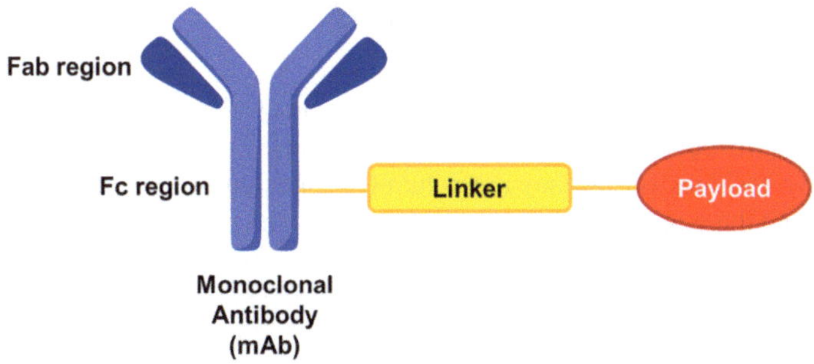

Fig. 1 General structure of Antibody Drug Conjugates (ADCs). Monoclonal antibody (mAb), payload and the linker form the three major components of an ADC

main reasons. First, the number of small molecules that could be conjugated on the antibody without affecting its binding to the relevant antigen was limited, and second the number of antigens on the target tissue were restricted and therefore sufficient cytotoxic drug accumulation in the target tissue was not achieved (McCombs and Owen 2015). By improving the linker chemistry and by using highly potent payloads some of these problems were solved and successful ADCs such as trastuzumab emtansine (Kadcyla®) (Phillips et al. 2008; Amiri-Kordestani et al. 2014) and brentuximab vedotin (Adcetris®) (Senter and Sievers 2012) were introduced.

Herein, conjugation chemistry, linkers, and different approaches used for conjugation of highly potent payloads to antibody scaffolds in ADC constructs are discussed. Conventional conjugation practices, as well as more recent conjugation approaches such as enzymatic ligation, click chemistry-based conjugation, and sequence specific conjugation will be discussed. Introduced in the linker selection, various linker properties, i.e. cleavable and non-cleavable linkers, are introduced. Finally, different payloads that have been successfully utilized in the context of ADCs are introduced.

Conjugation Chemistry

Where to Conjugate the Payload?

One of the most commonly used antibodies for drug conjugation is immunoglobulin G (IgG) with a molecular weight of 150 kDa. IgG has several native residues that can be utilized for the purpose of conjugation and further reactive sites such as aldehydes and ketones that serve as electrophilic handles can be introduced in its structure for conjugation practice (McCombs and Owen 2015). To minimize the interference of the conjugation on the targetability of mAbs, the conjugation usually takes place in the Fc region of the antibody (Ducry and Stump 2009).

Conjugation Through Native Residues

Some native amino acid residues such as lysine and cysteine are inherently reactive and can be utilized for conjugating various payloads. The main advantage of using such native residues is that they are naturally expressed in the antibody's primary structure and there is no need to genetically modify the protein to incorporate them. This facile expression is an important reason for their wide application for conjugation practice. In fact, FDA many approved ADCs such as Kadcyla® (Phillips et al. 2008; Amiri-Kordestani et al. 2014) and Adcetris® (Senter and Sievers 2012) are conjugated using their native residues. The problem associated with these residues however is that a wide range of variable and heterogenous products can be formed when using them as the conjugation point and such product heterogeneity may cause variable pharmacokinetic profile (McCombs and Owen 2015). Furthermore, conjugation on native residues can alter the electrostatic properties of the scaffold and result in a change in their isoelectric point. It can also affect the hydrophilicity/hydrophobicity of the residues and as a result affect the stability and pharmacokinetic profile of ADCs (McCombs and Owen 2015).

Lysine Based Coupling

Lysine has been widely used for the purpose of antibody drug conjugation due to the presence of a primary amino group in its structure. The primary amine moiety in lysine can react with a variety of reactive groups including different activated esters (Sakamoto and Hamachi 2018; Tsuchikama and An 2018). The most commonly used approach for lysine conjugation is to perform amide coupling between the primary amine of lysine and an active ester of the payload. This is usually achieved using succinimide reagents such as N-hydroxy succinimide (NHS) and sulfo-NHS where sulfo-NHS activated esters are associated with better water solubility (Fig. 1a) (Tsai et al. 2001; Lewis et al. 2001; Chen et al. 2015; Deguchi et al. 1986). Another approach is to use imidate esters that can react with lysine's amine to form amidine bonds (Fig. 2b). The amidine bond is protonated at physiological pH and therefore does not affect the native charges of the mAbs. Imidate esters are more resistant to hydrolysis than many other active esters such as NHS esters but are associated with a slower reaction rate and longer reaction times (Wilbur 1992; Batmanghelich et al. 1992). Traut's reagent or 2-iminothiolane is a cyclic thioimidate that can also form amidine bonds and is used to introduce thiol groups on the protein through its lysine residues where the introduced thiol groups can undergo thiol-based conjugation reactions (Fig. 2c). This is especially helpful if the protein of interest does not contain native thiols (Fig. 2c) (Ducry and Stump 2009; Traut et al. 1973; Xu et al. 2007; Mandler et al. 2000). Isothiocyanate mediated conjugation is another potential approach that can be used for antibody drug conjugation and has been widely used for fluorescence labeling of proteins results in the formation of thiourea linkages (Fig. 2d) (Cockrell et al. 2015; Zeng et al. 2018; Rath et al. 2017).

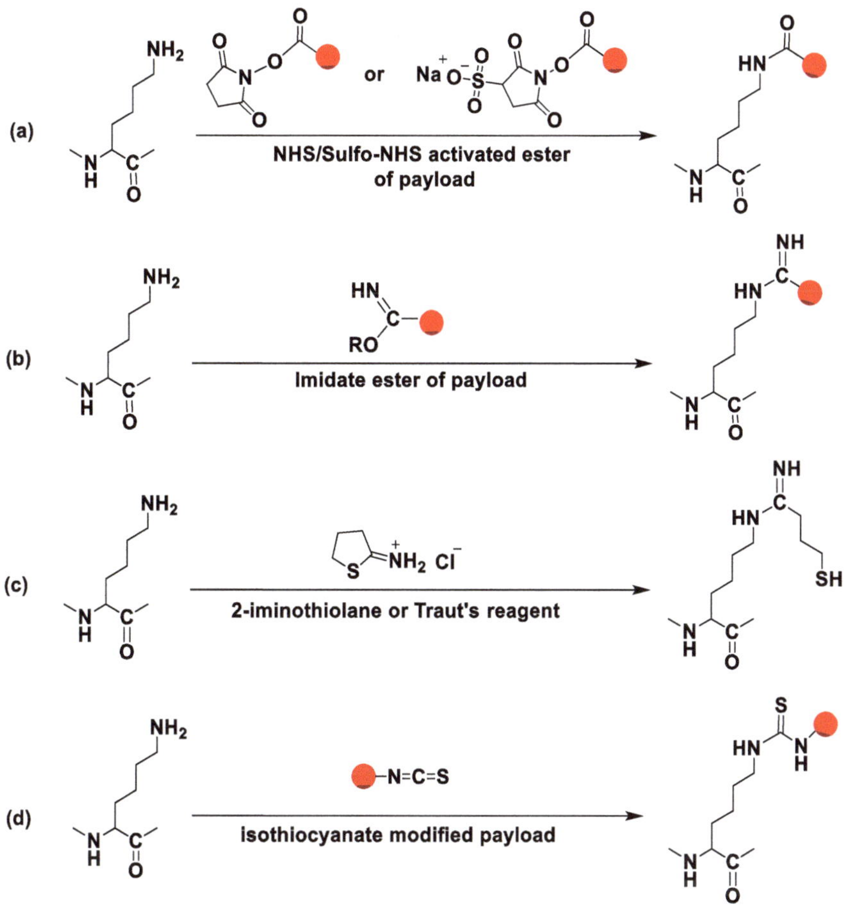

Fig. 2 Lysine conjugation chemistry. The primary amine moiety in lysine can react with a variety of reactive groups including (**a**) N-hydroxy succinimide (NHS) and sulfo-NHS activated ester payloads, (**b**) Imidate ester payloads (**c**) 2-iminothiolane or Traut's reagent and (**d**) isothiocyanate modified payload. The payload is showed as a red circle

Lysine is a very abundant residue and is highly present on the protein surface. For this reason it is particularly helpful for non-selective protein modification both in regard with the number and location of conjugation (Sakamoto and Hamachi 2018). For example, more than 80 lysine residues exist in IgG scaffold, from which more than 20 are in highly solvent accessible regions of the protein. Therefore, using lysine for conjugating the payload will result in conjugation in varying sites and provides a mixture of heterogenous products with variable Drug to Antibody Ratio (DAR) (McCombs and Owen 2015). Kadcyla® is an FDA approved lysine coupled ADC that is consisted of trastuzumab molecules covalently conjugated to a microtubule inhibitor DM1 through their lysine residues via a non-reducible thio-ether linker with an average molar ratio of ~3.5 (Fig. 3) (Kim et al. 2014). The

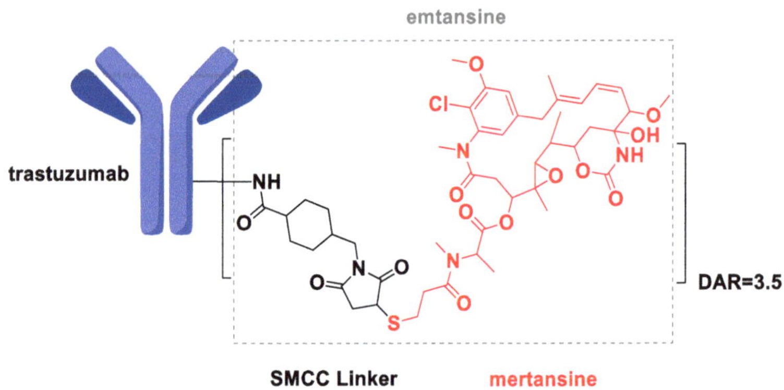

Fig. 3 Schematic structure of trastuzumab emtansine (Kadcyla®). The lysine residues on trastuzumab were modified with SMCC (succinimidyl-4-(N-maleimidomethyl) cyclohexane-1-carboxylate) linker and conjugated to the thiol group on mertansine with a drug to antibody ratio (DAR) of 3.5

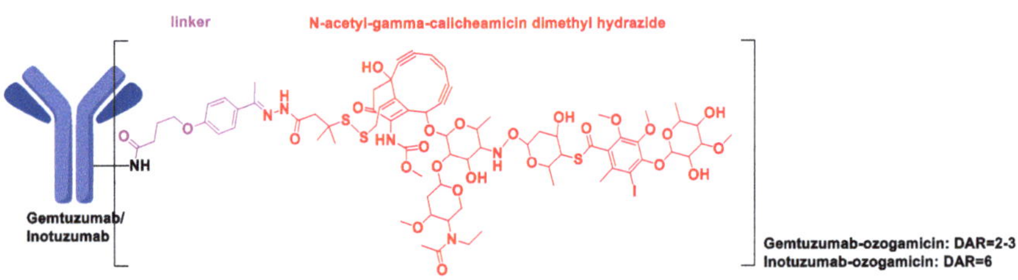

Fig. 4 Schematic structure of gemtuzumab-ozogamicin (Mylotarg®) and inotuzumab-ozogamicin (Besponsa®); In these molecules the antibody is attached to N-acetyl-gamma-calicheamicin dimethyl hydrazide through a bifunctional linker 4-(4′-acetylphenoxy) butanoic acid

resulting conjugate gives a favorable toxicity profile *in vivo* and *in vitro* and presents a potent activity selectively in HER-2 overexpressing tumor cells. This ADC is internalized by HER-2 overexpressing cells and is released inside cells due to intracellular proteolytic digestion (Phillips et al. 2008; Amiri-Kordestani et al. 2014).

Gemtuzumab ozogamicin (Mylotarg®) (Bross et al. 2001) and inotuzumab-ozogamicin (Besponsa®) (DiJoseph et al. 2006) are other FDA approved ADCs in which lysine residues are used for linking the payload to the antibody (Fig. 4). In Mylotarg®, ozogamicin, a calicheamicin derivative was conjugated to an anti-CD33 mAb through its lysine residues with variable DAR values with an average of four to six and was indicated for acute myeloid leukemia. In this ADC about half of the antibody is unconjugated. Unfortunately, according to the phase 3 clinical trials, Mylotarg® was withdrew from the market in 2010 (Senter and Sievers 2012; Bross et al. 2001) however its approval was revisited by FDA in 2017 and was approved for treating patients with acute myeloid leukemia with CD33 positive tumors (de la Torre and Albericio 2018).

Cysteine Based Coupling

Unlike lysine, cysteine is less abundant and may pose a better native residue for conjugation practice with an occurrence of less than 2% in proteins. Cysteine carries a highly reactive thiol group that shows robust nucleophilic activity (Sakamoto and Hamachi 2018). In IgG, 16 pairs of cysteines exist that 12 of them are involved in the formation of intra-chain disulfide bonds and 4 of them in inter-chain disulfide bond formation. The proteins that have cysteine groups that are engaged in disulfide bond formation can be reduced to form free thiol groups through the action of reagents such as β-mercaptoethanol and dithiothreitol (Wilbur 1992). The cysteines that are responsible for inter-chain disulfide bond formation are more of interest for the conjugation practice (McCombs and Owen 2015) as they are not extensively critical for IgG stability and more solvent accessible (Tsuchikama and An 2018). The reducing condition is of great importance here and needs to be carefully selected to prevent any negative impacts on the structural integrity of the antibody (McCombs and Owen 2015). Also, as mentioned earlier, lysine conjugation with Traut's reagent can be used to introduce thiols into a protein that do not contain native thiol groups (Traut et al. 1973).

Several electrophiles such as Michael acceptors (including maleimides), and α-haloketones can alkylate the thiol functional groups (Vinogradova et al. 2015; Kim et al. 2013). Maleimides react with thiol groups to form thioether bonds (Fig. 5a) (Wilbur 1992; Christie et al. 2015). Adcetris® that has received FDA approval for relapsed Hodgkin lymphoma and anaplastic large cell lymphoma is an ADC in which a maleimide functional group was used to conjugate the spacer to the mAb (Fig. 6) (Senter and Sievers 2012). In this ADC the cytotoxic agent is conjugated to the antibody through partial substitution of cysteines involved in interchain disulfide bond formation (Senter and Sievers 2012). α-haloketone groups (such as

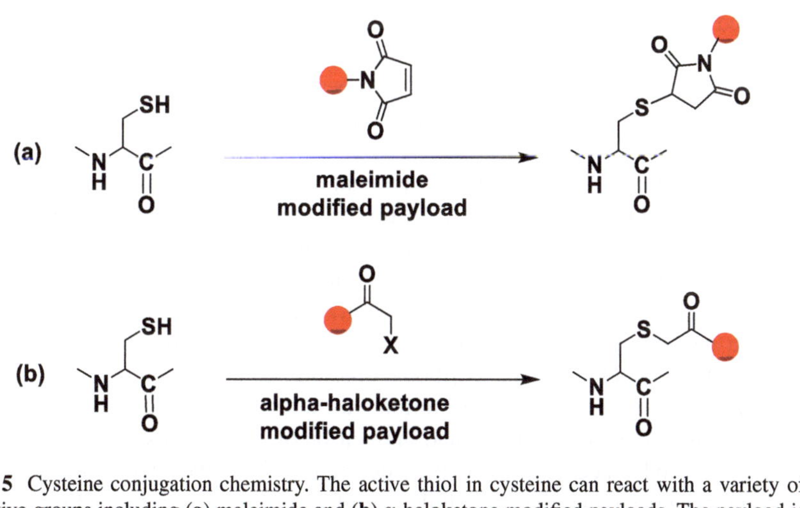

Fig. 5 Cysteine conjugation chemistry. The active thiol in cysteine can react with a variety of reactive groups including (**a**) maleimide and (**b**) α-haloketone modified payloads. The payload is showed as a red circle

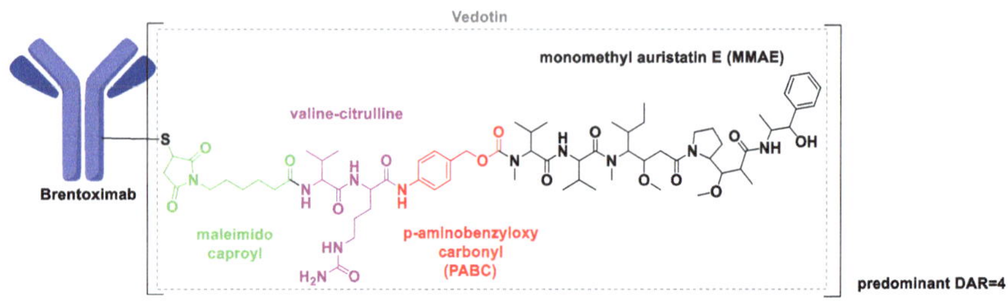

Fig. 6 Schematic structure of brentuximab-vedotin (Adcetris®). In this molecule the antibody is attached to monomethyl auristatin E (MMAE) through a linker consisting of maleimidocaproyl, valine-citrulline and p-amino benzyloxy carbonyl (PABC)

chloroacetyl, bromoacetyl and iodoacetyl) can also react with nucleophilic thiol groups of cysteines to form thioether linkages (Fig. 5b) (Wilbur 1992).

To evaluate the effect of DAR and drug loading on the therapeutic potential of the developed ADCs, Hamblett et al. evaluated monomethyl auristatin E (MMAE) conjugated anti-CD30 with different DARs. Various conjugated species with DAR values of 2, 4 and 8 were prepared and separated using hydrophobic interaction chromatography. When reacting all cysteines involved in interchain disulfide bond formation, a highly uniform ADC with a DAR of 8 was obtained and by controlling the concentration of the reducing agent, other conjugated species with lower DAR values were achieved. Although ADCs potency was directly correlated with their loading in cellular assays, *in vivo* studies showed that ADCs with a DAR of 8 showed similar therapeutic activities as ADCs with a DAR of 4 with the same mAb doses. ADCs with a DAR of 2 also showed therapeutic activity but required a higher mAb dose. ADCs with a DAR of 8 are more rapidly cleared from blood circulation than those with a DAR of 4 and for this reason ADCs with a DAR of 4 have a longer tumor exposure time that compensates for their reduced potency. The maximum tolerated dose was inversely correlated with DAR values where ADCs with a DAR of 8 showed the smallest maximum tolerated dose. It is therefore postulated that decreasing the DAR will increase the therapeutic index and highly loaded mAbs may not be the best choice with regards to potency and safety (Senter and Sievers 2012; Hamblett et al. 2004).

One of the main limitations of conventional cysteine conjugation strategies is the heterogeneity of the products where the product consists of a mixture of conjugated molecules with different DAR and different conjugation sites. Such species may show different pharmacokinetic profiles and may be different in respect with their safety and efficacy. To overcome this problem, Junutula et al. developed THIOMABs using a genetic engineering technology in which two additional cysteines were genetically incorporated in the antibody structure, on heavy and light chains. The suitable position for cysteine substitution was determined by phage display to not perturb immunoglobulin folding and antigen binding. These extra cysteines were utilized for payload conjugation to provide chemically defined conjugated products without modifying interchain disulfide bond formation. When conjugating MMAE to anti-MUC16 antibody using this method a nearly homogenous ADC showed up with more than 90% of the species showing a DAR of 2 (Junutula et al. 2008). In another attempt a maytansinoid was conjugated to genetically modified trastuzumab and similar results were observed where a DAR of 2 was achieved and the payload was selectively conjugated to the genetically engineered cysteines (Junutula et al. 2010).

McDonagh et al. used a different genetic engineering approach for decreasing product heterogeneity and to provide chemically defined ADCs. In their approach they simply mutated the unneeded solvent accessible cysteine residues to serine and reduced the number of possible reactive positions from eight cysteines to two or four cysteines. They were able to show that the binding affinity of the engineered antibodies remained intact following these mutations and the antitumor activity, pharmacokinetic and safety profile of the ADC with DAR 4 was comparable to the parental ADC (McDonagh et al. 2006).

One of the main problems associated with cysteine conjugation through maleimide coupling is the instability of ADC *in vivo*. The maleimide containing conjugates can have two different fates: (1) they can undergo a retro-Michael reaction that results in the loss of drug-linker component from the ADC. This can result in premature drug release before reaching to the target tissue (2). They can undergo hydrolysis on their thiosuccinimide ring which causes ring opening. The hydrolyzed product formed during ring opening is now stable towards drug-linker elimination (Lyon et al. 2014; Tumey et al. 2014). Lyon et al. designed an approach for improving the stability of ADCs towards the loss of conjugated drug by developing self-hydrolyzing maleimide derivatives. For this purpose, they incorporated a basic group, diaminopropionic acid, adjacent to the maleimide ring where the basic group promotes the thiosuccinimide ring opening and acts as an intramolecular catalyst. The hydrolyzed thiosuccinimide ring is resistant towards drug-linker elimination and in this way non-specific loss of drug is prevented (Lyon et al. 2014).

The traditional maleimide based coupling on reduced interchain cysteine residues only results in monoalkylation and leaves the interchain disulfide bonds unbridged, This can potentially cause physical and structural instability to the protein of interest (Bryant et al. 2015). To overcome this problem, cysteine rebridging strategy was introduced. Badescu et al. utilized a bis-sulfone reagent to conjugate MMAE to trastuzumab. Such bis-sulfone linkers not only incorporate the payload but also maintain the interchain disulfide bonds between a cysteine pair. In this case the bis-sulfone linker is incorporated between two reduced cysteines and result in the formation of conjugated antibody in which the two cysteines are linked through a three-carbon atom bridge that is covalently attached to the payload. The three-carbon atom bridge is flexible and small enough to not perturb the tertiary structure of the antibody. The conjugated antibody provided using this linker was more stable than maleimide conjugated antibodies in serum and in the presence of high concentrations of albumin. The authors further conjugated a Fab fragment derived from trastuzumab with MMAE using such bis-alkylation linkers and produced homogenous ADCs with a DAR of 1 (Bryant et al. 2015; Badescu et al. 2014). Other reagents such as dibromomaleimide linkers (Behrens et al. 2015; Bryden et al.

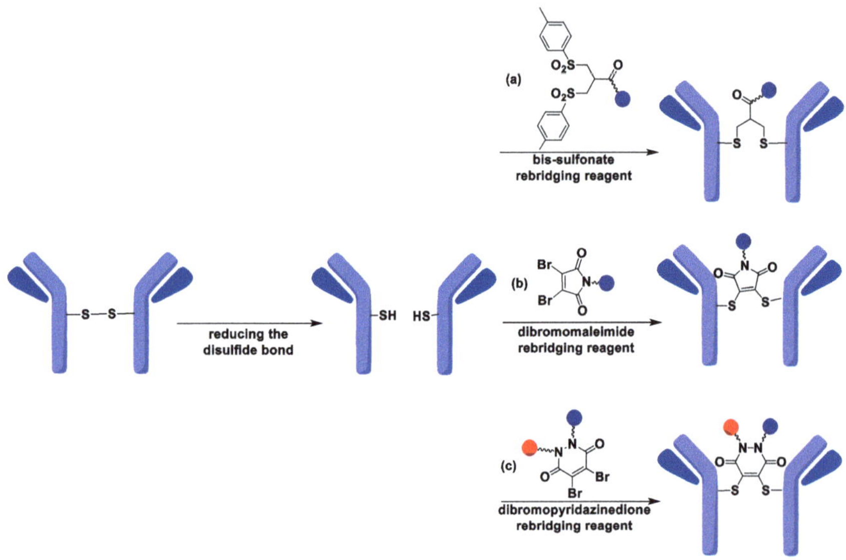

Fig. 7 Cysteine rebridging strategy using (**a**) bis-sulfonate rebridging reagent, (**b**) dibromomaleimide rebridging reagent and (**c**) dibromopyridazinedione rebridging reagent. The payload is showed as either a blue or red circle

2014; Schumacher et al. 2013), and dibromopyridazinedione linkers (Maruani et al. 2015) have also been utilized for payload incorporation along with cysteine rebridging (Fig. 7).

Vinogradova et al. introduced cysteine conjugation using aryl-palladium complexes that resulted in highly selective linker free cysteine conjugation. Aryl-palladium complex is prepared by mixing aryl-halides with palladium phosphine complexes and can cause thiol arylation on reduced cysteine residues. Using this approach, partially reduced trastuzumab was conjugated through its cysteine residues to a kinase inhibitor called vandetanib resulting in ADCs with a DAR of 4.4 where the resulting conjugated antibody was stable towards external thiols, acids, bases and oxidants and retained its binding. The main limitation of this approach is the risk of palladium toxicity which needs vigorous attention to make sure that any trace metal remnants has been removed in the final product (Vinogradova et al. 2015).

A promising strategy for site selective cysteine conjugation is to tune cysteine reactivity by modifying its local environment or the other residues surrounding it. In an interesting finding, Pentelute group found a four amino acid tag called "π-clamp' consisting of phe-cys-pro-phe that tunes cysteine's thiol reactivity towards perfluoroaromatic reagents. In this tag, the phenylalanine residues activate cysteine thiol to react with the perfluoroaryl group and proline helps cysteine and phenylalanine to position. π-clamp showed its reactivity regardless of its position in the antibody where it was efficiently conjugated at the N-terminus, C-terminus and the middle of a polypeptide chain. Incorporation of the π-clamp in the C-terminus of trastuzumab allowed for site selective cysteine conjugation with either biotin or monomethyl auristatin F (MMAF) linked to a perfluoroaryl group without affecting the endogenous cysteines of the antibody. The antibodies modified in this manner retained their binding affinity (Zhang et al. 2016). In an attempt to find other reactive sequences that undergo perfluorarylation in water, they later designed a doped library based on the π-clamp sequence to perform selection using mRNA display strategy and identified another 29-mer tag that undergoes chemo and regio-selective arylation on its cysteine residue (Evans and Pentelute 2017). In another effort, a seven-residue tag called "DBCO tag" was identified consisting of Leu-Cys-Tyr-Pro-Trp-Val-Tyr that can tune the reactivity of cysteine towards aza-dibenzocyclooctyne (DBCO) reagents. DBCO tag was fused to the C-terminus of trastuzumab heavy chains and was used for protein labeling through site selective cysteine thiol-yne reaction (Zhang et al. 2018).

Coupling Through Other Canonical Amino Acids

Conjugation and site selective protein modification on other canonical less abundant residues such as tyrosine and tryptophan can be helpful for producing uniform and homogenous products. Tyrosine has a unique reactivity due to the presence of an acidic proton on its phenol ring. Although tyrosine is still commonly found in proteins primary structure, surface accessible tyrosines are less abundant and provide a desirable point for bioconjugation. A click-like tyrosine reaction using phenyltriazolinediones (PTAD) derivatives has been used for conjugating aplaviroc (an HIV entry inhibitor) to trastuzumab and the resulting ADC exhibited potent HIV neutralizing activity comparable to that of unconjugated aplaviroc (Fig. 8a) (Ban et al. 2013).

Tryptophan is especially interesting in this context as it is the least abundant residue with about 1% natural occurrence in proteins. About 90% of native proteins have tryptophan in their primary structure which makes it a facile site for conjugation. Using metal catalysts for tryptophan conjugation is mainly limited by unfavorable biochemical conditions such as higher temperatures and organic solvents (Sakamoto and Hamachi 2018). Seki et al. reported a metal free tryptophan modification approach in which 9-azabicyclo [3.3.1] nonane-3-one-N-oxy in combination with $NaNO_2$ and acetic acid for tryptophan bioconjugation (Fig. 8b). This reaction did not cause any cross reactivity with other amino acid residues such as tyrosine and lysine and did not require the use of organic co-solvents. The tryptophan selective conjugation method was successfully applied for conjugating an anti-$A\beta_{1-16}$ antibody with fluorescein and showed that conjugation has taken place in both heavy and light chains while the functionality of the antibody has remained unaffected (Seki et al. 2016).

Methionine is another less abundant residue that is rarely present on the surface of a protein due to its hydrophobic nature and is the second least abundant residue after tryptophan. Like cysteine, methionine is also a sulfur containing amino acid but is associated with a weaker nucleophilicity. Successful methionine conjugation for site selective protein modification has been reported by Lin et al. where they

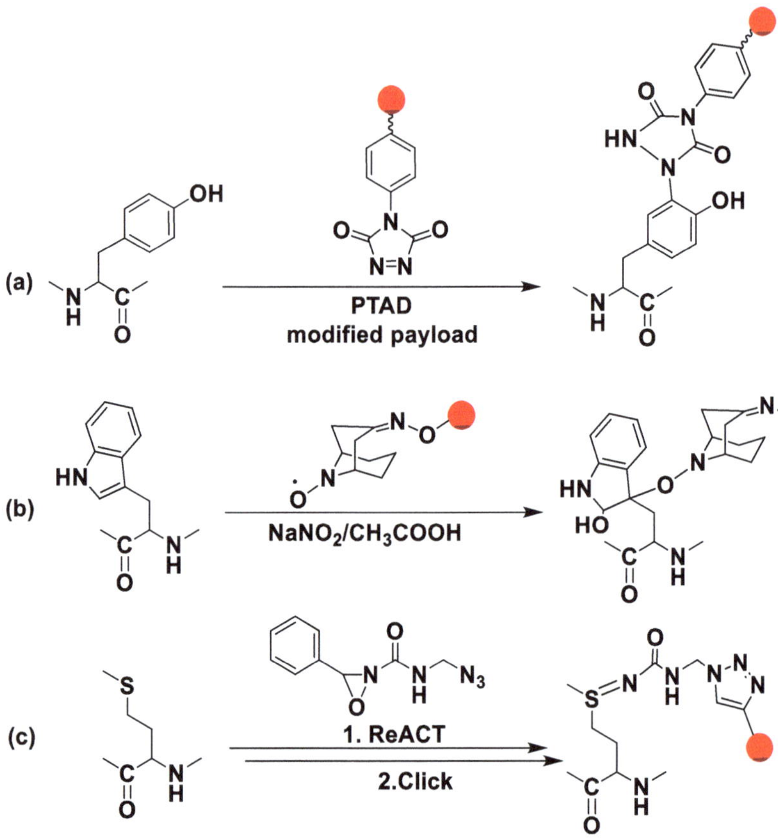

Fig. 8 Conjugation methods for tyrosine, tryptophan and methionine. (**a**) Tyrosine conjugation using phenyltriazolinediones (PTAD) derivatives of the payload (**b**) Tryptophan conjugation using a payload modified with 9-azabicyclo [3.3.1] nonane-3-one-N-oxy derivatives (**c**) Methionine conjugation using azide or alkyne modified oxaziridine followed by click reaction. The payload is showed as a red circle

developed a redox-activated chemical tagging technology (ReACT) in which methionine specific labeling was achieved using oxaziridine under biocompatible reaction conditions and the approach was successfully applied for antibody drug conjugation. For this study, a Fab antibody fragment to green fluorescent protein was used that carries one methionine residue in its light chain and two methionine residues in its heavy chain. They observed that no bioconjugation is achieved using ReACT strategy even at high oxaziridine concentrations. The reason is that these native methionine residues are buried in the hydrophobic core of the Fab and are not solvent accessible. In the next step, extra methionine residues were genetically engineered in this Fab and were modified using ReACT strategy with an azide group that can be used for further functionalization through click chemistry (Fig. 8c). In a similar approach, trastuzumab Fab was genetically engineered to incorporate one or two surface accessible methionine residues that were conjugated to MMAE to provide a uniform ADC with a defined DAR while native methionine residues of trastuzumab Fab remained intact (Lin et al. 2017).

Site-Specific Conjugation Through Genetically Engineered Unnatural Residues

Introducing new building blocks to protein sequence provides a new platform for protein modification and antibody drug conjugation (Wang and Schultz 2005; Axup et al. 2012). Unnatural residues that carry bioorthogonal reaction handles are incorporated into proteins through genetic codon expansion. Bioorthogonal reactive groups such as alkyne, azide, alkene, tetrazine and ketone can be introduced to protein structure in this manner (Sakamoto and Hamachi 2018). Genetic code is consisted of 64 codons where 61 code for 20 canonical amino acids and three specify stop codons to terminate translation. To incorporate unnatural amino acids, new tRNA and aminoacyl tRNA synthetase are expressed in live cells to charge a certain unnatural residue on the tRNA. The unnatural amino acid charged tRNA incorporates the unnatural amino acid in response to a unique codon such as amber stop codon UAG (Wang 2017).

By genetic codon expansion, unnatural amino acid p-acetylphenylalanine was incorporated into the trastuzumab full-length IgG and Fab fragment in one or two sites, respectively and was utilized to conjugate the antibody to an auristatin derivative. p-acetylphenylalanine can selectively react with a terminal alkoxyamine through its keto group which results in the formation of a stable oxime bond while allowing for meticulous control of the site of conjugation as well as its stoichiometry. Incorporated p-acetylphenylalanine was conjugated to alkoxyamine auristatin without deteriorating antibody binding properties and the resulting conjugates showed selective and potent cytotoxicity in HER-2 expressing cells and HER-2 positive xenografts (Axup et al. 2012).

Incorporation of selenocysteine for regiospecific protein modification is another approach used for antibody drug conjugation. Selenocysteine is in fact a rare natural residue with a pKa of 5.2 and unique nucleophilic reactivity. Selenocysteine's high chemical reactivity allows for its regiospecific conjugation in the presence of other natural residues including cysteine. Selenocysteine was incorporated into an IgG derived Fc fragment through UGA nonsense codon. Selenocysteine was then conjugated to a maleimide modified peptidomimetic with picomolar affinity for integrin $\alpha 4\beta 1$ to form monomeric Fc conjugates with improved pharmacokinetic properties (Hofer et al. 2008). selenocysteine incorporation has been also used in the context of alternative antibody formats such as Fab fragment to conjugate biotin, fluorescein and polyethylene glycol without disrupting the antigen binding properties of the antibodies (Hofer et al. 2009).

Enzymatic Conjugation

Sortase a Mediated Peptide Coupling

Enzymatic conjugation approaches are considered an interesting alternative to chemical conjugation strategies and have the potential to site specifically conjugate the payload on the mAb with a defined DAR. Sortase mediated transpeptidation is one of the available approaches that is considered a site specific and versatile conjugation approach for different payloads and results in high yields (Guimaraes et al. 2013).

Sortases are membrane associated transpeptidase enzymes in Gram positive bacteria that have a role in protein attachment to peptidoglycan cell wall and cell wall biosynthesis (Guimaraes et al. 2013; Theile et al. 2013). Sortase A is an enzyme that is able to recognize a sortase recognition motif at or near the C-terminus of the protein substrate consisting of Leu-Pro-X-Thr-Gly (LPXTG) where X can be any amino acid (Guimaraes et al. 2013). Sortase A mediates a transpeptidation reaction by nucleophilic attack of active cysteine's thiol group to the sortase recognition motif and subsequently cleaving the bond between threonine and glycine. The active cysteine in the enzyme then forms a bond with the threonine carbonyl residue which results in the formation of a thioacyl intermediate and the release of the terminal glycine of the recognition motif. The LPXT-thioacyl-enzyme intermediate later undergoes an attack by a nucleophilic primary amino moiety of an oligoglycine modified substrate to form an amide bond between the threonine within sequence and the incoming glycine of the conjugation partner (Fig. 9a) (Guimaraes et al. 2013). Sortases can be derived from different microorganisms such as *Staphylococcus aureus* or *Streptococcus pyogenes*. *Staphylococcus aureus* sortase requires the presence of calcium ions for sortagging reaction but sortase from *Streptococcus pyogenes* and some mutant forms of those from *Staphylococcus aureus* are calcium independent (Guimaraes et al. 2013).

This enzymatic activity can be utilized for protein conjugation with different payloads. The protein substrates utilized for this type of modification should either carry a sortase recognition motif or have exposed stretches of glycine for nucleophilic attack. These reaction sites can be introduced into the protein through genetic engineering. The conjugation partner can also be a payload carrying peptide that carries a stretch consisting of 1 to 5 exposed glycine residues at its N-terminus (with a single glycine residue being enough in most cases) or a peptide consisting of recognition motif (Guimaraes et al. 2013). Sortase A ligation has been used for antibody drug conjugation for conjugating various payloads such as small cytotoxic agents (Beerli et al. 2015; Lai et al. 2018), peptides and proteins (Kornberger and Skerra 2014; Swee et al. 2013), radioactive agents (Paterson et al. 2014; Rashidian et al. 2015) and various other tags (Madej et al. 2012).

Madej et al. modified a single chain antibody (sc528) that binds to the extracellular domain of epidermal growth factor receptor by a sortase recognition tag sequence and conjugated it with an oligo-glycine modified biotin through sortase action (Madej et al. 2012). In another study, an Fab for HER-2 was conjugated to

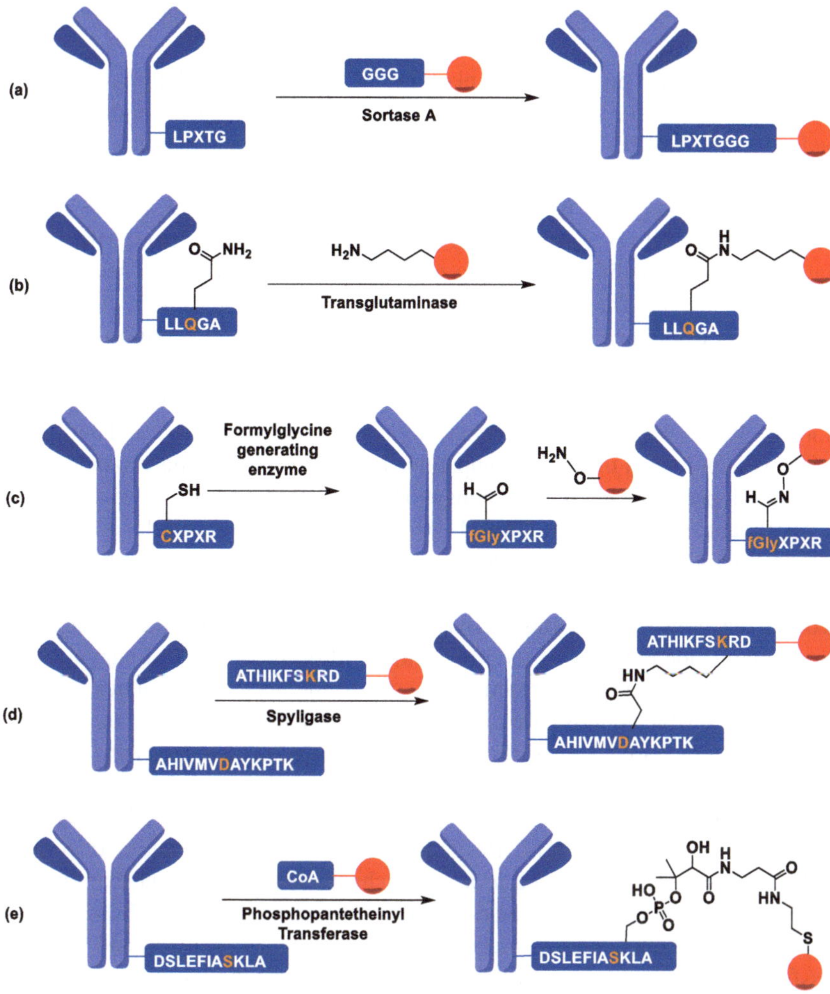

Fig. 9 Various strategies used for enzymatic conjugation; (**a**) Sortase mediated conjugation; (**b**) Transglutaminase mediated conjugation; (**c**) Formylglycine-Generating Enzymes mediated coupling; (**d**) Spyligase mediated conjugation and (**e**) Phosphopantetheinyl transferase mediated conjugation. The payload is showed as a red circle

gelonin, a 30 kDa protein plant toxin through the action of sortase A. In this case the Fab was modified at its C-terminus with the recognition motif and a penta-glycine stretch was added to the N-terminus of gelonin. The binding affinity of the Fab was not affected by the conjugation practice and the cellular toxicity was improved by a 1000-fold when compared to the free gelonin which is suggested to be associated with the cellular targeting function of the developed immunoconjugates (Kornberger and Skerra 2014). Swee et al. conjugated αDEC205 mAb heavy chain with peptide and protein antigens using sortase activity by introducing LPXTG motif to the C-terminus of the mAb with the goal of delivering these antigens to dendritic cells and monitor the resulting immune responses to develop protein-based vaccines (Swee et al. 2013).

Beerli et al. later presented this platform as "SMAC-Technology" or "Sortase-mediated antibody conjugation Technology" for site specifically conjugating payloads to mAbs. SMAC Technology was used for sortagging antibodies with toxins (Beerli et al. 2015). In this study, sortase A recognition motif was added to the C-terminus of heavy and light chains of immunoglobulins and *S. aureus* derived sortase A was used for transpeptidation reaction. On the other hand, MMAE and maytansine were modified by adding a penta-glycine stretch which serves as the nucleophilic group. The initial conjugation of immunoglobulin light chain to the payload was associated with poor efficiency but adding a five amino acid spacer between the C-terminus and the recognition motif improved the efficiency and allowed for concomitant conjugation of both light and heavy chains of the immunoglobulin. In this way, several enzymatically conjugated anti-HER-2 and anti-CD30 mAbs were produced that were comparable to their corresponding classical conjugates developed using chemical conjugation (Beerli et al. 2015).

Möhlmann et al. tried to modify a Fab fragment for the extracellular domain B of fibronectin with a fluorescently modified oligo-glycine probe via sortagging when using limited concentration of the oligo-glycine probe. They were able to recognize two side reactions for the sortase activity with oligo-glycine limited concentrations. One of these side reactions is the irreversible hydrolysis of the acyl-enzyme intermediate and the second side reaction was due to the nucleophilic attack of lysine amino residue which in this case resulted in the covalent interaction of the light and heavy chains of Fab. Using sufficient concentrations of oligo-glycine probe was able to prevent both side reactions (Möhlmann et al. 2011).

Sortagging has also proved to be helpful in nuclear chemistry and for diagnostic purposes (Paterson et al. 2014; Rashidian et al. 2015). For instance, radioactive [^{64}Cu] complex was conjugated to a single-chain variable fragment that targets a glycoprotein receptor on activated platelets through sortagging. The developed radiolabeled immunoconjugates selectively bind to activated platelets and were successfully used for positron emission tomography imaging in a carotid artery thrombosis (Paterson et al. 2014).

Transglutaminases Mediated Coupling

Bacterial transglutaminases are another group of enzymes that can be exploited for antibody drug conjugation. These enzymes catalyze the covalent conjugation of the side chains of glutamine residues to primary amine groups such as lysine's side (McCombs and Owen 2015). Through catalyzing an acyl transfer reaction, transglutaminases result in the formation of an amide bond between the amine donor and the glutamine. For transglutaminase activity, a glutamine residue on the target protein is required but the amine containing donor does not need to be a peptide and can even be a small molecule such as alkylamine or oligoamine as long as it contains a primary amino group (Fig. 9b). Transglutaminase enzymes can be derived from eukaryotic sources such as guinea pig liver or can have a prokaryotic source. Transglutaminases that are derived from *Streptomyces mobaraensis* act independently from calcium ions and are more favorable (Falck and Müller 2018). Although the primary amine substrates for transglutaminase can be very variable and are used promiscuously, the glutamine substrate should have certain properties. In fact, neighboring amino acids, antibody glycosylation status and secondary and tertiary structures of the protein affect the choice of glutamine substrate (Ebenig et al. 2019). Some proteins can be directly used as glutamine substrates for transglutaminase activity (Buchardt et al. 2010) but most transglutaminases are not able to recognize natural glutamines in the glycosylated antibodies and this discrimination between glutamines allowed the design of a specific "glutamine tag" as the recognition motif for transglutaminase enzymes (Strop et al. 2013). The transglutaminase however is able to recognize Q295 in the deglycosylated IgG and provide conjugated IgG with a DAR of 2. When N297 is further mutated into glutamine, the transglutaminase recognizes two conjugation sites in the resulting mutated IgG and provides conjugates with a DAR of 4 (Jeger et al. 2010; Lhospice et al. 2015). In addition to the solely enzymatic approach, chemo-enzymatic conjugation approach has also been used in which a reactive spacer is first introduced to the antibody using transglutaminase enzyme and is then reacted with the payload. Using this approach, MMAE was successfully conjugated to trastuzumab with a DAR of 2 where 32 times less molar excess of toxin to antibody was needed to provide a homogenous mixture of ADCs (Dennler et al. 2014).

Strop et al., engineered a glutamine tag with LLQG sequence and found 12 sites on an anti-EGFR antibody for incorporating this glutamine tag that were effectively undergoing conjugation. In this study, various payloads such as an amine containing fluorophores and cytotoxic agents were conjugated to the antibody through the action of transglutaminase enzyme. Anti-M1S1 antibody C16 and anti-HER-2 antibody were also conjugated by introducing the above glutamine tag in their structure (Strop et al. 2013). The same group later utilized mass spectrometry to characterize transglutaminase mediated conjugation more extensively and found out that the presence of small amounts of unglycosylated antibodies in the starting material will result in a small level of off-target conjugation on the exposed glutamines (Farias et al. 2014).

Ebenig et al. have recently reported new recognition motifs for transglutaminase and successfully used it for antibody drug conjugation (Ebenig et al. 2019; Siegmund et al. 2015). These recognition motifs were derived from the natural substrate of transglutaminase, streptomyces papain inhibitor (SPI_P) (Ebenig et al. 2019) and dispase autolysis-inducing protein (DAIP) (Siegmund et al. 2015) and allowed for successful transglutaminase mediated antibody payload conjugation. When used as transglutaminase recognition motif, SPI7G motif (DIPIGQGMTG) which is derived from SPI_P provided a conjugation reaction with improved kinetics that can minimize the reaction time which is especially useful when the ADC is prone to aggregation (Ebenig et al. 2019). DAIP based engineered recognition motif (GECTYFQAYGCTE) on the other hand is a conformationally locked motif that contains an intramolecular disulfide bond and provides a better kinetic rate when compared to its linear counterpart (Siegmund et al. 2015).

Nucleic acid based therapeutics have also been subjected to transglutaminase mediated conjugation using a chemoenzymatic approach, Huggins et al. conjugated an azide carrying peptide to the C-terminus of an anti-CD33 IgG4 mAb via transglutaminase enzyme and subsequently conjugated it to a DBCO carrying SiRNA using strain promoted azide alkyne cycloaddition (Huggins et al. 2019).

Formylglycine-Generating Enzymes (FGEs) Mediated Coupling

Formylglycine-Generating Enzymes (FGEs) can convert cysteine residues into formylglycine and therefore present an aldehyde group that can be used for further modification (Fig. 9c) (Falck and Müller 2018; Berteau et al. 2006). The natural substrate of FGEs are sulfatases which carry a sulfatase motif consisting of about 13 residues. This motif carries a CXPXR submotif where X is usually alanine, serine, threonine or glycine (Carrico et al. 2007). FGEs that are found in both eukaryotic and prokaryotic organisms can convert the cysteine residue of this motif to formylglycine (Falck and Müller 2018). Bertozzi group designed an "aldehyde tag" (LCTPSR) based on sulfatase motif and presented it as a small peptide tag that can undergo conjugation with molecules carrying aminooxy or hydrazide functional groups for site specific protein and antibody labeling (Carrico et al. 2007; Wu et al. 2009) where the motif can be installed at either C or N-terminus (Carrico et al. 2007; Wu et al. 2009) or the interior positions of the protein (Carrico et al. 2007). Aldehyde reaction with aminooxy and hydrazide reagents will result in the formation of oximes and hydrazones, respectively (Wu et al. 2009; Rabuka et al. 2012). Later investigations suggested that a variety of other sequences can also be used as aldehyde tags and can get recognized by FGEs (Rush and Bertozzi 2008). By conducting a systematic evaluation using FGE based bioconjugation platform, Rabuka et al. demonstrated that the site of conjugation has a significant impact on the pharmacokinetic profile and *in vivo* efficacy of ADCs (Drake et al. 2014).

Interestingly, concomitant expression of FGE with aldehyde-tagged protein substrates leads to expression of proteins containing formylglycine in their sequence (York et al. 2016). Rabuka et al. found that by optimizing the culture conditions and supplementing it with copper ions, the activity of copper dependent FGE can be maximized (York et al. 2016).

In an interesting application, Dickgiesser et al. conjugated both nucleic acid-based aptamers and functionalized fluorophores to IgG1 Fc fragments using two different enzymatic based conjugation approaches. In this study, sortase was used to conjugate peptide nucleic acids (PNA) to the C-terminus of an Fc fragment and FGE conjugation was used to ligate an aminooxy functionalized fluorophore. The PNA was further hybridized with a complementary DNA aptamer that binds to a hepatocyte growth factor receptor and can target the cells of interest (Dickgiesser et al. 2015).

Spyligase Mediated Coupling

Cell surface proteins of gram-positive bacteria can form intramolecular isopeptide bonds that contribute to their outer-membrane stability (Kang et al. 2007). Using these isopeptide bonds, Zakeri et al., designed a "SpyTag" that spontaneously forms isopeptide bonds with its protein partner in various pH, temperature and buffer conditions (Zakeri et al. 2012; Zakeri and Howarth 2010). SpyTag was designed based on the fibronectin adhesion protein CnaB2 domain of *Streptococcus pyogenes* and can be installed internally or at either N or C-terminus. SpyTag forms a covalent bond between its aspartate residue and the lysine residue of its protein partner (Zakeri et al. 2012). Later, the system was redesigned by introducing two peptide tags called SpyTag (AHIVMVDAYKPTK) and K-Tag (ATHIKFSKRD) where a protein domain called SpyLigase, conjugated these two peptide tags together (Fig. 9d) (Fierer et al. 2014). Siegemund et al. applied this strategy to develop ADCs. In this study, MMAE was chemically linked to a K-Tag through solid phase peptide synthesis and an SpyTag was installed on the C-terminus of cetuximab and the conjugation resulted in the formation of an ADC with sub-nanomolar IC_{50} Siegmund et al. 2016).

Phosphopantetheinyl Transferases (PPTases) Mediated Coupling

PPTases catalyze the transfer and covalent attachment of phosphopantethein group from coenzyme A to a certain serine residue in the target protein via a phosphodiester bond (Falck and Müller 2018). Phosphopantetheinyl transferase derived from *Bacillus subtilis* can recognize an 11-mer motif called "ybbR-tag" (DSLEFIASKLA) at either termini or in the internal flexible loops of the target protein (Fig. 9e). *Bacillus subtilis* PPTase can accept a variety of small molecule substrates conjugated to coenzyme A including fluorophores, sugars, biotin and so on (Yin et al. 2005; Yin et al. 2006). Two 12 residue recognition motifs called S6 (GDSLSWLLRLLN) and A1 (GDSLDMLEWSLM) were also identified that can be recognized and modified by *Bacillus subtilis* and *Escherichia coli* PPTases, respectively. These recognition motifs can also be inserted in either termini of the target protein (Zhou et al. 2007). Grünewald used PPTases for developing ADCs where they used 11 and 12-mer recognition motifs at different loop positions in the constant region of trastuzumab. They were able to locate 63 positions that were effectively conjugated with the payload, auristatin toxin, through the action of PPTases and prepared 95 homogenous ADCs. Various conjugated products displayed different pharmacokinetic profiles where ADCs that were conjugated in CH1 domain were associated with favorable pharmacokinetic profiles but the ones that were conjugated to the CH2 domain were associated with rapid clearance due to thermal destabilization of this domain. PPTase catalyzed conjugation was most efficient with S6 recognition motif and no conjugation was detected on native residues in the protein (Grünewald et al. 2015).

Other Enzymatic Approaches

There are several other enzymatic approaches that can be used for developing ADCs. For example, *E.coli* lipoic acid ligase that is an enzyme that catalyzes the conjugation of lipoic acid to a 13-mer recognition motif called "LAP" can be used for this purpose. Mutations in the enzymes binding pocket allows the ligation of a variety of other substrates via lipoic acid ligase enzymatic activity. Lipoic acid ligase has been successfully used for conjugating aryl-aldehyde and aryl-hydrazine substrates to LAP tagged proteins such as streptavidin, neurexin-1b and low-density lipoprotein receptor. These substrates can then be derivatized with hydrazine and aldehyde probes respectively (Cohen et al. 2012). In another instance, Best et al. used a mutant lipoic acid ligase to conjugate a norbornene reagent to LAP and later modified this functional group using inverse-electron-demand Diels–Alder cycloaddition. This approach was further successfully used on protein models such as dihydrofolate reductase (Best et al. 2015) and has the potential to be used for antibody drug conjugation.

Glyco-Conjugation

Glyco-conjugation or conjugating payload on the carbohydrate branches of glycosylated mAbs is another approach that can be used for developing ADCs. Glyco-conjugation can result in ADCs with a limited DAR of about 2. For Glyco-conjugation the antibody should be minimally treated before conjugation practice (McCombs and Owen 2015). For example, mild oxidation of terminal carbohydrates using periodate has been used for providing aldehyde functional groups that can be used for conjugation. In an earlier experience, the conjugation of calicheamicin was directed on oxidized carbohydrates of the anti-CD33 mAb. Although successfully used in this case, there are certain antibodies that cannot tolerate even mild oxidizing conditions. For example, CDR-grafted humanized P67.6 antibody showed sensitivity to periodate oxidation and its conjugation using this approach was not sucessful (Hamann et al. 2002; Rodwell et al. 1986).

In a different strategy, the carbohydrate groups on the antibody were subjected to galactosyl and sialyl transferases enzymes to add sialic acid groups to the antibody. These sialic acid groups were then treated with periodate to provide aldehyde functional groups and underwent oxime ligation with the aminooxy functionalized payload. This approach resulted in a DAR of about 1.6 (Zhou et al. 2014).

Boeggeman et al., used glycoengineering for developing ADCs from four commonly used mAbs bevacizumab, rituximab, infliximab, and trastuzumab. In human IgG, variable glycosylation levels exist that result in a mixture of glycoforms: G0 with no terminal galactoses and G1 and G2 with 1 and 2 terminal galactoses, respectively. These mAbs were subjected to *Streptococcus pneumoniae* β-1,4 galactosidase treatment to convert into G0 glycoform. The G0 glycoform was then galactosylated with a ketone or azide carrying galactose using enzymatic approaches and the functional handle on the introduced modified galactose was used for glycoconjugation (Boeggeman et al. 2009). The same strategy was also used for replacing native fucose with thiolated fucose using fucosyltransferase and the thiolated carbohydrate was subjected to maleimide based conjugation practice (Okeley et al. 2013).

Linkers

The payload can be either directly conjugated to the antibody or through a linker. Many drugs and payloads can be directly conjugated to antibodies but using a heterobifunctional linker may facilitate the ADC development. In addition to providing a reactive group for facilitating the conjugation reaction, linkers provide several other advantages such as improving the solubility and stability of the ADCs as well as the controlled release of drug in the action site. A stable linker can prevent premature drug release before reaching the destination hence preventing damage to healthy tissues. In fact, a critical consideration in the choice of linkers is the mechanism of drug release where the linkers are divided into three groups of non-cleavable linkers, chemically cleavable linkers and enzymatically cleavable linkers. Non-cleavable linkers do not release the active ingredient by chemical and enzymatic means and the release happens when the scaffold is degraded within the lysosomes. Chemically cleavable linkers allow drug release in response to various stimuli such as pH or glutathione concentrations and enzymatic cleavable linkers undergo enzymatic degradation (McCombs and Owen 2015).

Non-cleavable Linkers

Non-cleavable linkers provide the advantage of an increased stability in blood stream and by limiting drug release before reaching to the target tissue, prevent unwanted side effects in healthy body tissues (Fig. 10). In the design of FDA approved Kadcyla®, a non-cleavable thioether bond was used. In this case the drug release occurs when the ADC gets internalized through receptor mediated transcytosis and following mAb degradation within the lysosomes. In this case, succinimidyl-4-(N-maleimidomethyl) cyclohexane-1-carboxylate (SMCC) which is a bifunctional linker was used for payload conjugation to trastuzumab where the linker introduced a maleimide functional group on the antibody and the payload was conjugated on trastuzumab using a non-reducible thioeter bond (Fig. 3) (Erickson et al. 2006; Lambert and Chari 2014). After mAb degradation, the released maytansinoid remains attached to a lysine residue through the linker but this modification does not negatively impact its cytotoxic activity. However, this charged metabolite cannot exert cytotoxic activity on the neighboring cells as it is not able to diffuse across the plasma membrane of the cells that are not expressing the target antigen due to its hydrophilicity. This lack of bystander effect narrows the efficacy of the

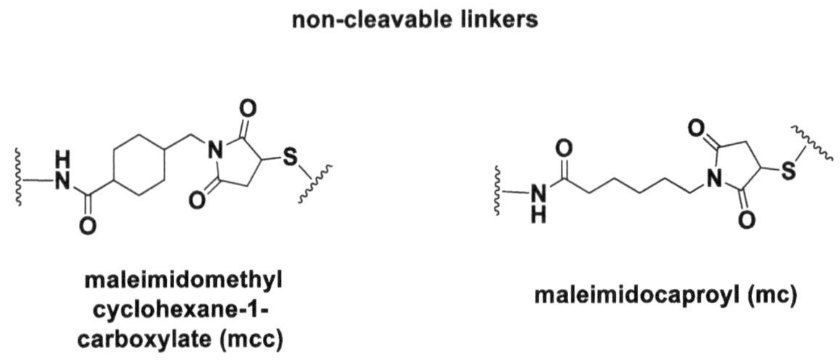

Fig. 10 Non-cleavable linkers maleimidomethyl cyclohexane carboxylate (mcc) and maleimidocaproyl (mc)

non-cleavable linker-based ADCs to some extent (Erickson et al. 2006; Lambert and Chari 2014; Kovtun and Goldmacher 2007). In addition to the bystander effect, other reasons also contribute to a narrower efficacy and higher dependence on lysosomal trafficking observed with non-cleavable linkers when compared with cleavable linker. Both ADCs with non-cleavable and cleavable linkers get activated after internalization, but cleavable linker-based ADCs may be active even when they are poorly internalized where extracellular cleavage assists free drug permeation into the target cells. However, this is not true about non-cleavable linkers where the activation only happens when the ADC is internalized and degraded within the cells (Ducry and Stump 2009; Polson et al. 2009). Nevertheless, non-cleavable linker ADCs are safer choices as they are associated with lower drug release outside the target cells. As a result, non-cleavable linker ADCs may present a good choice for improving the therapeutic index (Polson et al. 2009).

The challenge associated with this release mechanism is that any differences between the parent drug and ADC metabolites needs to be considered. Some cytotoxic agents retain their activity in the form of a derivatized metabolite while many lose their activity as a result of derivatization. MMAF molecules linked to cysteine residues of an antibody using maleimidocaproyl (mc) non-cleavable linker is another example of non-cleavable linker-based ADCs. This alkyl-maleimide ADC must be internalized within the cells to be degraded by the lysozymes. As a result of mAb degradation due to cellular internalization MMAF is released in the form of a conjugated metabolite to the linker and a cysteine residue. This chemical modification does not diminish the *in vivo* activity of MMAF. MMAE, however, was not active when conjugated in this manner (Doronina et al. 2006). Replacing mc linker with bromoacetamidecaproyl linker for conjugating MMAF to antibody not only resulted in similar activity but also improved plasma stability by preventing a retro-Michael reaction in plasma that can result in transfer and covalent conjugation of the payload to albumin's cysteines which could lead to premature release of the drug in the blood stream (Alley et al. 2008).

Chemically Degradable Linkers

Selective linker cleavage was achieved through utilizing the different characteristics of the blood stream and cytoplasmic environment. The most commonly used chemically degradable linkers are acid sensitive linkers such as hydrazones and silyl ethers (Fig. 11). The plasma half-life of hydrazones in neutral pH is about 183 hours but it is as low as 4.4 hours at an acidic pH of 5. Therefore, such linkers are cleaved in the acidic environment of the lysosomes (McCombs and Owen 2015). Disulfide bridges present another group of chemically cleavable linkers that are cleaved in the reducing environment of the cells and can release the payload after cellular internalization (McCombs and Owen 2015; Ducry and Stump 2009). Mylotarg® is one of the earliest FDA approved ADCs in which N-acetyl-gamma-calicheamicin was attached to anti-CD33 antibody using a bifunctional linker carrying a hydrazone bond that is susceptible to acid cleavage and will act as a hydrolytic site (Hamann et al. 2002). After internalization the hydrazone bond is cleaved and the enediyne agent is activated through cleavage of disulfide bond in reducing intracellular environment. The disulfide bond in this linker is stabilized by two adjacent methyl groups to prevent the nonspecific release of the agent by circulating reducing agents such as glutathione. The challenge with such linkers however is that there are other parts of human body that are associated with acidic pH and can cause premature and non-specific drug release outside the target tissue. The original withdrawal of Mylotarg® from US market due to toxicity is also associated with the poor stability of hydrazone linkers in plasma as a significant amount of drug is released in the systemic circulation before reaching the target (McCombs and Owen 2015; Ducry and Stump 2009). Inotuzumab-ozogamicin (Besponsa®) is another FDA approved ADC that takes advantage of chemically cleavable linkers. In this ADC,

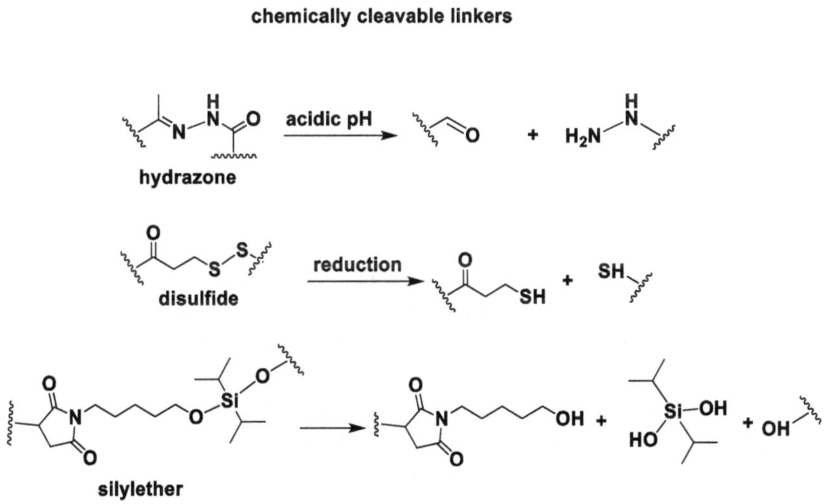

Fig. 11 Chemically cleavable linkers including hydrazone, disulfide and silylether linkers

N-acetyl-gamma-calicheamicin is attached to an anti CD22 antibody through a hydrazone linkage. Inotuzumab-ozogamicin is structurally very close to Mylotarg® but has showed desirable clinical activity and good serum and plasma stability (DiJoseph et al. 2006; Ricart 2011).

Disulfide bond linkers cleavage is dependent on the high glutathione concentration within cells. Glutathione concentration in blood stream is in micromolar range but in the cytoplasm a 1000-fold higher concentration in millimolar range is present. Furthermore, in tumor microenvironment, the increased activity of reducing enzymes results in an even higher glutathione concentration (Ducry and Stump 2009). N-succinimidyl 4-(2-pyridyldithio)butanoate (SPDB), N-succinimidyl 3-(2-pyridyldithio)propanoate (SPDP), N-succinimidyl 4-methyl-4-(2-pyridyldithio)pentanoate (SMPP) and N-succinimidyl 4-(2-pyridyldithio)pentanoate (SPP) are examples of disulfide linkers (Kellogg et al. 2011). Maytansinoid ADCs that had a SPDB linker generated lysine conjugated metabolites after lysosomal degradation but this metabolite was further reduced and S-methylated through disulfide exchange with reducing agents such as glutathione which resulted in the formation of an active lipophilic metabolite that was diffusible through plasma membranes allowing the exertion of bystander effect (Erickson et al. 2006; Kovtun et al. 2006). SPDB, SPDP and SPP differ in the level of steric hindrance around their disulfide bridge that affects their stability. A higher level of steric hindrance is associated with a more stable disulfide bond. The SPDB conjugates with intermediate disulfide bond stability showed the highest efficacy (Kellogg et al. 2011).

Silylether linkers are another group of chemically cleavable linkers that are degraded in acidic environment. Finniss et al. conjugated gemcitabine to trastuzumab via a silylether maleimide linker. The linkage was pretty stable in neutral pH where minimal release was observed after 72 hours but in an acidic pH of 5 about 80% of the payload was released during this time (Finniss et al. 2014). Wang et al. also conjugated MMAE to an anti-HER-2 antibody using a silyl ether linker that contained a p-hydroxybenzyl alcohol group which resulted in improved stability and a half-life of 7 days at 37 °C (Wang et al. 2019).

Enzymatically Degradable Linkers

Enzymatically degradable linkers are among the most promising linkers that are commonly used in ADC design and offer a better control over payload release. Such linkers are cleaved due to the action of lysosomal enzymes such as cathepsin or plasmin where these enzymes exist with a higher concentration in tumor tissues (Fig. 12) (Ducry and Stump 2009). The most prevalently used linker of this group are dipeptide valine-citrulline linkers that are combined with p-aminobenzyl alcohol (PAB). The cleavage of this amide linked PAB results in 1,6 elimination of a carbon dioxide and release of the parent drug in the amine form. Burke et al. used

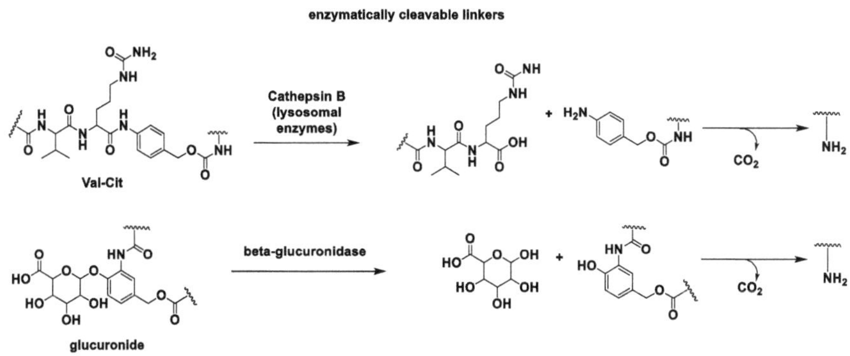

Fig. 12 Enzymatically cleavable linkers such as dipeptide valine-citrulline (Val-Cit) and glucuronide linkers

such dipeptide linkers for conjugating aniline containing camptothecin analogues to various antibodies which allowed drug release following ADC internalization and lysosomal degradation. The designed ADCs were highly potent and efficacious at desired doses in a renal cell carcinoma xenograft model (Burke et al. 2009). Dubowchik et al. have screened various dipeptide linkers to determine the impact of the linker on the release rate by lysosomal proteases and found the Phe-Lys was cleaved the fastest ($t_{1/2} = 8$ min), followed by Val-Lys ($t_{1/2} = 9$ min). Val-Cit however was very stable ($t_{1/2} = 2$ h). The presence of self-immolative PAB group increased the cleavage rate probably due to the steric constraints in the active site of the enzyme (Dubowchik et al. 2002a; Dubowchik et al. 2002b).

MMAE and MMAF conjugation to antibodies using dipeptide linkers have resulted in promising outcomes. Val-Cit and Phe-Lys both showed a higher stability when compared to acid labile hydrazone linkers. Val-Cit linkers provided better stability than Phe-Lys with an estimated half-life of 30 days in mouse plasma and 230 days in human plasma. As a result, a higher level of specificity and lower toxicity was achieved with Val-Cit conjugated ADCs along with great results in tumor xenograft models (Doronina et al. 2003). In FDA approved Adcetris® MMAE is conjugated to cAC10 mAb through Val-Cit linker and MMAE is released by proteolysis of the dipeptide linker, followed by elimination of the self-immolative p-aminobenzyloxycarbonyl (PABC) group (Senter and Sievers 2012).

Another type of enzymatically cleavable linkers are glucuronide linkers that are cleaved by the lysosomal enzyme β-glucuronidase. Jeffrey et al. conjugated MMAE, MMAF and a doxorubicin derivative to mAbs using glucuronide linkers and reported high plasma stability and a desirable safety profile (Jeffrey et al. 2006). Burke et al. compared dipeptide and glucuronide linkers that contained a self immolative spacer group and observed a higher level of aggregation with dipeptide linkers when compared with the more hydrophilic glucuronide linkers (Burke et al. 2009).

Payloads

Some of the highly toxic payloads that cannot be used as anticancer agents due to their unfavorable safety profile may present promising candidates for designing ADCs (Ducry and Stump 2009). Many different classes of anticancer agents have been subjected to antibody conjugation including maytansinoids, taxanes, duocarmycins, calicheamicins and auristatins (Hamann 2005). The choice of an appropriate payload for ADC development depends on several parameters including potency, solubility, stability and having appropriate functional handles. The potency is a critical parameter as minimal concentration of payload may reach to the target tissue and in order to be effective the drug should be potent enough to provide enough cytotoxic activity with such low concentrations (McCombs and Owen 2015). An IC50 in sub nanomolar range can indicate a desirable candidate for ADC development (Ducry and Stump 2009; Lambert and Berkenblit 2018). Solubility is a two-edged sword where the payload should be sufficiently lipophilic in order to get out of the lysosomes after release but must also be soluble enough to be amenable to conjugation in aqueous buffer as most antibodies are sensitive to high organic solvent concentrations and their folding may get negatively affected in such environments (McCombs and Owen 2015). The right choice of the linker with respect to lipophilicity can compensate for undesirable partition coefficient of the payload. For example, hydrophilic linkers that contain polyethylene glycol or sulfonate groups are more hydrophilic and may present a better choice for conjugating hydrophobic payloads such as maytansinoids while lowering the risk of aggregation and losing binding affinity (Zhao et al. 2011).

Another important aspect when selecting payloads for antibody drug conjugation is their amenability for conjugation and having functional groups that can undergo covalent interactions. If the drug does not have such handles, it may be necessary to modify the drug to incorporate required functional groups and this may negatively affect drug's activity. Furthermore, some ADCs do not release the payload as the parent molecule and instead release certain metabolites that may or may not be as functional as the parent molecule and this should be considered when selecting the appropriate linkers for ADC design. Having self immolative groups such as PAB and PABC has the advantage of releasing the drug in its original form (McCombs and Owen 2015).

The stability of the payload should also be considered as many of ADCs are internalized and released through lysosomal degradation, if a payload is specifically sensitive to acidic environment such as drug molecules that contain epoxide, disulfide or alkene bonds in their structure the drug itself can get degraded in the lysosomal compartment (McCombs and Owen 2015).

As mentioned earlier several anticancer drugs have been subjected to antibody conjugation however, only a few of these drugs have been regularly evaluated in clinical settings. These leading drugs belong to two families: Tubulin inhibitors that interfere with tubulin formation and DNA reactive agents.

Tubulin Inhibitors

Auristatins and maytansinoids are two major groups of tubulin inhibitors that have been successfully used for developing ADCs. Dolastatin 10, MMAE and MMAF belong to the family of auristatins and act as highly potent tubulin polymerization inhibitors (Fig. 13). MMAE is a synthetic pentapeptide designed based on the naturally occurring dolastatin-10 (Doronina et al. 2003). MMAE is consisted of four amino acid residues monomethylvaline, valine, dolaisoleuine and dolaproine and has a 2-amino-1-phenylpropan-1-ol (norephedrine) group added to its C-terminus (Waight et al. 2016; Adair et al. 2012). MMAE has nanomolar to subnanomolar activity in tumor cell lines and is the payload used for Adcetris® (Adair et al. 2012; Gualberto 2012). An analogue of MMAE, MMAF (Monomethyl auristatin F), with a phenylalanine replacing the norephedrine group in the C-terminus exhibits comparable potency to MMAE when delivered to cells as an ADC (Adair et al. 2012).

Maytansinoids present another class of tubulin inhibitors that were derived from a natural product called maytansine which is extracted from the bark of the African shrub *Maytenus ovatus* and exhibit picomolar activity and have been successfully used in the context of ADCs (Fig. 14) (Amiri-Kordestani et al. 2014; Lambert and Chari 2014; Adair et al. 2012; Bargh et al. 2019). Maytansine analogs including DM1 (mertansine), DM3 and DM4 are widely studied as potential payloads for ADC design. In addition to the higher hydrophobicity of DM3 and DM4, the thiol

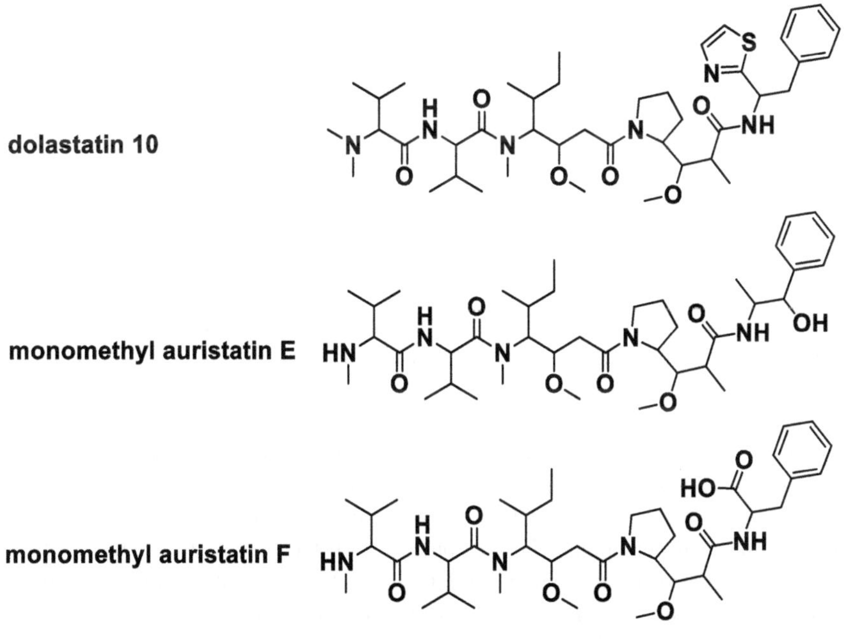

Fig. 13 Auristatin family of tubulin polymerization inhibitors including dolastatin 10, monomethyl auristatin E (MMAE) and monomethyl auristatin F (MMAF)

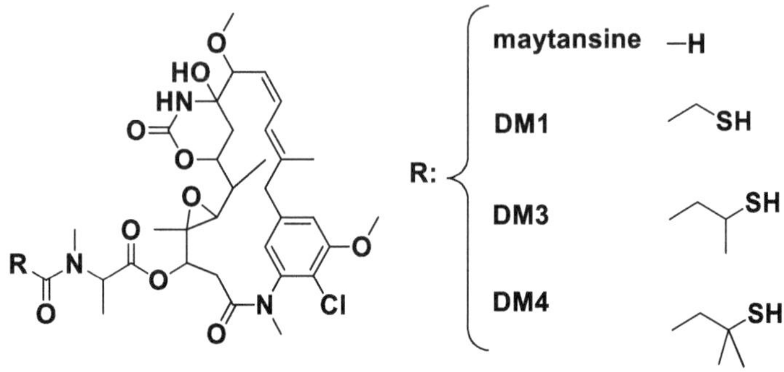

Fig. 14 Maytansinoid family of tubulin polymerization inhibitors including maytansine, DM1, DM3 and DM4

group in the structure of DM3 and DM4 is more hindered compared to DM1 and therefore a more pronounced bystander effect is achieved with these analogues. Furthermore, DM3 and DM4 metabolites are more effectively S-methylated than DM1 which again contributes to the greater bystander effect observed with DM3 and DM4 (Kellogg et al. 2011).

DNA Reactive Agents

Calicheamicins (Fig. 15a), duocarmycins (Fig. 15b) and pyrrolobenzodiazepines (PBDs) (Fig. 15c) are DNA reactive agents that are widely explored in the field of ADCs. Calicheamicins are a class of highly potent antitumor antibiotics that belong to enediyne family and were originally extracted from *Micromonospora echinospora ssp. calichensis* which is a soil microorganism. These agents are associated with the minor groove of DNA and when activated by reducing glutathione can generate diradical species and cause DNA damage (Hamann et al. 2002; Adair et al. 2012). Due to toxic effects, these agents are not suitable to be exploited as single agents but present promising agents for targeted drug delivery (Hamann et al. 2002). Mylotarg® and Besponsa® are ADCs that are designed based on calicheamicin family and have been approved by FDA to be used in clinical settings (DiJoseph et al. 2006; Ricart 2011).

CC-1065 and duocarmycins belong to another group of naturally occurring antitumor antibiotics that can bind to the minor groove of DNA in adenine thymine rich regions and alkylate the adenine nucleotides via their cyclopropyl group (Hamann 2005; Elgersma et al. 2015). CC-1065 was isolated in 1970s from *Streptomyces* bacteria and was very potent *in vitro* but was not that successful *in vivo* specially due to its delayed hepatotoxicity (Elgersma et al. 2015). Seco-duocarmycins are halogen containing derivatives of duocarmycins that act as an inactive prodrug and require a ring closure to become active. This inactive form of ducoarmycin was used for generating SYD985 as an ADC with cleavable Val-Cit linker and is currently under investigation in clinical stages (Elgersma et al. 2015).

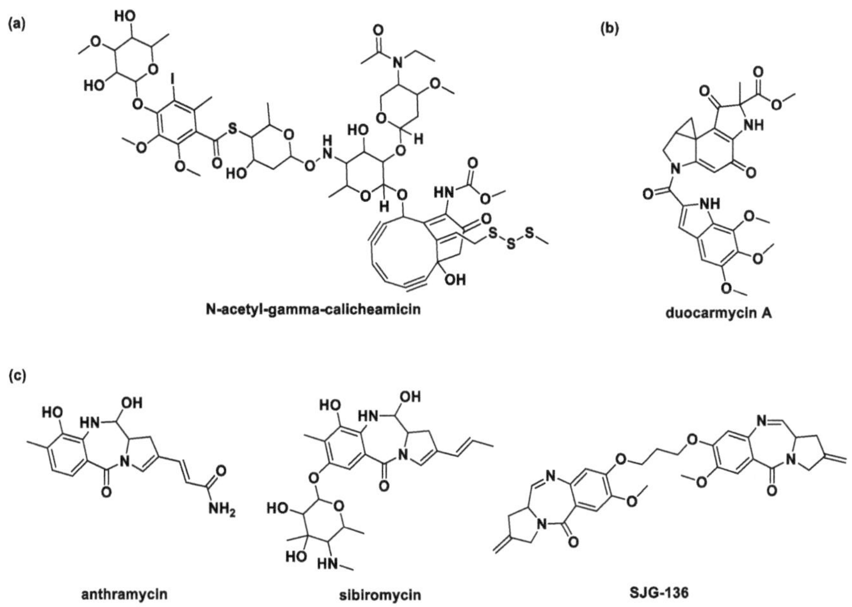

Fig. 15 DNA reactive agents (**a**) N-acetyl-gamma-calicheamicin (**b**) duocarmycin A (**c**) pyrrolobenzodiazepines (PBDs) including mono-pyrrolobenzodiazepines (anthramycin and sibiromycin) and a pyrrolobenzodiazepines dimer (SJG-136)

PBDs are another class of natural antitumor antibiotics that are found in *Streptomyces* species. PBD monomers such as anthramycin and sibiromycin can bind to the minor groove of DNA in a sequence-selective manner at 5′-purine-guanine-purine motifs and form a covalent bond with the guanine bases. PBD dimers designed by linking two PBD monomers together via a flexible linker can cause DNA strands crosslinking and are highly cytotoxic. Many PBD dimeric agents are associated with sub-nanomolar and even picomolar activity in cancer cell lines (Adair et al. 2012; Hartley et al. 2010). PBD dimers are specially promising for ADC design and many such ADCs have been evaluated in clinical and preclinical stages (Mantaj et al. 2017). For example, Vadastuximab talirine (Stein et al. 2018) and Rovalpituzumab tesirine ("Rova-T") (Rudin et al. 2017) are ADCs that contain a PBD dimer.

Other Drug Payloads

Traditional anticancer agents such as antifolates, vinca alkaloids and anthracyclines have also been subjected to antibody conjugation in several instances. Methotrexate is an antifolate that can inhibit dihydrofolate reductase and interferes with nucleotide synthesis and cellular reproduction and has been historically conjugated to antibodies specially using hydrazide linkers (Johnson et al. 1991; Umemoto et al. 1992). Vinca alkaloids such as vincristine and vinblastine are antitubulin agents that inhibit microtubule production and interfere with cellular division. This group and specially desacetylvinblastine were also subjected to antibody conjugation (Apelgren et al. 1993; Starling et al. 1991). Anthracyclines are another group of payloads used in the context of ADCs. These intercalating agents disrupt DNA function and result in its cleavage through the action of topoisomerase II enzyme (Hermentin et al. 1990). BR96-DOX is a doxorubicin conjugated to an anti-Lewis antibody that was evaluated in clinical trial and was designed using a hydrazide linker (Trail et al. 1993; Smith 2001). This ADC showed a potency comparable to doxorubicin itself but was not successful when used as a single agent (Saleh et al. 2000; Tolcher et al. 1999). Other anthracyclines such as cyano morpholino anthracyclines with a modified amine group have also been used for designing ADCs (King et al. 2003). Taxanes such as paclitaxel and docetaxel that stabilize tubules and can cause cellular arrest have also been tested specially because of the several challenges associated with their formulation, nonspecific action and multi drug resistance. Many groups have tried to design ADCs based on taxanes (Miller et al. 2004; Ojima et al. 2002) and in some cases the IC50 of the conjugated taxoids was 60 time less than the IC50 values of the parent drug (Miller et al. 2004). In a more recent study, docetaxel was conjugated to cetuximab and panitumumab using a heterobifunctional linker which shows the potential of these traditional anticancer agents to be further investigated in the format of ADCs (Glatt et al. 2018).

Conclusion

Antibody drug conjugation is a promising approach to improve the clinical outcomes of cancer therapy. ADCs allow for targeted delivery of cytotoxic agents to the tumor cells and can diminish dose limiting toxicity that is mostly associated with the off-target toxic effects of cancer therapeutics on normal cells. The conjugation chemistry is a major aspect when considering ADCs that can have a profound impact on the potency and safety of the developed agents. The linker technology and the choice of payload are other important features that have critical impact on the outcomes. In the current chapter we tried to cover an extensive evaluation of the chemistry adapted for conjugation practice including conjugation through native residues such as lysine, and cysteine and other less frequently used canonical amino acids, followed by discussing unnatural amino acids application for conjugation practice and enzymatic approaches used for developing ADCs. Various linkers that are commonly used for ADC development including non-cleavable and chemically or enzymatically cleavable linkers were further discussed and various payload that have been subjected to antibody drug conjugation were introduced.

References

Adair, J.R., Howard, P.W., Hartley, J.A., Williams, D.G., Chester, K.A.: Antibody–drug conjugates–a perfect synergy. Expert. Opin. Biol. Ther. **12**(9), 1191–1206 (2012)

Alley, S.C., Benjamin, D.R., Jeffrey, S.C., Okeley, N.M., Meyer, D.L., Sanderson, R.J., Senter, P.D.: Contribution of linker stability to the activities of anticancer immunoconjugates. Bioconjug. Chem. **19**(3), 759–765 (2008)

Amiri-Kordestani, L., Blumenthal, G.M., Xu, Q.C., Zhang, L., Tang, S.W., Ha, L., Weinberg, W.C., Chi, B., Candau-Chacon, R., Hughes, P.: FDA approval: ado-trastuzumab emtansine for the treatment of patients with HER2-positive metastatic breast cancer. Clin. Cancer Res. **20**(17), 4436–4441 (2014)

Apelgren, L., Bailey, D., Briggs, S., Barton, R., Guttman-Carlisle, D., Koppel, G., Nichols, C., Scott, W., Lindstrom, T.: Chemoimmunoconjugate development for ovarian carcinoma therapy: preclinical studies with vinca alkaloid-monoclonal antibody constructs. Bioconjug. Chem. **4**(2), 121–126 (1993)

Axup, J.Y., Bajjuri, K.M., Ritland, M., Hutchins, B.M., Kim, C.H., Kazane, S.A., Halder, R., Forsyth, J.S., Santidrian, A.F., Stafin, K.: Synthesis of site-specific antibody-drug conjugates using unnatural amino acids. Proc. Natl. Acad. Sci. **109**(40), 16101–16106 (2012)

Badescu, G., Bryant, P., Bird, M., Henseleit, K., Swierkosz, J., Parekh, V., Tommasi, R., Pawlisz, E., Jurlewicz, K., Farys, M.: Bridging disulfides for stable and defined antibody drug conjugates. Bioconjug. Chem. **25**(6), 1124–1136 (2014)

Ban, H., Nagano, M., Gavrilyuk, J., Hakamata, W., Inokuma, T., Barbas III, C.F.: Facile and stabile linkages through tyrosine: bioconjugation strategies with the tyrosine-click reaction. Bioconjug. Chem. **24**(4), 520–532 (2013)

Bargh, J.D., Isidro-Llobet, A., Parker, J.S., Spring, D.R.: Cleavable linkers in antibody–drug conjugates. Chem. Soc. Rev. **48**(16), 4361–4374 (2019)

Batmanghelich, S., Brown, R.C., Woodhead, J.S., Weeks, I., Smith, K.: Preparation of a chemiluminescent imidoester for the non-radioactive labelling of proteins. J. Photochem. Photobiol. B Biol. **12**(2), 193–201 (1992)

Beerli, R.R., Hell, T., Merkel, A.S., Grawunder, U.: Sortase enzyme-mediated generation of site-specifically conjugated antibody drug conjugates with high in vitro and in vivo potency. PLoS One. **10**(7), e0131177 (2015)

Behrens, C.R., Ha, E.H., Chinn, L.L., Bowers, S., Probst, G., Fitch-Bruhns, M., Monteon, J., Valdiosera, A., Bermudez, A., Liao-Chan, S.: Antibody–drug conjugates (ADCs) derived from interchain cysteine cross-linking demonstrate improved homogeneity and other pharmacological properties over conventional heterogeneous ADCs. Mol. Pharm. **12**(11), 3986–3998 (2015)

Berteau, O., Guillot, A., Benjdia, A., Rabot, S.: A new type of bacterial sulfatase reveals a novel maturation pathway in prokaryotes. J. Biol. Chem. **281**(32), 22464–22470 (2006)

Best, M., Degen, A., Baalmann, M., Schmidt, T.T., Wombacher, R.: Two-step protein labeling by using lipoic acid ligase with norbornene substrates and subsequent inverse-electron demand diels–alder reaction. Chembiochem. **16**(8), 1158–1162 (2015)

Boeggeman, E., Ramakrishnan, B., Pasek, M., Manzoni, M., Puri, A., Loomis, K.H., Waybright, T.J., Qasba, P.K.: Site specific conjugation of fluoroprobes to the remodeled fc N-glycans of monoclonal antibodies using mutant glycosyltransferases: application for cell surface antigen detection. Bioconjug. Chem. **20**(6), 1228–1236 (2009)

Bross, P.F., Beitz, J., Chen, G., Chen, X.H., Duffy, E., Kieffer, L., Roy, S., Sridhara, R., Rahman, A., Williams, G.: Approval summary: gemtuzumab ozogamicin in relapsed acute myeloid leukemia. Clin. Cancer Res. **7**(6), 1490–1496 (2001)

Bryant, P., Pabst, M., Badescu, G., Bird, M., McDowell, W., Jamieson, E., Swierkosz, J., Jurlewicz, K., Tommasi, R., Henseleit, K.: In vitro and in vivo evaluation of cysteine rebridged trastuzumab–MMAE antibody drug conjugates with defined drug-to-antibody ratios. Mol. Pharm. **12**(6), 1872–1879 (2015)

Bryden, F., Maruani, A., Savoie, H., Chudasama, V., Smith, M.E., Caddick, S., Boyle, R.W.: Regioselective and stoichiometrically controlled conjugation of photodynamic sensitizers to a HER2 targeting antibody fragment. Bioconjug. Chem. **25**(3), 611–617 (2014)

Buchardt, J., Selvig, H., Nielsen, P.F., Johansen, N.L.: Transglutaminase-mediated methods for site-selective modification of human growth hormone. Pept. Sci. **94**(2), 229–235 (2010)

Burke, P.J., Senter, P.D., Meyer, D.W., Miyamoto, J.B., Anderson, M., Toki, B.E., Manikumar, G., Wani, M.C., Kroll, D.J., Jeffrey, S.C.: Design, synthesis, and biological evaluation of antibody– drug conjugates comprised of potent camptothecin analogues. Bioconjug. Chem. **20**(6), 1242–1250 (2009)

Carrico, I.S., Carlson, B.L., Bertozzi, C.R.: Introducing genetically encoded aldehydes into proteins. Nat. Chem. Biol. **3**(6), 321–322 (2007)

Chen, Y., Liu, G., Guo, L., Wang, H., Fu, Y., Luo, Y.: Enhancement of tumor uptake and therapeutic efficacy of EGFR-targeted antibody cetuximab and antibody–drug conjugates by cholesterol sequestration. Int. J. Cancer. **136**(1), 182–194 (2015)

Christie, R.J., Fleming, R., Bezabeh, B., Woods, R., Mao, S., Harper, J., Joseph, A., Wang, Q., Xu, Z.-Q., Wu, H.: Stabilization of cysteine-linked antibody drug conjugates with N-aryl maleimides. J. Control. Release. **220**, 660–670 (2015)

Cockrell, G.M., Wolfe, M.S., Wolfe, J.L., Schöneich, C.: Photoinduced aggregation of a model antibody–drug conjugate. Mol. Pharm. **12**(6), 1784–1797 (2015)

Cohen, J.D., Zou, P., Ting, A.Y.: Site-specific protein modification using lipoic acid ligase and bis-aryl hydrazone formation. Chembiochem. **13**(6), 888–894 (2012)

de la Torre, B., Albericio, F.: The pharmaceutical industry in 2017. An analysis of fda drug approvals from the perspective of molecules. Molecules. **23**(3), 533 (2018)

Deguchi, T., Chu, T.M., Leong, S.S., Horoszewicz, J.S., Lee, C.-L.: Effect of methotrexate-monoclonal anti-prostatic acid phosphatase antibody conjugate on human prostate tumor. Cancer Res. **46**(8), 3751–3755 (1986)

Dennler, P., Chiotellis, A., Fischer, E., Brégeon, D., Belmant, C., Gauthier, L., Lhospice, F., Romagne, F.O., Schibli, R.: Transglutaminase-based chemo-enzymatic conjugation approach yields homogeneous antibody–drug conjugates. Bioconjug. Chem. **25**(3), 569–578 (2014)

Dickgiesser, S., Rasche, N., Nasu, D., Middel, S., Hörner, S., Avrutina, O., Diederichsen, U., Kolmar, H.: Self-assembled hybrid aptamer-fc conjugates for targeted delivery: a modular chemoenzymatic approach. ACS Chem. Biol. **10**(9), 2158–2165 (2015)

DiJoseph, J.F., Dougher, M.M., Kalyandrug, L.B., Armellino, D.C., Boghaert, E.R., Hamann, P.R., Moran, J.K., Damle, N.K.: Antitumor efficacy of a combination of CMC-544 (inotuzumab ozogamicin), a CD22-targeted cytotoxic immunoconjugate of calicheamicin, and rituximab against non-Hodgkin's B-cell lymphoma. Clin. Cancer Res. **12**(1), 242–249 (2006)

Doronina, S.O., Toki, B.E., Torgov, M.Y., Mendelsohn, B.A., Cerveny, C.G., Chace, D.F., DeBlanc, R.L., Gearing, R.P., Bovee, T.D., Siegall, C.B.: Development of potent monoclonal antibody auristatin conjugates for cancer therapy. Nat. Biotechnol. **21**(7), 778–784 (2003)

Doronina, S.O., Mendelsohn, B.A., Bovee, T.D., Cerveny, C.G., Alley, S.C., Meyer, D.L., Oflazoglu, E., Toki, B.E., Sanderson, R.J., Zabinski, R.F.: Enhanced activity of monomethylauristatin F through monoclonal antibody delivery: effects of linker technology on efficacy and toxicity. Bioconjug. Chem. **17**(1), 114–124 (2006)

Drake, P.M., Albers, A.E., Baker, J., Banas, S., Barfield, R.M., Bhat, A.S., de Hart, G.W., Garofalo, A.W., Holder, P., Jones, L.C.: Aldehyde tag coupled with HIPS chemistry enables the production of ADCs conjugated site-specifically to different antibody regions with distinct in vivo efficacy and PK outcomes. Bioconjug. Chem. **25**(7), 1331–1341 (2014)

Dubowchik, G.M., Firestone, R.A., Padilla, L., Willner, D., Hofstead, S.J., Mosure, K., Knipe, J.O., Lasch, S.J., Trail, P.A.: Cathepsin B-labile dipeptide linkers for lysosomal release of doxorubicin from internalizing immunoconjugates: model studies of enzymatic drug release and antigen-specific in vitro anticancer activity. Bioconjug. Chem. **13**(4), 855–869 (2002a)

Dubowchik, G.M., Radia, S., Mastalerz, H., Walker, M.A., Firestone, R.A., King, H.D., Hofstead, S.J., Willner, D., Lasch, S.J., Trail, P.A.: Doxorubicin immunoconjugates containing bivalent, lysosomally-cleavable dipeptide linkages. Bioorg. Med. Chem. Lett. **12**(11), 1529–1532 (2002b)

Ducry, L., Stump, B.: Antibody– drug conjugates: linking cytotoxic payloads to monoclonal antibodies. Bioconjug. Chem. **21**(1), 5–13 (2009)

Ebenig, A., Juettner, N.E., Deweid, L., Avrutina, O., Fuchsbauer, H.-L., Kolmar, H.: Efficient site-specific antibody-drug conjugation by engineering of a nature-derived recognition tag for microbial transglutaminase. Chembiochem. **20**, 2411–2419 (2019)

Elgersma, R.C., Coumans, R.G., Huijbregts, T., Menge, W.M., Joosten, J.A., Spijker, H.J., de Groot, F.M., van der Lee, M.M., Ubink, R., van den Dobbelsteen, D.J.: Design, synthesis, and evaluation of linker-duocarmycin payloads: toward selection of HER2-targeting antibody–drug conjugate SYD985. Mol. Pharm. **12**(6), 1813–1835 (2015)

Erickson, H.K., Park, P.U., Widdison, W.C., Kovtun, Y.V., Garrett, L.M., Hoffman, K., Lutz, R.J., Goldmacher, V.S., Blättler, W.A.: Antibody-maytansinoid conjugates are activated in targeted cancer cells by lysosomal degradation and linker-dependent intracellular processing. Cancer Res. **66**(8), 4426–4433 (2006)

Evans, E.D., Pentelute, B.L.: Discovery of a 29-amino-acid reactive abiotic peptide for selective cysteine arylation. ACS Chem. Biol. **13**(3), 527–532 (2017)

Falck, G., Müller, K.M.: Enzyme-based labeling strategies for antibody–drug conjugates and antibody mimetics. Antibodies. **7**(1), 4 (2018)

Farias, S.E., Strop, P., Delaria, K., Galindo Casas, M., Dorywalska, M., Shelton, D.L., Pons, J., Rajpal, A.: Mass spectrometric characterization of transglutaminase based site-specific antibody–drug conjugates. Bioconjug. Chem. **25**(2), 240–250 (2014)

Fierer, J.O., Veggiani, G., Howarth, M.: SpyLigase peptide–peptide ligation polymerizes affibodies to enhance magnetic cancer cell capture. Proc. Natl. Acad. Sci. **111**(13), E1176–E1181 (2014)

Finniss, M.C., Chu, K.S., Bowerman, C.J., Luft, J.C., Haroon, Z.A., DeSimone, J.M.: A versatile acid-labile linker for antibody–drug conjugates. MedChemComm. **5**(9), 1355–1358 (2014)

Glatt, D.M., Beckford Vera, D.R., Prabhu, S.S., Mumper, R.J., Luft, J.C., Benhabbour, S.R., Parrott, M.C.: Synthesis and Characterization of Cetuximab–Docetaxel and Panitumumab–Docetaxel Antibody–Drug Conjugates for EGFR-Overexpressing Cancer Therapy. Mol. Pharm. **15**(11), 5089–5102 (2018)

Grünewald, J., Klock, H.E., Cellitti, S.E., Bursulaya, B., McMullan, D., Jones, D.H., Chiu, H.-P., Wang, X., Patterson, P., Zhou, H.: Efficient preparation of site-specific antibody–drug conjugates using phosphopantetheinyl transferases. Bioconjug. Chem. **26**(12), 2554–2562 (2015)

Gualberto, A.: Brentuximab vedotin (SGN-35), an antibody–drug conjugate for the treatment of CD30-positive malignancies. Expert Opin. Investig. Drugs. **21**(2), 205–216 (2012)

Guimaraes, C.P., Witte, M.D., Theile, C.S., Bozkurt, G., Kundrat, L., Blom, A.E., Ploegh, H.L.: Site-specific C-terminal and internal loop labeling of proteins using sortase-mediated reactions. Nat. Protoc. **8**(9), 1787–1799 (2013)

Hamann, P.R.: Monoclonal antibody–drug conjugates. Expert Opin. Ther. Pat. **15**(9), 1087–1103 (2005)

Hamann, P.R., Hinman, L.M., Hollander, I., Beyer, C.F., Lindh, D., Holcomb, R., Hallett, W., Tsou, H.-R., Upeslacis, J., Shochat, D.: Gemtuzumab ozogamicin, a potent and selective anti-CD33 antibody– calicheamicin conjugate for treatment of acute myeloid leukemia. Bioconjug. Chem. **13**(1), 47–58 (2002)

Hamblett, K.J., Senter, P.D., Chace, D.F., Sun, M.M., Lenox, J., Cerveny, C.G., Kissler, K.M., Bernhardt, S.X., Kopcha, A.K., Zabinski, R.F.: Effects of drug loading on the antitumor activity of a monoclonal antibody drug conjugate. Clin. Cancer Res. **10**(20), 7063–7070 (2004)

Hartley, J.A., Hamaguchi, A., Coffils, M., Martin, C.R., Suggitt, M., Chen, Z., Gregson, S.J., Masterson, L.A., Tiberghien, A.C., Hartley, J.M.: SG2285, a novel C2-aryl-substituted pyrrolobenzodiazepine dimer prodrug that cross-links DNA and exerts highly potent antitumor activity. Cancer Res. **70**(17), 6849–6858 (2010)

Hermentin, P., Doenges, R., Gronski, P., Bosslet, K., Kraemer, H., Hoffmann, D., Zilg, H., Steinstraesser, A., Schwarz, A.: Attachment of rhodosaminylanthracyclinone-type anthracyclines to the hinge-region of monoclonal antibodies. Bioconjug. Chem. **1**(2), 100–107 (1990)

Hofer, T., Thomas, J.D., Burke, T.R., Rader, C.: An engineered selenocysteine defines a unique class of antibody derivatives. Proc. Natl. Acad. Sci. **105**(34), 12451–12456 (2008)

Hofer, T., Skeffington, L.R., Chapman, C.M., Rader, C.: Molecularly defined antibody conjugation through a selenocysteine interface. Biochemistry. **48**(50), 12047–12057 (2009)
Huggins, I.J., Medina, C.A., Springer, A.D., van den Berg, A., Jadhav, S., Cui, X., Dowdy, S.F.: Site selective antibody-oligonucleotide conjugation via microbial transglutaminase. Molecules. **24**(18), 3287 (2019)
Jeffrey, S.C., Andreyka, J.B., Bernhardt, S.X., Kissler, K.M., Kline, T., Lenox, J.S., Moser, R.F., Nguyen, M.T., Okeley, N.M., Stone, I.J.: Development and properties of β-glucuronide linkers for monoclonal antibody– drug conjugates. Bioconjug. Chem. **17**(3), 831–840 (2006)
Jeger, S., Zimmermann, K., Blanc, A., Grünberg, J., Honer, M., Hunziker, P., Struthers, H., Schibli, R.: Site-specific and stoichiometric modification of antibodies by bacterial transglutaminase. Angew. Chem. Int. Ed. **49**(51), 9995–9997 (2010)
Johnson, D.A., Laguzza, B.C., Scott, W.L.: Cytotoxic antibody conjugates of hydrazide derivatized methotrexate analogs via simple organic linkers. Google Patents. (1991)
Junutula, J.R., Raab, H., Clark, S., Bhakta, S., Leipold, D.D., Weir, S., Chen, Y., Simpson, M., Tsai, S.P., Dennis, M.S.: Site-specific conjugation of a cytotoxic drug to an antibody improves the therapeutic index. Nat. Biotechnol. **26**(8), 925–932 (2008)
Junutula, J.R., Flagella, K.M., Graham, R.A., Parsons, K.L., Ha, E., Raab, H., Bhakta, S., Nguyen, T., Dugger, D.L., Li, G.: Engineered thio-trastuzumab-DM1 conjugate with an improved therapeutic index to target human epidermal growth factor receptor 2–positive breast cancer. Clin. Cancer Res. **16**(19), 4769–4778 (2010)
Kang, H.J., Coulibaly, F., Clow, F., Proft, T., Baker, E.N.: Stabilizing isopeptide bonds revealed in gram-positive bacterial pilus structure. Science. **318**(5856), 1625–1628 (2007)
Kellogg, B.A., Garrett, L., Kovtun, Y., Lai, K.C., Leece, B., Miller, M., Payne, G., Steeves, R., Whiteman, K.R., Widdison, W.: Disulfide-linked antibody– maytansinoid conjugates: optimization of in vivo activity by varying the steric hindrance at carbon atoms adjacent to the disulfide linkage. Bioconjug. Chem. **22**(4), 717–727 (2011)
Kim, C.H., Axup, J.Y., Schultz, P.G.: Protein conjugation with genetically encoded unnatural amino acids. Curr. Opin. Chem. Biol. **17**(3), 412–419 (2013)
Kim, M.T., Chen, Y., Marhoul, J., Jacobson, F.: Statistical modeling of the drug load distribution on trastuzumab emtansine (Kadcyla), a lysine-linked antibody drug conjugate. Bioconjug. Chem. **25**(7), 1223–1232 (2014)
King, H.D., Staab, A.J., Pham-Kaplita, K., Yurgaitis, D., Firestone, R.A., Lasch, S.J., Trail, P.A.: BR96 conjugates of highly potent anthracyclines. Bioorg. Med. Chem. Lett. **13**(13), 2119–2122 (2003)
Kornberger, P., Skerra, A.: In Sortase-catalyzed in vitro functionalization of a HER2-specific recombinant Fab for tumor targeting of the plant cytotoxin gelonin, MAbs, pp. 354–366. Taylor & Francis (2014)
Kovtun, Y.V., Goldmacher, V.S.: Cell killing by antibody–drug conjugates. Cancer Lett. **255**(2), 232–240 (2007)
Kovtun, Y.V., Audette, C.A., Ye, Y., Xie, H., Ruberti, M.F., Phinney, S.J., Leece, B.A., Chittenden, T., Blättler, W.A., Goldmacher, V.S.: Antibody-drug conjugates designed to eradicate tumors with homogeneous and heterogeneous expression of the target antigen. Cancer Res. **66**(6), 3214–3221 (2006)
Lai, J., Wang, Y., Wu, S.-S., Ding, D., Sun, Z.-Y., Zhang, Y., Zhou, J., Zhou, Z., Xu, Y.-C., Pan, L.-Q.: Elimination of melanoma by sortase A-generated TCR-like antibody-drug conjugates (TL-ADCs) targeting intracellular melanoma antigen MART-1. Biomaterials. **178**, 158–169 (2018)
Lambert, J.M., Berkenblit, A.: Antibody–drug conjugates for cancer treatment. Annu. Rev. Med. **69**, 191–207 (2018)
Lambert, J.M., Chari, R.V.: Ado-Trastuzumab Emtansine (T-DM1): an Antibody–Drug Conjugate (ADC) for HER2-Positive Breast Cancer, vol. 57, pp. 6949–6964. ACS Publications (2014)
Lewis, M.R., Kao, J.Y., Anderson, A.-L.J., Shively, J.E., Raubitschek, A.: An improved method for conjugating monoclonal antibodies with N-hydroxysulfosuccinimidyl DOTA. Bioconjug. Chem. **12**(2), 320–324 (2001)

Lhospice, F., Bregeon, D., Belmant, C., Dennler, P., Chiotellis, A., Fischer, E., Gauthier, L., Boedec, A., Rispaud, H., Savard-Chambard, S.: Site-specific conjugation of monomethyl auristatin E to anti-CD30 antibodies improves their pharmacokinetics and therapeutic index in rodent models. Mol. Pharm. **12**(6), 1863–1871 (2015)

Lin, S., Yang, X., Jia, S., Weeks, A.M., Hornsby, M., Lee, P.S., Nichiporuk, R.V., Iavarone, A.T., Wells, J.A., Toste, F.D.: Redox-based reagents for chemoselective methionine bioconjugation. Science. **355**(6325), 597–602 (2017)

Lyon, R.P., Setter, J.R., Bovee, T.D., Doronina, S.O., Hunter, J.H., Anderson, M.E., Balasubramanian, C.L., Duniho, S.M., Leiske, C.I., Li, F.: Self-hydrolyzing maleimides improve the stability and pharmacological properties of antibody-drug conjugates. Nat. Biotechnol. **32**(10), 1059–1062 (2014)

Madej, M.P., Coia, G., Williams, C.C., Caine, J.M., Pearce, L.A., Attwood, R., Bartone, N.A., Dolezal, O., Nisbet, R.M., Nuttall, S.D.: Engineering of an anti-epidermal growth factor receptor antibody to single chain format and labeling by sortase A-mediated protein ligation. Biotechnol. Bioeng. **109**(6), 1461–1470 (2012)

Mandler, R., Wu, C., Sausville, E.A., Roettinger, A.J., Newman, D.J., Ho, D.K., King, C.R., Yang, D., Lippman, M.E., Landolfi, N.F.: Immunoconjugates of geldanamycin and anti-HER2 monoclonal antibodies: antiproliferative activity on human breast carcinoma cell lines. J. Natl. Cancer Inst. **92**(19), 1573–1581 (2000)

Mantaj, J., Jackson, P.J., Rahman, K.M., Thurston, D.E.: From anthramycin to Pyrrolobenzodiazepine (PBD)-Containing Antibody–Drug Conjugates (ADCs). Angew. Chem. Int. Ed. **56**(2), 462–488 (2017)

Maruani, A., Smith, M.E., Miranda, E., Chester, K.A., Chudasama, V., Caddick, S.: A plug-and-play approach to antibody-based therapeutics via a chemoselective dual click strategy. Nat. Commun. **6**, 6645 (2015)

McCombs, J.R., Owen, S.C.: Antibody drug conjugates: design and selection of linker, payload and conjugation chemistry. AAPS J. **17**(2), 339–351 (2015)

McDonagh, C.F., Turcott, E., Westendorf, L., Webster, J.B., Alley, S.C., Kim, K., Andreyka, J., Stone, I., Hamblett, K.J., Francisco, J.A.: Engineered antibody–drug conjugates with defined sites and stoichiometries of drug attachment. Protein Eng. Des. Sel. **19**(7), 299–307 (2006)

Miller, M.L., Roller, E.E., Wu, X., Leece, B.A., Goldmacher, V.S., Chari, R.V., Ojima, I.: Synthesis of potent taxoids for tumor-specific delivery using monoclonal antibodies. Bioorg. Med. Chem. Lett. **14**(15), 4079–4082 (2004)

Möhlmann, S., Mahlert, C., Greven, S., Scholz, P., Harrenga, A.: In vitro sortagging of an antibody fab fragment: overcoming unproductive reactions of sortase with water and lysine side chains. Chembiochem. **12**(11), 1774–1780 (2011)

Ojima, I., Geng, X., Wu, X., Qu, C., Borella, C.P., Xie, H., Wilhelm, S.D., Leece, B.A., Bartle, L.M., Goldmacher, V.S.: Tumor-specific novel taxoid– monoclonal antibody conjugates. J. Med. Chem. **45**(26), 5620–5623 (2002)

Okeley, N.M., Toki, B.E., Zhang, X., Jeffrey, S.C., Burke, P.J., Alley, S.C., Senter, P.D.: Metabolic engineering of monoclonal antibody carbohydrates for antibody–drug conjugation. Bioconjug. Chem. **24**(10), 1650–1655 (2013)

Paterson, B.M., Alt, K., Jeffery, C.M., Price, R.I., Jagdale, S., Rigby, S., Williams, C.C., Peter, K., Hagemeyer, C.E., Donnelly, P.S.: Enzyme-mediated site-specific bioconjugation of metal complexes to proteins: Sortase-mediated coupling of copper-64 to a single-chain antibody. Angew. Chem. Int. Ed. **53**(24), 6115–6119 (2014)

Phillips, G.D.L., Li, G., Dugger, D.L., Crocker, L.M., Parsons, K.L., Mai, E., Blättler, W.A., Lambert, J.M., Chari, R.V., Lutz, R.J.: Targeting HER2-positive breast cancer with trastuzumab-DM1, an antibody–cytotoxic drug conjugate. Cancer Res. **68**(22), 9280–9290 (2008)

Polson, A.G., Calemine-Fenaux, J., Chan, P., Chang, W., Christensen, E., Clark, S., de Sauvage, F.J., Eaton, D., Elkins, K., Elliott, J.M.: Antibody-drug conjugates for the treatment of non–hodgkin's lymphoma: target and linker-drug selection. Cancer Res. **69**(6), 2358–2364 (2009)

Rabuka, D., Rush, J.S., Gregory, W.D., Wu, P., Bertozzi, C.R.: Site-specific chemical protein conjugation using genetically encoded aldehyde tags. Nat. Protoc. **7**(6), 1052–1067 (2012)

Rashidian, M., Keliher, E.J., Dougan, M., Juras, P.K., Cavallari, M., Wojtkiewicz, G.R., Jacobsen, J.T., Edens, J.G., Tas, J.M., Victora, G.: Use of 18F-2-fluorodeoxyglucose to label antibody fragments for immuno-positron emission tomography of pancreatic cancer. ACS Central Sci. **1**(3), 142–147 (2015)

Rath, T., Bojarski, C., Neurath, M.F., Atreya, R.: Molecular imaging of mucosal α4β7 integrin expression with the fluorescent anti-adhesion antibody vedolizumab in Crohn's disease. Gastrointest. Endosc. **86**(2), 406–408 (2017)

Ricart, A.D.: Antibody-drug conjugates of calicheamicin derivative: gemtuzumab ozogamicin and inotuzumab ozogamicin. Clin. Cancer Res. **17**(20), 6417–6427 (2011)

Rodwell, J.D., Alvarez, V.L., Lee, C., Lopes, A.D., Goers, J., King, H.D., Powsner, H.J., McKearn, T.J.: Site-specific covalent modification of monoclonal antibodies: in vitro and in vivo evaluations. Proc. Natl. Acad. Sci. **83**(8), 2632–2636 (1986)

Rudin, C.M., Pietanza, M.C., Bauer, T.M., Ready, N., Morgensztern, D., Glisson, B.S., Byers, L.A., Johnson, M.L., Burris III, H.A., Robert, F.: Rovalpituzumab tesirine, a DLL3-targeted antibody-drug conjugate, in recurrent small-cell lung cancer: a first-in-human, first-in-class, open-label, phase 1 study. Lancet Oncol. **18**(1), 42–51 (2017)

Rush, J.S., Bertozzi, C.R.: New aldehyde tag sequences identified by screening formylglycine generating enzymes in vitro and in vivo. J. Am. Chem. Soc. **130**(37), 12240–12241 (2008)

Sakamoto, S., Hamachi, I.: Recent progress in chemical modification of proteins. Anal. Sci., 18R003 (2018)

Saleh, M.N., Sugarman, S., Murray, J., Ostroff, J.B., Healey, D., Jones, D., Daniel, C.R., LeBherz, D., Brewer, H., Onetto, N.: Phase I trial of the Anti–Lewis Y Drug immunoconjugate BR96-doxorubicin in patients with Lewis Y–expressing epithelial tumors. J. Clin. Oncol. **18**(11), 2282–2292 (2000)

Schumacher, F.F., Sanchania, V.A., Tolner, B., Wright, Z.V., Ryan, C.P., Smith, M.E., Ward, J.M., Caddick, S., Kay, C.W., Aeppli, G.: Homogeneous antibody fragment conjugation by disulfide bridging introduces 'spinostics'. Sci. Rep. **3**, 1525 (2013)

Seki, Y., Ishiyama, T., Sasaki, D., Abe, J., Sohma, Y., Oisaki, K., Kanai, M.: Transition metal-free tryptophan-selective bioconjugation of proteins. J. Am. Chem. Soc. **138**(34), 10798–10801 (2016)

Senter, P.D., Sievers, E.L.: The discovery and development of brentuximab vedotin for use in relapsed Hodgkin lymphoma and systemic anaplastic large cell lymphoma. Nat. Biotechnol. **30**(7), 631–637 (2012)

Siegmund, V., Schmelz, S., Dickgiesser, S., Beck, J., Ebenig, A., Fittler, H., Frauendorf, H., Piater, B., Betz, U.A., Avrutina, O.: Locked by design: a conformationally constrained transglutaminase tag enables efficient site-specific conjugation. Angew. Chem. Int. Ed. **54**(45), 13420–13424 (2015)

Siegmund, V., Piater, B., Zakeri, B., Eichhorn, T., Fischer, F., Deutsch, C., Becker, S., Toleikis, L., Hock, B., Betz, U.A.: Spontaneous isopeptide bond formation as a powerful tool for engineering site-specific antibody-drug conjugates. Sci. Rep. **6**, 39291 (2016)

Smith, S.: Technology evaluation: SGN-15, Seattle genetics Inc. Curr. Opin. Mol. Ther. **3**(3), 295–302 (2001)

Starling, J.J., Maciak, R.S., Law, K.L., Hinson, N.A., Briggs, S.L., Laguzza, B.C., Johnson, D.A.: In vivo antitumor activity of a monoclonal antibody-vinca alkaloid immunoconjugate directed against a solid tumor membrane antigen characterized by heterogeneous expression and noninternalization of antibody-antigen complexes. Cancer Res. **51**(11), 2965–2972 (1991)

Stein, E.M., Walter, R.B., Erba, H.P., Fathi, A.T., Advani, A.S., Lancet, J.E., Ravandi, F., Kovacsovics, T., DeAngelo, D.J., Bixby, D.: A phase 1 trial of vadastuximab talirine as monotherapy in patients with CD33-positive acute myeloid leukemia. Blood. **131**(4), 387–396 (2018)

Strop, P., Liu, S.-H., Dorywalska, M., Delaria, K., Dushin, R.G., Tran, T.-T., Ho, W.-H., Farias, S., Casas, M.G., Abdiche, Y.: Location matters: site of conjugation modulates stability and pharmacokinetics of antibody drug conjugates. Chem. Biol. **20**(2), 161–167 (2013)

Swee, L.K., Guimaraes, C.P., Sehrawat, S., Spooner, E., Barrasa, M.I., Ploegh, H.L.: Sortase-mediated modification of αDEC205 affords optimization of antigen presentation and immunization against a set of viral epitopes. Proc. Natl. Acad. Sci. **110**(4), 1428–1433 (2013)
Theile, C.S., Witte, M.D., Blom, A.E., Kundrat, L., Ploegh, H.L., Guimaraes, C.P.: Site-specific N-terminal labeling of proteins using sortase-mediated reactions. Nat. Protoc. **8**(9), 1800–1807 (2013)
Tolcher, A.W., Sugarman, S., Gelmon, K.A., Cohen, R., Saleh, M., Isaacs, C., Young, L., Healey, D., Onetto, N., Slichenmyer, W.: Randomized phase II study of BR96-doxorubicin conjugate in patients with metastatic breast cancer. J. Clin. Oncol. **17**(2), 478–484 (1999)
Trail, P., Willner, D., Lasch, S., Henderson, A., Hofstead, S., Casazza, D.A., Firestone, R., Hellstrom, I., Hellstrom, K.: Cure of xenografted human carcinomas by BR96-doxorubicin immunoconjugates. Science. **261**(5118), 212–215 (1993)
Traut, R.R., Bollen, A., Sun, T.-T., Hershey, J.W., Sundberg, J., Pierce, L.R.: Methyl 4-mercaptobutyrimidate as a cleavable crosslinking reagent and its application to the Escherichia coli 30S ribosome. Biochemistry. **12**(17), 3266–3273 (1973)
Tsai, S., Li, L., Williams, L., Anderson, A.-L., Raubitschek, A., Shively, J.: Metabolism and renal clearance of 111In-labeled DOTA-conjugated antibody fragments. Bioconjug. Chem. **12**(2), 264–270 (2001)
Tsuchikama, K., An, Z.: Antibody-drug conjugates: recent advances in conjugation and linker chemistries. Protein Cell. **9**(1), 33–46 (2018)
Tumey, L.N., Charati, M., He, T., Sousa, E., Ma, D., Han, X., Clark, T., Casavant, J., Loganzo, F., Barletta, F.: Mild method for succinimide hydrolysis on ADCs: impact on ADC potency, stability, exposure, and efficacy. Bioconjug. Chem. **25**(10), 1871–1880 (2014)
Umemoto, N., Kato, Y., Hara, T.: Molecular design of methotrexate-antibody conjugates for targeted cancer treatment. J. Bioact. Compat. Polym. **7**(2), 191–219 (1992)
Vinogradova, E.V., Zhang, C., Spokoyny, A.M., Pentelute, B.L., Buchwald, S.L.: Organometallic palladium reagents for cysteine bioconjugation. Nature. **526**(7575), 687–691 (2015)
Waight, A.B., Bargsten, K., Doronina, S., Steinmetz, M.O., Sussman, D., Prota, A.E.: Structural basis of microtubule destabilization by potent auristatin anti-mitotics. PLoS One. **11**(8), e0160890 (2016)
Wang, L.: Engineering the genetic code in cells and animals: biological considerations and impacts. Acc. Chem. Res. **50**(11), 2767–2775 (2017)
Wang, L., Schultz, P.G.: Expanding the genetic code. Angew. Chem. Int. Ed. **44**(1), 34–66 (2005)
Wang, Y., Fan, S., Xiao, D., Xie, F., Li, W., Zhong, W., Zhou, X.: Novel silyl ether-based acid-cleavable antibody-MMAE conjugates with appropriate stability and efficacy. Cancers. **11**(7), 957 (2019)
Wilbur, D.S.: Radiohalogenation of proteins: an overview of radionuclides, labeling methods and reagents for conjugate labeling. Bioconjug. Chem. **3**(6), 433–470 (1992)
Wu, P., Shui, W., Carlson, B.L., Hu, N., Rabuka, D., Lee, J., Bertozzi, C.R.: Site-specific chemical modification of recombinant proteins produced in mammalian cells by using the genetically encoded aldehyde tag. Proc. Natl. Acad. Sci. **106**(9), 3000–3005 (2009)
Xu, H., Baidoo, K., Gunn, A.J., Boswell, C.A., Milenic, D.E., Choyke, P.L., Brechbiel, M.W.: Design, synthesis, and characterization of a dual modality positron emission tomography and fluorescence imaging agent for monoclonal antibody tumor-targeted imaging. J. Med. Chem. **50**(19), 4759–4765 (2007)
Yin, J., Straight, P.D., McLoughlin, S.M., Zhou, Z., Lin, A.J., Golan, D.E., Kelleher, N.L., Kolter, R., Walsh, C.T.: Genetically encoded short peptide tag for versatile protein labeling by Sfp phosphopantetheinyl transferase. Proc. Natl. Acad. Sci. **102**(44), 15815–15820 (2005)
Yin, J., Lin, A.J., Golan, D.E., Walsh, C.T.: Site-specific protein labeling by Sfp phosphopantetheinyl transferase. Nat. Protoc. **1**(1), 280–285 (2006)
York, D., Baker, J., Holder, P.G., Jones, L.C., Drake, P.M., Barfield, R.M., Bleck, G.T., Rabuka, D.: Generating aldehyde-tagged antibodies with high titers and high formylglycine yields by supplementing culture media with copper (II). BMC Biotechnol. **16**(1), 23 (2016)

Zakeri, B., Howarth, M.: Spontaneous intermolecular amide bond formation between side chains for irreversible peptide targeting. J. Am. Chem. Soc. **132**(13), 4526–4527 (2010)
Zakeri, B., Fierer, J.O., Celik, E., Chittock, E.C., Schwarz-Linek, U., Moy, V.T., Howarth, M.: Peptide tag forming a rapid covalent bond to a protein, through engineering a bacterial adhesin. Proc. Natl. Acad. Sci. **109**(12), E690–E697 (2012)
Zeng, H., Zhang, D., Zhai, X., Wang, S., Liu, Q.: Enhancing the immunofluorescent sensitivity for detection of Acidovorax citrulli using fluorescein isothiocyanate labeled antigen and antibody. Anal. Bioanal. Chem. **410**(1), 71–77 (2018)
Zhang, C., Welborn, M., Zhu, T., Yang, N.J., Santos, M.S., Van Voorhis, T., Pentelute, B.L.: Π-Clamp-mediated cysteine conjugation. Nat. Chem. **8**(2), 120–128 (2016)
Zhang, C., Dai, P., Vinogradov, A.A., Gates, Z.P., Pentelute, B.L.: Site-selective cysteine–cyclooctyne conjugation. Angew. Chem. **130**(22), 6569–6573 (2018)
Zhao, R.Y., Wilhelm, S.D., Audette, C., Jones, G., Leece, B.A., Lazar, A.C., Goldmacher, V.S., Singh, R., Kovtun, Y., Widdison, W.C.: Synthesis and evaluation of hydrophilic linkers for antibody–maytansinoid conjugates. J. Med. Chem. **54**(10), 3606–3623 (2011)
Zhou, Z., Cironi, P., Lin, A.J., Xu, Y., Hrvatin, S., Golan, D.E., Silver, P.A., Walsh, C.T., Yin, J.: Genetically encoded short peptide tags for orthogonal protein labeling by Sfp and AcpS phosphopantetheinyl transferases. ACS Chem. Biol. **2**(5), 337–346 (2007)
Zhou, Q., Stefano, J.E., Manning, C., Kyazike, J., Chen, B., Gianolio, D.A., Park, A., Busch, M., Bird, J., Zheng, X.: Site-specific antibody–drug conjugation through glycoengineering. Bioconjug. Chem. **25**(3), 510–520 (2014)

Clinical Pharmacology Considerations in Antibody-Drug Conjugates Development

Chandrasekar Durairaj, Dawei Xuan, and Diane Wang

Abstract Antibody-drug conjugates (ADCs) combine monoclonal antibodies and chemotherapeutics to target cancer cells. The clinical development of these complex drugs presents pharmacology challenges. This review analyzes the six FDA-approved ADCs to summarize lessons for optimizing first-in-human dose selection, study design, bioanalytical methods, drug interaction evaluations, modeling approaches, and immunogenicity assessments. Implementing integrated clinical pharmacology strategies for ADCs can enable regulatory approval and ideal use of these emerging targeted cancer therapies. Analysis of approved ADCs reveal key considerations for future ADC programs regarding data integration, modeling, bioanalysis, early clinical trials, dose rationale, and immunogenicity.

Keywords Antibody-drug conjugation · Bioconjugation · Conjugation chemistry · Linkers · Payloads · Cytotoxic drugs · Monoclonal antibodies · Antibody engineering · Targeted delivery · Drug delivery systems

---

C. Durairaj (✉)
Clinical Pharmacology, Global Product Development, Pfizer Inc., San Diego, CA, USA

Quantitative Clinical Pharmacology, Takeda Pharmaceuticals, Cambridge, MA, USA

D. Xuan
Clinical Pharmacology, Turning Point Therapeutics, San Diego, CA, USA

Clinical Pharmacology, Erasca Inc., San Diego, CA, USA

D. Wang
Clinical Pharmacology, Global Product Development, Pfizer Inc., San Diego, CA, USA

Clinical Pharmacology, BlossomHill Therapeutics, Inc., San Diego, CA, USA

H. J. Gukasyan, A. Kulkarni (eds.), *Antibody Drug Conjugates*, AAPS Advances in the Pharmaceutical Sciences Series 52, https://doi.org/10.1007/978-3-031-89319-3_5

Introduction

Antibody drug conjugate (ADC) is a unique class of therapeutic agent that combines a monoclonal antibody (mAb) targeting a specific antigen linked to a small molecule drug designed for release in the target cells. Since the first approval of ADC gemtuzumab ozogamicin (Mylotarg®) in 2000, ADCs have emerged as a novel modality mainly in the treatment of various cancers. At the time of this chapter's authorship, six ADCs had been approved by the Unites States Food and Drug Administration (FDA): gemtuzumab ozogamicin (Mylotarg®), brentuximab vedotin (Adcetris®), trastuzumab emtansine (Kadcyla®), inotuzumab ozogamicin (Besponsa®), polatuzumab-vedotin (Polivy®) and trastuzumab deruxtecan (Enhertu®). In addition, there are several new ADCs that have been recently approved and many others currently at various stages of preclinical and clinical development. Given the intricate structure of ADCs which comprise a macromolecule covalently linked to a cytotoxic drug, characterization of ADCs is challenging from a clinical pharmacology perspective due to the complex interplay of pharmacokinetic (PK) and pharmacodynamic (PD) properties of large and small molecule components in the ADC. Thus, the clinical pharmacology strategy should incorporate the relevant aspects of mAb and small molecule drug development to fully characterize the ADC. Based on the lessons learned from the six approved ADCs, this chapter focuses on the clinical pharmacology components to be considered during clinical development of an ADC. Also, the strategies utilized during clinical development of these ADCs are discussed from a clinical pharmacology perspective in this chapter.

Overview of Recently Approved ADCs

Gemtuzumab ozogamicin (GO) is an ADC composed of CD33-directed mAb (hP67.6; recombinant humanized immunoglobulin (Ig) G4, kappa antibody) that is covalently linked to the cytotoxic agent N-acetyl-γ calicheamicin (NAGC) by a bifunctional linker 4-(4-acetylphenoxy)butanoic acid (Wyeth Pharmaceuticals LLC 2018b). GO was granted accelerated approval in May 2000 as a monotherapy for the treatment of patients with CD33 positive acute myeloid leukemia (AML) followed by its withdrawal in November 2011 due to lack of additional efficacy to the backbone therapy (daunorubicin or cytarabine) and higher induction toxicity (FDA 2017a). However, using a fractionated dosing approach at a lower dose level, GO was subsequently reapproved in 2017 in combination with daunorubicin and cytarabine for the treatment of newly diagnosed CD33-positive AML in adults and as a single agent for treatment of relapsed or refractory CD33-positive AML in adults and in pediatric patients 2 years and older.

Brentuximab vedotin (BV) is a CD30-directed ADC consisting of a chimeric IgG1 antibody cAC10 which is covalently linked to the microtubule disrupting agent monomethyl auristatin E (MMAE) through a protease-cleavable linker. BV was approved in August 2011 for the treatment of patients with Hodgkin lymphoma and for the treatment of patients with systemic anaplastic large cell lymphoma (sALCL) (Seattle Genetics, Inc. 2019).

Trastuzumab emtansine (T-DM1) is a human epidermal growth factor receptor 2 (HER2)-targeted ADC which contains the humanized anti-HER2 IgG1 trastuzumab covalently linked to the microtubule inhibitory drug DM1 (a maytansine derivative) via the stable thioether linker 4-[N-maleimidomethyl] cyclohexane-1-carboxylate (MCC). Emtansine refers to the MCC-DM1 complex. T-DM1 was approved in February 2013 for the treatment of patients with HER2-positive, metastatic breast cancer (Genentech, Inc. 2019a).

Inotuzumab ozogamicin (IO) is a CD22-directed ADC consisting of a recombinant humanized IgG4 kappa antibody inotuzumab covalently linked to NAGC by an acid-cleavable linker composed of the condensation product of 4-(4′-acetylphenoxy)-butanoic acid and 3-methyl-3-mercaptobutane hydrazide (known as dimethylhydrazide). IO was approved in August 2017 for the treatment of adults with relapsed or refractory B-cell precursor acute lymphoblastic leukemia (ALL) (Wyeth Pharmaceuticals LLC 2018a).

Polatuzumab vedotin (PV) is a CD79b-directed ADC consisting of a humanized IgG1 mAb covalently linked to MMAE through a protease-cleavable linker maleimidocaproyl-valine-citrulline-p-aminobenzyloxycarbonyl. PV received accelerated approval in June 2019 and is indicated in combination with bendamustine and rituximab for the treatment of adult patients with relapsed or refractory diffuse large B-cell lymphoma, after at least two prior therapies (Genentech, Inc. 2019b).

Trastuzumab deruxtecan (TDXd) is a HER2-directed antibody and topoisomerase inhibitor conjugate that received accelerated approval recently in December 2019. TDXd comprises a humanized anti-HER2 IgG1 mAb, covalently linked to DXd (a topoisomerase inhibitor) via a protease-cleavable maleimide tetrapeptide linker. Deruxtecan refers to the protease-cleavable maleimide tetrapeptide linker and DXd. TDXd is indicated for the treatment of adult patients with unresectable or metastatic HER2-positive breast cancer who have received two or more prior anti-HER2-based regimens in the metastatic setting (Daiichi Sankyo, Inc. 2019).

The list of six approved ADCs by the FDA are presented in Table 1 along with the targets, structural components and the drug-to-antibody ratio (DAR). Majority of the approved ADCs contain mAb of IgG1 isotype. Protease cleavable linkers were used in three of the ADCs; BV, PV and TDXd. While acid cleavable linkers were used in GO and IO, a noncleavable alkyl thioether linker was used for conjugation in T-DM1. The average DAR was in the range of approximately 3.5–8.

Bioanalytical Strategies in Characterization of ADCs

Selection of right analytes is essential to fully characterize the PK profile of an ADC. As ADCs are designed to release their payload in the target cells and remain relatively nontoxic to normal cells, multiple analytes are usually measured during early development in order to understand their disposition and metabolic fate. Therefore, quantification of various components of the ADC including total

Table 1 List of six antibody-drug conjugates approved by the FDA

Antibody-drug conjugate	Target	Antibody type	Drug	Linker	Drug to antibody ratio
Gemtuzumab ozogamicin	CD33	IgG4	N-acetyl-gamma calicheamicin	4-(4-Acetylphenoxy) butanoic acid [Acbut]	0–6
Brentuximab vedotin	CD30	IgG1	Monomethyl auristatin E	Valine-citrulline	4
Trastuzumab emtansine	HER2	IgG1	DM1 (a maytansine derivative)	4-(N-maleimidomethyl) cyclohexane-1-carboxylate	3.5
Inotuzumab ozogamicin	CD22	IgG4	N-acetyl-gamma calicheamicin	Condensation product of AcBut and 3-methyl-3-mercaptobutane hydrazide	6
Polatuzumab vedotin	CD79b	IgG1	Monomethyl auristatin E	Maleimidocaproyl-valine-citrulline-p-aminobenzyloxycarbonyl	3.5
Trastuzumab deruxtecan	HER2	IgG1	DXd	Maleimide tetrapeptide	8

FDA Food and Drug Administration, *HER2* human epidermal growth factor receptor 2, *IgG* immunoglobulin G

antibody (conjugated and unconjugated), ADC (conjugated antibody), conjugated drug, unconjugated antibody and unconjugated drug were considered during early development to assess the in vivo stability and integrity of the ADC (Gorovits et al. 2013). Enzyme linked immunosorbent assay (ELISA) methods are generally used for the measurement of total antibody and ADC. Depending on the assay configuration, these assays can be DAR sensitive or DAR insensitive (i.e., cannot differentiate ADC moieties with different DARs). Measuring conjugated drug provides the ability to quantify the average concentration of drug associated with an antibody but is less commonly used in the assay of clinical samples. Measuring unconjugated or free drug concentrations can be accomplished using a liquid chromatography tandem mass spectrometry (LC-MS/MS) or ELISA method. A detailed review of the various bioanalytical techniques and its utility in ADC characterization have been published extensively (Gorovits et al. 2013; Kumar et al. 2015).

GO was the first developed ADC and the analytes characterized in clinical studies included total antibody, total calicheamicin and unconjugated calicheamicin (Sievers et al. 1999; Dowell et al. 2001). Clinical characterization of IO also included similar analytes (total calicheamicin and unconjugated calicheamicin) in addition to total antibody and ADC (FDA 2017b). However, with the recent advancements in understanding the exposure-response relationship of ADC and the cumulative knowledge gained over this many years of ADC development, the field has narrowed down to the measurement of three key analytes: ADC, total antibody, and unconjugated (free) drug as evident in the case of approved ADCs including BV, T-DM1, PV and TDXd (FDA 2011, 2012, 2019a, b). Measurement of these three analytes in systemic circulation is considered to be adequate to characterize the PK profile of an ADC.

Clinical Pharmacology Considerations During Early Clinical Development

Selection of First in Human Dose

Due to the unique structure of ADCs that comprise a small molecule and an antibody, a holistic approach should be utilized to identify a tolerable starting dose for the first in human (FIH) study. As there is no guidance currently available to select the FIH dose for ADCs, common approaches used for small molecules and mAbs can be utilized for selecting a safe starting dose (FDA 2005; ICH Expert Working Group 2009). In general, the current practice for ADCs is to select 1/10th of the severely toxic dose (STD10) if rodent is the relevant species or to select 1/6th of the highest non-severely toxic dose (HNSTD) if nonrodent is considered more relevant followed by extrapolation to human based on the body weight (mg/kg) or body surface area (mg/m^2). Unlike small molecules, caution should be exercised in selecting a FIH dose for ADCs. If the selected FIH dose is too low, it would result in investigation of multiple non-informative doses with limited contribution to overall safety, and an undue increase in the study duration before attaining the dose level where clinical benefit can be seen. In contrast, if FIH dose selected is too high, any observed toxic effect in the nonclinical species would persist longer in humans due to the extended circulation half-life. Another challenge to consider when utilizing the STD10 or HNSTD data from nonclinical species to estimate a safe FIH dose for humans is the lack of cross-reactivity for some humanized mAbs in these species. Thus, rational selection of FIH dose for ADCs should encompass all relevant pharmacology, efficacy and safety data from nonclinical evaluation integrated with PK/PD modeling and simulation to project the safe yet not over conservative FIH dose.

A recent review conducted by the FDA on FIH dose selection based on the nonclinical safety data submitted for 20 investigational new drug applications indicated that selection of a FIH dose that is 1/6th HNSTD in cynomolgus monkeys or 1/10th STD10 in rodents scaled according to body surface area (BSA) generally resulted in an acceptable balance of safety and efficient dose-escalation in a Phase 1 trial (Saber and Leighton 2015). Table 2 summarizes the doses investigated during clinical development and the approved doses for the six ADCs.

In case of the six approved ADCs, the FIH dose was estimated from nonrodent studies. Monkey was the sensitive toxicological species for five of the ADCs compared to rodents which is expected for large molecules. Of interest is the selection of FIH dose for PV as the antibody binds only to human CD79b. To combat this issue, a surrogate ADC that binds to cynomolgus monkey CD79b with similar affinity was developed and used in the toxicology studies which enabled FIH dose selection (Li et al. 2019). Therefore, as a conservative approach, a safety factor of 30 was considered when estimating the FIH dose for PV. For TDXd, in addition to the nonclinical pharmacology data, PK simulations were conducted to support the FIH dose and dosing regimen selection. Assuming similar PK in humans and cynomolgus monkeys, the simulation results indicated that the selected FIH dose of 0.8 mg/

Table 2 Doses investigated during clinical development of the six approved antibody-drug conjugates

Antibody-drug conjugate	First-in-human dose	Dose levels investigated in phase 1 escalation study	Maximum tolerated dose	Approved clinical dose
Gemtuzumab ozogamicin	0.25 mg/m^2 (1/2 of well tolerated dose in monkey)	0.25, 0.5, 1, 2, 4, 5, 6, and 9 mg/m^2	Not observed	3 mg/m^2 on days 1, 4 and 7 (monotherapy)
Brentuximab vedotin	0.1 mg/kg (1/10th of HNSTD in monkey)	0.1, 0.2, 0.4, 0.6, 0.8, 1.2, 1.8, 2.7, and 3.6 mg/kg	1.8 mg/kg	1.8 mg/kg every 3 weeks (monotherapy)
Trastuzumab emtansine	0.3 mg/kg (1/12th of HNSTD in monkey)	0.3, 0.6, 1.2, 2.4, 3.6, and 4.8 mg/kg	3.6 mg/kg	3.6 mg/kg every 3 weeks
Inotuzumab ozogamicin	0.4 mg/m^2 (1/10th of highest dose with no dose limiting toxicities or life-threatening toxicity in monkey and rat)	0.4, 0.8, 1.34, 1.8, and 2.4 mg/m^2	1.8 mg/m^2	0.8 mg/m^2 on day 1 and 0.5 mg/m^2 on days 8 and 15
Polatuzumab vedotin	0.1 mg/kg (1/30th of HNSTD based on surrogate antibody-drug conjugate in monkey)	0.1, 0.25, 0.5, 1, 1.8, and 2.4 mg/kg	Not observed	1.8 mg/kg every 3 weeks
Trastuzumab deruxtecan	0.8 mg/kg (1/12th of HNSTD in monkey)	0.8, 1.6, 3.2, 5.4, 6.4, and 8 mg/kg	Not observed	5.4 mg/kg every 3 weeks

HNSTD highest non-severely toxic dose

kg would achieve trough concentrations close to the minimum effective concentration when administered once every 3 weeks (Q3W) in humans (Doi et al. 2017). Such a combined approach provides confidence that the selected FIH dose is not conservatively too low. While majority of the ADCs used a body-weight based human equivalent dose, BSA based normalization was used for GO and IO.

Considerations in Early Study Designs

Early Phase 1 clinical trials are conducted to characterize the PK and safety of the ADC and to determine the maximum tolerated dose (MTD) in human. Often, in oncology trials, antitumor activity is also explored during early studies. In addition, the relationship between drug exposure and biomarkers are often explored and/or exposure-response analyses for key efficacy and safety endpoints are conducted to inform the selection of an optimal recommended phase 2 dose (RP2D) for further evaluation in late stage trials. Therefore, due importance should be given to the clinical pharmacology components when designing a Phase 1 trial to ensure that PK objectives are adequately addressed in the design (Iasonos and O'Quigley 2013).

The commonly used Phase 1 study designs for dose-escalation of ADCs include a typical 3 + 3 design with or without accelerated titration (GO, BV, IO, and PV) (Sievers et al. 1999; Palanca-Wessels et al. 2015; Advani et al. 2010; Younes et al. 2010), a modified Fibonacci sequence (T-DM1) (Krop et al. 2010) and a continuous modified reassessment method using a Bayesian logistic regression model (TDXd) (Doi et al. 2017). The FIH starting dose and dose titration scheme determines the number of dose escalation arms evaluated prior to reaching the MTD. As shown in Table 2, MTD was not observed for GO, PV and TDXd at the dose levels investigated in the Phase 1 dose escalation studies. The knowledge gained from prior clinical development of the ADCs also contribute to more efficient design of studies and improved confidence in dose selections. This can be inferred from the review of actual number of dose levels investigated in the Phase 1 dose escalation studies to reach the MTD or maximum administered dose (MAD); a maximum of six dose levels were screened prior to attaining the MTD or MAD for the recently approved ADCs (T-DM1, IO, PV and TDXd) while eight to nine dose levels were used for previously approved ADCs (GO and BV).

During early clinical development, importance should be given for collection of robust PK/PD data which is central for further development of the compound. Such data collected including various endpoints such as PK, PD, efficacy and safety are integrated to develop population PK, PK/PD and exposure-response models to address some of the key questions on optimal dose and dosing schedule for attainment of appropriate exposure in humans. This information is critical to guide the selection of dose and appropriate regimen and to optimize the design of future clinical trials. The robustness of a model is dependent on the quality of data collected from early studies. From a clinical pharmacology perspective, the PK/PD components collected in early clinical studies should include adequate data to characterize the PK, enable assessment of covariates that accounts for the variability in exposure, and exposure-response relationship in terms of efficacy and safety in the target patient population. In case of FIH study, PK and PD characterization is generally included as one of the key objectives. Thus, importance should be given in the selection of sampling times that would allow characterization of complete PK profile, robust estimation of PK parameters of interest, and PD changes post treatment. As multiple doses are evaluated in the dose escalation part, PK samples collected at different dose levels will aid in developing a robust exposure-response relationship.

Characterization of Distribution, Metabolism and Elimination Profiles of ADCs

As ADCs are generally administered as an intravenous infusion, no studies are required to characterize the absorption. Distribution studies for ADCs are generally limited to *in vitro* evaluation of the plasma protein binding of the small molecule drug. In addition to the noncompartmental methods, the volume of distribution of ADCs can also be obtained using the population PK modeling approach.

Metabolism of ADCs first involve deconjugation where the drug is released from the ADC by enzymatic or chemical processes. The mAb component of the ADC undergoes proteolytic metabolism to form amino acids which are further recycled into other proteins. The released small molecule drug is generally metabolized by phase I and or phase II metabolizing enzymes. Guidance for assessment of CYP-mediated and transporter-mediated drug-drug interactions (DDI) for small molecules should be followed to evaluate the DDI potential of the small molecule drug (FDA 2020). The four small molecule drugs in the list of six approved ADCs (Table 1) are primarily metabolized by CYP3A4 except NAGC which undergoes non-enzymatic reduction. All the four drugs are also substrate of the transporter, P-glycoprotein (P-gp). Clinical evaluation of the DDI potential for these drugs are discussed in later section.

Similar to mAb, several factors contribute to the elimination of ADC including target-mediated uptake followed by degradation in the lysosomal compartment, nonspecific proteolysis (uptake via pinocytosis and catabolism), recycle by the neonatal F_c receptors, etcetera (Kraynov et al. 2016). Due to the challenges involved in conducting a radiolabeled mass balance study for ADCs in human, these studies are generally conducted in animals to elucidate the excretion pathways (Roberts et al. 2013). However, a radiolabeled mass balance study can be performed for the small molecule drug component of the ADC in humans. In addition, due to the cytotoxic nature of the small molecules used in oncology, such studies cannot be conducted in healthy subjects. Also, any available data from previously conducted studies using same drug or linker drug as part of another mAb can be used to elucidate the disposition of an ADC. Another approach is to include the metabolite characterization as part of other clinical studies where additional matrices (e.g., urine, feces) can be collected to aid in metabolite quantification. Similar approach was followed for BV, where excreted metabolites of MMAE were measured in urine and feces sample collected as part of a DDI study (Lu et al. 2013a).

Considerations in Selection of Dose and Dosing Regimen

The current approach to administer an ADC is based on body surface area or body weight. While GO and IO are the only two approved ADCs administered based on BSA (mg/m^2), other ADCs are administered based on the body weight (mg/kg). This approach of body size-based dosing is largely driven by extrapolation of the animal dose to humans based on the assumption that PK of a drug is proportional to body size. However, it is increasingly recognized now for large molecules like mAbs and therapeutic peptides, body size only accounts for a portion of the PK variability between patients and flat dosing performs similarly to body-sized based approaches in minimizing the between subject variability in drug exposure (Wang et al. 2009; Zhang et al. 2012). Since the mAb portion of the ADC mainly dictates the PK properties of the molecule, similar flat dosing-based approach remains attractive for ADCs as it minimizes the dose calculation errors and is easy for preparation and administration. Due to the limited number of ADCs available when compared to mAbs and therapeutic proteins, no such inference can be derived for ADCs in general. Based on the power coefficient (α) for the effect of body weight on clearance, body weight-based dosing was expected to provide lower variability in exposure for three ADCs (BV, GO, PV; α value > 0.5) while flat dosing was expected to provide lower variability for TDXd (α value < 0.5). In case of T-DM1 and for a group of eight ADCs containing MMAE under development (Kagedal et al. 2017), the estimated power coefficient value was approximately 0.5 indicating that both flat and body weight based dosing approaches should result in similar between subject variability for these compounds. Thus, the utility of flat dosing approach still needs to be validated for ADCs when more data becomes available.

The selection of optimal dosing regimen is critical for achieving the desired efficacy while minimizing adverse events and can be accomplished by considering the efficacy and safety data in conjunction with PK properties of the ADC. Despite the relatively shorter half-lives of the approved ADCs (4–12 days) compared to regular mAbs, as shown in Table 2, most of the ADCs are administered Q3W which is similar to majority of the standard of care chemotherapies to allow recovery of patients from toxicities before proceeding to subsequent treatment. An alternative approach is to administer lower doses of an ADC at more frequent intervals (Simon and Norton 2006). Such a fractionated dosing approach could delay the onset and lessen the severity of acute toxicities thereby improving the tolerability of the administered ADC while maintaining efficacy. This is evident in the case of GO where a fractionated lower dose of 3 mg/m^2 administered on Days 1, 4, and 7 (in combination with daunorubicin and cytarabine) showed an improved clinical benefit when compared to the previously approved dose of 9 mg/m^2 as a monotherapy resulting in the reapproval of GO in combination with daunorubicin and cytarabine for de novo AML and as a single agent for relapsed AML (Gbadamosi et al. 2018). However, a post-marketing commitment (PMC) to evaluate the efficacy, safety and PK of the fractionated dosing regimen (3 mg/m^2 on Days 1, 4, and 7 for induction) as a monotherapy was recommended by the FDA and is further discussed in the context of exposure-response analysis in subsequent sections (FDA 2017a). As IO shares the same cytotoxic drug (NAGC) as GO, a fractionated dosing regimen of the MTD 1.8 mg/m^2 divided into three weekly doses of 0.8 mg/m^2 on Day 1 and 0.5 mg/m^2 on Days 8 and 15 in a 21-day cycle was investigated in clinical studies which formed the basis for currently approved dosing regimen of IO (Wyeth Pharmaceuticals LLC 2018a). However, due to the increased rate of venous-occlusive disorder (VOD) in patients treated with IO and based on prior experience with NAGC containing ADC (GO) exhibiting higher VOD, the FDA requested a dose finding study as post-marketing requirement (PMR) for IO (FDA 2017b). The agency indicated that the submitted trial was not robust to identify an optimal dose as a narrow range of doses (1.2, 1.6, and 1.8 mg/m^2) were evaluated with limited number of patients at the lowest dose and recommended to conduct a PMR study with at least two dose levels. The sponsor was conducting a Phase 4 study to compare a lower starting dose of 1.2 mg/m^2

administered in three divided doses per cycle to the currently approved dose of 1.8 mg/m^2 (ClinicalTrials.gov Identifier: NCT03677596). This highlights the importance of selecting dose levels in a dose finding study and the prerequisite to include adequate number of patients at each dose level to enable optimal dose selection (Le Tourneau et al. 2009).

Considerations in Assessment of Drug-Drug Interactions

The assessment for potential DDI of an ADC should include both mAb and small molecule drug. Unlike the small molecules, limited data is available on the DDI for large molecules such as mAbs and there is a consensus that classical biotransformation studies are not required for mAbs. As mAbs are not metabolized by CYP or other drug metabolizing enzymes, it is unlikely to have a DDI with small molecule drugs (Schrieber et al. 2019). However, if a mAb has potential to modulate CYP enzyme levels (e.g., cytokines or cytokine modulators), then the need for in vitro or in vivo studies of ADC has to be considered on a case-by-case basis. Therefore, in vitro assessment should be conducted to determine the inhibition or induction potential of CYP enzymes using the ADC in addition to the small molecule drug. For the approved six ADCs, no metabolism studies were conducted using the intact ADC except IO which included in vitro assessments for the ADC to assess its potential to inhibit or induce CYP enzymes.

In case of the small molecule drug portion of the ADCs, relevant DDI guidance from agencies should be considered in determining the need for a clinical DDI study (FDA 2020). Although the released drug from an ADC may achieve lower systemic concentrations when compared to the drug administered alone, given the narrow therapeutic index of the ADCs, a relatively mild to moderate change in the systemic concentration may increase the risk for any untoward effects of the ADC. Hence, a thorough DDI risk assessment is necessary to evaluate the need for a dedicated clinical study. Table 3 summarizes the metabolism studies conducted to support DDI assessment and its impact on the product label for the six approved ADCs. As calicheamicin mainly undergoes non-enzymatic reduction and does not inhibit or induce CYP enzymes, no clinical DDI studies were conducted for GO and IO. In case of T-DM1, in vitro studies indicated that DM1 is mainly metabolized by CYP3A4 and due to the lack of clinical DDI study results, the concomitant use of strong CYP3A4 inhibitors should be avoided (Genentech, Inc. 2019a). However, if a clinical DDI study was conducted for another ADC (e.g., BV) with same small molecule drug (MMAE) (Han et al. 2013a), physiologically-based pharmacokinetic (PBPK) modeling approach can be utilized to predict the exposure and thereby support any label claim on the potential for or lack of any DDI (e.g., PV) (FDA 2019b; Chen et al. 2015).

Table 3 Metabolism and drug-drug interaction assessments conducted for the six approved antibody-drug conjugates

Antibody-drug conjugate	In vitro metabolism studies	Clinical DDI studies	DDI specific recommendations in product label
Gemtuzumab ozogamicin	Conducted only for calicheamicin. Undergoes non-enzymatic reduction. Substrate of P-gp. Low potential for inhibition or induction of CYP enzymes.	No formal studies conducted.	None
Brentuximab vedotin	Conducted only for MMAE. Substrate of CYP3A4 and P-gp. Inhibitor of CYP3A4.	Conducted DDI study. Coadministration with ketoconazole (a strong CYP3A4 inhibitor) increased MMAE exposure by ~34%. Coadministration with rifampin (a strong CYP3A4 inducer) reduced MMAE exposure by ~46%. Coadministration with midazolam (a CYP3A4 probe substrate) did not affect midazolam exposure.	Closely monitor for adverse reactions when given concomitantly with strong CYP3A4 inhibitors.
Trastuzumab emtansine	Conducted for DM1. Substrate of CYP3A4 and P-gp.	No formal studies conducted.	Avoid concomitant use of strong CYP3A4 inhibitors. If unavoidable and treatment cannot be delayed, closely monitor for adverse reactions.
Inotuzumab ozogamicin	Conducted for both ADC and unconjugated calicheamicin. *Calicheamicin*: Substrate of CYP3A4 and P-gp. *ADC*: Not an inhibitor or inducer of CYP enzymes.	No formal studies conducted.	None

(continued)

Table 3 (continued)

Antibody-drug conjugate	In vitro metabolism studies	Clinical DDI studies	DDI specific recommendations in product label
Polatuzumab vedotin	Conducted only for MMAE. Substrate of CYP3A4 and P-gp. Inhibitor of CYP3A4.	No formal studies conducted. PBPK modeling used for DDI predictions: MMAE exposure increased by 45% when coadministered with ketoconazole. MMAE exposure decreased by 63% when coadministered with rifampin. Midazolam exposure unaltered with concomitant use of MMAE.	Monitor for signs of toxicity when given concomitantly with strong CYP3A4 inhibitors.
Trastuzumab deruxtecan	Conducted only for DXd. Substrate of CYP3A4, and transporters P-gp, MRP1, BCRP, OATP1B1, OATP1B3, and MATE2-K.	Conducted DDI study. Coadministration with itraconazole (a strong CYP3A4 inhibitor) increased ADC exposure by 11% and DXd exposure by 18%. Coadministration with ritonavir (a dual inhibitor of OATP1B and CYP3A) ADC exposure by 19% and DXd exposure by 22%.	None

ADC antibody-drug conjugate, *BCRP* breast cancer resistance protein, *CYP* cytochrome P450, *CYP3A4* cytochrome P450 3A4 isoenzyme, *DDI* drug-drug interaction, *MATE2-K* multidrug and toxin extrusion 2-K protein, *MMAE* monomethyl auristatin E, *MRP1* multidrug resistance-associated protein 1, *OATP* organic anion transporting polypeptide, *PBPK* physiologically-based pharmacokinetic, *P-gp* P-glycoprotein

Organ Impairment Studies

Renal Dysfunction

The need for assessment of renal dysfunction on the PK and safety of an ADC is mainly dictated by the elimination route of the small molecule drug. Typically, large molecules (>69 kDa) are not cleared in the kidneys by glomerular filtration and thus a dedicated renal impairment study is not expected for mAbs. The extent of renal clearance for the ADC or small molecule drug can be obtained from the animal or human (if available) mass balance study and from existing data if a prior study was conducted for the drug. In addition, the effect of renal impairment on the PK of ADC can also be characterized using population PK analysis which is described under the pharmacometric strategies.

Based on the results from animal mass balance studies that indicated renal excretion was a minor route of elimination for GO and IO (13% and 9%, respectively), no dedicated renal impairment studies were conducted for these ADCs. This is also consistent with the population PK analyses which indicated that renal impairment had no effect on the PK of these ADCs (Garrett et al. 2019; Hibma and Knight 2019). Excretion of MMAE after administration of BV was quantified as part of a clinical DDI study where 17.2% and 6.3% of MMAE were recovered in feces and urine, respectively. As patients in early clinical trials had only normal renal function, a dedicated renal impairment study was conducted for BV in patients with CD30-positive hematologic malignancy and with varying degrees of renal impairment (Zhao et al. 2016). While mild and moderate renal impairment had no impact on the PK of ADC and MMAE, exposure of MMAE increased 1.9-fold in patients with severe renal impairment and thus, BV is not recommended for patients with severe renal impairment (Seattle Genetics, Inc. 2019). Another recently approved ADC, PV, shares the same small molecule drug (MMAE) as BV. Unlike BV, no dedicated renal impairment study was conducted, and the effect of renal impairment was investigated in the population PK analysis using creatinine clearance as a marker for renal function (Lu et al. 2020). Though limited ($n = 3$) number of patients with severe renal impairment were included in the analysis dataset, no dose adjustment was recommended for patients with any degree of renal impairment as renal elimination is not a significant clearance pathway for PV or MMAE (Genentech, Inc. 2019b). Similar approach of using population PK analysis to characterize the impact of renal impairment was used for T-DM1 and TDXd (FDA 2019a; Lu et al. 2013b).

Hepatic Dysfunction

As hepatic metabolism plays a major role in the elimination of majority of the small molecule drugs that are part of ADCs, the effect of hepatic dysfunction on the PK and safety of ADCs need to be characterized as part of clinical development. Similar to renal impairment assessment, population PK analysis can also be used to characterize the effect of hepatic impairment on the PK of ADCs and small molecule drugs. Out of the six approved ADCs, only BV conducted a dedicated hepatic impairment study (Zhao et al. 2016) while population PK analysis was used for other ADCs to assess the impact of hepatic impairment.

Strategies for Clinical Evaluation of QTc Interval Prolongation

Similar to small molecule drugs, the effect of ADC on the potential to prolong QT/QTc interval has to be determined during its clinical development. Although there is lack of evidence for large molecules such as mAbs to induce delayed cardiac repolarization, the risk exists for the small molecule drug component of the ADC which may be prematurely released into circulation before reaching the target site. As per guidance, a thorough QT (TQT) study is recommended which includes evaluating a supratherapeutic dose and inclusion of a positive control to detect small changes in QTc. However, based on the risk for QT interval prolongation, alternative methods to assess the QT interval prolongation can be used and should be justified on a case-by-case basis (Rock et al. 2009).

In general, results from the in vitro hERG assay and preclinical ECG evaluations will inform the extent of ECG monitoring required in early phase clinical studies. Adequate ECG monitoring should be included in the FIH study that includes triplicate measurements at multiple timepoints including baseline, around the peak drug concentration (C_{max}), at steady-state and during follow-up visit. In addition, intensive ECG measurements are recommended in single and multiple dose escalation studies where QT effects can be evaluated at steady-state and at higher than clinically recommended doses. This approach would enable robust QTc assessment across entire dose range and concentration ranges higher than those achieved at anticipated therapeutic dose and should be adequate to determine the need for any additional ECG monitoring in later clinical trials.

Calicheamicin is the small molecule drug component of ADCs, GO and IO. In the INO-VATE ALL clinical trial, QTcF changes ≥60 ms from baseline were observed in 3% of patients treated with IO (FDA 2017b). Therefore, concomitant use of IO is not recommended with drugs known to prolong QT interval and additional caution is advised for use in patients with history of or predisposition for QTc prolongation (Wyeth Pharmaceuticals LLC 2018a). Although, a dedicated QTc study was not conducted for both ADCs, GO label includes the recommendation for additional ECG and electrolytes assessment in patients based on the clinical experience with IO (Wyeth Pharmaceuticals LLC 2018b). An alternate to TQT study was conducted to evaluate the effect of BV on cardiac ventricular repolarization in patients with CD30-positive hematologic malignancies and no clinically meaningful changes in the duration of ventricular repolarization was observed in patients as measured by QTcF (Han et al. 2013b). Similar approach of alternate to TQT study design was utilized for assessing the effect of T-DM1 (Gupta et al. 2013) and recently approved TDXd on QTc interval in breast cancer patients (FDA 2019a). The effect of PV on the QTc interval was evaluated based on triplicate ECG data from two open-label studies in patients with previously treated B-cell malignancies (FDA 2019b).

Pharmacometric Strategies to Characterize ADCs

Considerations in Population Pharmacokinetic Analysis

Population PK approach is widely adapted in the clinical development of ADCs to characterize the PK and to evaluate the effect of covariates on the PK of ADC and its components (Singh et al. 2015). As development of a population PK model involves pooling of PK data across studies, the importance of PK samples collection (intense sampling in early trials and at least sparse samples in the later phases) cannot be emphasized more as estimation of robust model parameters depends on the quality of data. As the final population PK model is used for simulating drug exposure which is subsequently used in exposure-response analyses, importance should be given for model selection criteria including goodness-of-fit metrics, diagnostic plots and model validation in order to select a stable model. The covariates selected for evaluation should have clinical significance and need to take into consideration the metabolic mechanisms of the ADC and its components. Some of the potential covariates of interest include demographic variables (age, body weight, gender and race), laboratory test variables (to assess the impact of hepatic and renal functions), formulations used in clinical trials or manufacturing process (to assess PK comparability), concomitant medications (to assess DDI), and disease characteristics (tumor type, tumor size, disease score, etc.). Of note, although population PK analysis can be used to assess the effect of hepatic and renal impairments, if the study population did not include sufficient number of patients under each severity category, and if the range of laboratory values that represent hepatic and renal functions are not broad enough to cover a wider range, then meaningful conclusions cannot be drawn from such analysis. Similarly, when addressing a DDI concern using a population PK approach, the analysis should consider the timing of administration of concomitant medications relative to the ADC dosing and PK sampling times to derive clinically meaningful conclusions.

Table 4 summarizes the population PK models along with significant covariates for the six approved ADCs. Population PK models were developed for both intact ADC and unconjugated drug for all ADCs except T-DM1 and IO, where model was developed only for the whole ADC. While a sequential modeling approach was utilized by GO, BV and TDXd during model development for the ADC and unconjugated drug, an integrated approach was used in PV by simultaneously fitting the ADC and unconjugated drug concentrations (Lu et al. 2020). Structural models developed for ADCs varied in their complexity from a simple empirical model with first order elimination (e.g., T-DM1) to models with multiple clearance pathways that include time varying clearance and Michaelis-Menten elimination kinetics (e.g., PV). In majority of the models, baseline body weight (or BSA) was a significant covariate on the clearance and/or volume of distribution of the ADC. Among other demographic factors, gender was found to be a significant covariate on the volume of distribution of ADCs in majority of the models. Another important laboratory variable that appeared significant in majority of the population PK models

Table 4 Summary of population PK models for approved antibody-drug conjugates

Antibody-drug conjugate	Number of studies (patients)	Analytes	Structural model	Significant covariates	References
Gemtuzumab ozogamicin	7 (407)	Separate models for ADC and unconjugated calicheamicin	*ADC*: two compartment model with linear and time-dependent clearances *Calicheamicin*: two compartment model with linear clearance and input rate of formation based on total antibody eliminated	Dose, baseline albumin and body weight (fixed) on clearance of ADC. Dose, baseline albumin, gender and body weight (fixed) on central volume of distribution of ADC. Baseline bone marrow blast percentage, baseline peripheral blast percentage and combination treatment on decay coefficient of the time-dependent clearance of ADC. Baseline body weight (fixed) on clearance and volume of calicheamicin.	Hibma and Knight (2019)
Brentuximab vedotin	5 (314)	Separate models for ADC and MMAE	*ADC*: three compartment model with zero order input and first order elimination *MMAE*: semi-mechanistic two compartment model with first order elimination and rate of formation based on total antibody elimination rate	Body weight on clearance and inter-compartmental clearances of ADC. Body weight on volume of central and peripheral compartments of ADC. Age on central compartment volume of ADC. Body weight on clearance and inter-compartmental clearance of MMAE. Body weight on volume of central and peripheral compartment of MMAE.	Li et al. (2017)
Trastuzumab emtansine	5 (671)	Model only for ADC	Two compartment model with first order elimination	Baseline body weight, baseline serum human epidermal growth factor receptor 2 shed extracellular domain concentration, serum albumin, baseline sum of longest dimension of target lesions, baseline trastuzumab concentration and baseline aspartate aminotransferase on clearance. Baseline body weight on volume of distribution in central compartment.	Lu et al. (2013b)

Inotuzumab ozogamicin	11 (765)	Model only for ADC	Two compartment model with linear and time dependent clearance	Baseline body surface area on linear and time-dependent clearances. Baseline body surface area on volume of distribution in central compartment. Baseline percentage of blasts in the peripheral blood on the decay coefficient of the time-dependent clearance. Concomitant rituximab treatment was a covariate of the linear clearance component.	Garrett et al. (2019)
Polatuzumab vedotin	4 (460)	Integrated model for ADC and MMAE	*ADC*: two compartments with three clearance pathways (nonspecific time-dependent linear clearance, a linear time-dependent exponentially declining clearance and a Michealis-Menten clearance). *MMAE*: two compartment model.	Baseline bodyweight, sex, serum albumin, B-cell count, tumor sum of the product of perpendicular dimensions, and combination therapy on the nonspecific linear clearance pathway of ADC. Baseline body weight, sex, race, and treatment-naive status on volume of distribution of central compartment of ADC. Prior treatment and combination therapy effects on rate constant of linear time-dependent exponentially declining clearance of ADC. Treatment-naive status, B-cell count, threshold of B-cells, and tumor sum of the product of perpendicular dimensions on linear time-dependent exponentially declining clearance of ADC. Baseline bodyweight, sex, serum albumin, treatment-naive status, combination therapy, hepatic impairment and ECOG performance status on relative conversion fraction for nonspecific elimination of MMAE.	Lu et al. (2020)

(continued)

Table 4 (continued)

Antibody-drug conjugate	Number of studies (patients)	Analytes	Structural model	Significant covariates	References
Trastuzumab deruxtecan	5 (639)	Separate models for ADC and DXd	*ADC*: two compartment model with linear elimination *DXd*: one compartment model with time varying input	Baseline body weight, baseline albumin, country (Japan vs. non-Japan), sex, and tumor size on clearance of ADC. Baseline body weight and sex on volume of distribution of the central compartment of ADC. Country (Japan vs. non-Japan) on volume of distribution of the peripheral compartment of ADC. Baseline body weight, baseline aspartate aminotransferase, total bilirubin, cycle effect and use of ritonavir or itraconazole on clearance of DXd. Age and formulation on volume of distribution of DXd.	FDA (2019a)

ADC antibody-drug conjugate, *ECOG* eastern cooperative oncology group, *MMAE* monomethyl auristatin E, *PK* pharmacokinetic was the effect of albumin levels on clearance of ADC and/or unconjugated drug. This is consistent with the observation that ADCs are generally cleared faster in patients with lower levels of albumin compared to those with normal levels, though the exact mechanism remains unknown (Gibiansky et al. 2016). Overall, the population PK model is extensively used to evaluate the effect of extrinsic and intrinsic factors on the PK of ADC and its components and thereby support the label statements.

Considerations in Exposure-Response Analyses

Exposure-response analyses plays a major role in determining the attribution of each component of an ADC to the efficacy and safety outcomes and to inform dosing recommendations. In oncology drug development, usually only one dose level (which is the RP2D) is evaluated in the Phase 2/3 studies and often there is limited time to explore additional doses and different dosing regimens in order to accelerate the approval timeline so that lifesaving medicines can be delivered to patients sooner. Most approved doses for ADC are the MTD or close to MTD which provides a narrow therapeutic index. Consequently, exposure-response analysis is key for dose justification and is an integral part of ADC submissions.

Given the importance of exposure-response modeling in selecting the optimal dose, efficacy and safety endpoints for analyses should be cautiously selected so that the results would translate into clinical benefits. For GO, the exposure-response analysis for efficacy included a combined endpoint of proportion of patients who achieved complete remission and complete remission without full recovery of platelets (FDA 2017a). However, the later was not considered as an adequate efficacy endpoint and the analysis results were not accepted by the agency. A revised analysis that included patients with complete remission and full platelet recovery only along with the exposure-response analysis of safety endpoints supported the fractionated dosing regimen for monotherapy with a lower dose resulting in improved risk/benefit profile.

As multiple analytes are measured for ADCs, the choice of analyte to derive exposure metrics for analysis is important in successful development of exposure-response relationship. Exploratory visual analysis, correlation analysis or univariate analysis would help in identifying the exposure metric and the most relevant analyte for model development. In case of BV, plasma concentrations of total antibody, ADC and unconjugated MMAE were measured in early clinical studies. Since the concentrations of total antibody and ADC were highly correlated, ADC and MMAE concentrations were used for exposure-response analyses (FDA 2011).

Another important factor to consider is how the exposure metrics used in exposure-response analyses are derived. Area under the concentration-time curve (AUC) is a commonly used exposure metric for exposure-response analyses. In case of T-DM1, AUC values estimated by noncompartmental analysis (NCA) was used for developing exposure-response relationships which may not be reliable due to limited sampling in those patients (FDA 2012). However, the AUC derived from population PK model was found to be more reliable based on the validation analyses performed by the agency. While no exposure-response relationship was observed for efficacy endpoints (overall survival and progression free survival) using NCA estimated AUC for T-DM1 and total trastuzumab, revised analysis using population PK predicted AUC indicated that higher T-DM1 exposure led to greater improvement in those efficacy endpoints. Based on these results, a PMC was requested to conduct exposure-response analyses for T-DM1 efficacy and safety endpoints using updated data from two trials (TH3RESA and EMILIA) and then to determine the need for a post-marketing trial to optimize the dose of T-DM1.

Immunogenicity Assessment

Since development of immunogenicity may alter the PK of ADC and compromise its efficacy and safety profile, immunogenic potential of an ADC should be assessed as part of clinical development. Unlike mAbs, antidrug antibody (ADA) for an ADC may develop against the mAb, linker and linker-cytotoxic drug portion of the ADC. The FDA guidance suggests a multi-tiered approach for evaluation of the immunogenicity which can be utilized for ADCs (Watson et al. 2017). It starts with screening of samples for ADA-positive (APA) samples and patients, further evaluation of positive samples to determine APA titer, and evaluation of the neutralizing capacity of the ADA. Also, the relationship between ADA formation and PK, efficacy and safety should be assessed. In order to characterize the onset and duration of antibody formation, paired PK and ADA samples are recommended to be collected at approximately every 16 weeks (Shankar et al. 2014).

Incidence rates of ADAs for the approved ADCs were 1.1% (182 patients), 30% (156 patients), 5.3% (836 patients), 3% (236 patients), 6% (134 patients), and 0.6% (640 patients) for GO, BV, T-DM1, IO, PV, and TDXd, respectively. The higher incidence rate of ADAs in BV compared to other ADCs can be attributed to the chimeric origin of the antibody in BV. Although clearance of BV was 18% higher in ADA positive patients, this was not considered to be clinically relevant and ADA was not considered to affect the PK or pharmacodynamics of BV (Li et al. 2017).

Summary and Conclusion

This chapter summarizes the clinical pharmacology strategies utilized in the development of ADCs with a focus on lessons learned from the development of six approved ADCs. The ADC field has evolved in the last two decades since the first approval of GO in 2000 with several novel ADCs currently being evaluated across various phases of preclinical and clinical development. Based on the experience gained over years of ADC development, progress has been made in efficiently designing clinical pharmacology components in clinical studies from FIH dose selection to advancements in pharmacometric approaches to support regulatory approvals. As there are new initiatives to design more efficient ADCs with variety of payloads, use of single chain fragment variable regions rather than mAbs to achieve greater tumor penetration and the increasing interest in clinical development of combination approaches of ADCs with other agents, these strategies will continue to evolve and adapt to the changes in this field.

References

Advani, A., et al.: Safety, pharmacokinetics, and preliminary clinical activity of inotuzumab ozogamicin, a novel immunoconjugate for the treatment of B-cell non-Hodgkin's lymphoma: results of a phase I study. J. Clin. Oncol. **28**(12), 2085–2093 (2010)

Chen, Y., et al.: Physiologically based pharmacokinetic modeling as a tool to predict drug interactions for antibody-drug conjugates. Clin. Pharmacokinet. **54**(1), 81–93 (2015)

Daiichi Sankyo, Inc.: ENHERTU® (fam-trastuzumab deruxtecan-nxki). Highlights of prescribing information (Dec 2019) [cited 2020 Jan 22]. Available from: https://dsi.com/prescribing-information-portlet/getPIContent?productName=Enhertu&inline=true

Doi, T., et al.: Safety, pharmacokinetics, and antitumour activity of trastuzumab deruxtecan (DS-8201), a HER2-targeting antibody-drug conjugate, in patients with advanced breast and gastric or gastro-oesophageal tumours: a phase 1 dose-escalation study. Lancet Oncol. **18**(11), 1512–1522 (2017)

Dowell, J.A., et al.: Pharmacokinetics of gemtuzumab ozogamicin, an antibody-targeted chemotherapy agent for the treatment of patients with acute myeloid leukemia in first relapse. J. Clin. Pharmacol. **41**(11), 1206–1214 (2001)

FDA: Guidance for industry: estimating the maximum safe starting dose in initial clinical trials for therapeutics in adult healthy volunteers (2005). Available from: https://www.fda.gov/regulatory-information/search-fda-guidance-documents/estimating-maximum-safe-starting-dose-initial-clinical-trials-therapeutics-adult-healthy-volunteers

FDA: ADCETRIS (brentuximab vedotin). Clinical pharmacology and biopharmaceutics review (Application number: 125388Orig1s000) (2011) [cited 2019 July 25]. Available from: https://www.accessdata.fda.gov/drugsatfda_docs/nda/2011/125388Orig1s000ClinPharmR.pdf

FDA: KADCYLA (trastuzumab emtansine). Clinical pharmacology and biopharmaceutics review(s) (Application number: 125427Orig1s000) (2012) [cited 2019 July 25]. Available from: https://www.accessdata.fda.gov/drugsatfda_docs/nda/2013/125427Orig1s000ClinPharmR.pdf

FDA: MYLOTARG (gemtuzumab ozogamicin). Clinical pharmacology and biopharmaceutics review(s) (Application number: 761060Orig1s000 761060Orig2s000) (2017a) [cited 2019 July 25]. Available from: https://www.accessdata.fda.gov/drugsatfda_docs/nda/2017/761060Orig1s000Orig1Orig2s000ClinPharmR.pdf

FDA: BESPONSA (inotuzumab ozogamicin). Multi-discipline review (Application number: 761040Orig1s000) (2017b) [cited 2019 July 25]. Available from: https://www.accessdata.fda.gov/drugsatfda_docs/nda/2017/761040Orig1s000MultidisciplineR.pdf

FDA: ENHERTU (fam-trastuzumab deruxtecan-nxki). Multi-discipline review (Application number: 761139Orig1s000) (2019a) [cited 2020 Jan 27]. Available from: https://www.accessdata.fda.gov/drugsatfda_docs/nda/2019/761139Orig1s000MultidisciplineR.pdf

FDA: POLIVY (polatuzumab vedotin). Clinical pharmacology and biopharmaceutics review(s) (Application number: 761121Orig1s000) (2019b) [cited 2019 Sept 28]. Available from: https://www.accessdata.fda.gov/drugsatfda_docs/nda/2019/761121Orig1s000ClinPharmR.pdf

FDA: Clinical drug interaction studies—cytochrome P450 enzyme- and transporter-mediated drug interactions guidance for industry (2020) [cited 2020 Feb 1]. Available from: https://www.fda.gov/regulatory-information/search-fda-guidance-documents/clinical-drug-interaction-studies-cytochrome-p450-enzyme-and-transporter-mediated-drug-interactions

Garrett, M., et al.: Population pharmacokinetics of inotuzumab ozogamicin in relapsed/refractory acute lymphoblastic leukemia and non-Hodgkin lymphoma. J. Pharmacokinet. Pharmacodyn. **46**(3), 211–222 (2019)

Gbadamosi, M., Meshinchi, S., Lamba, J.K.: Gemtuzumab ozogamicin for treatment of newly diagnosed CD33-positive acute myeloid leukemia. Future Oncol. **14**(30), 3199–3213 (2018)

Genentech, Inc.: KADCYLA® (ado-trastuzumab emtansine). Highlights of prescribing information (May 2019a) [cited 2020 Jan 22]. Available from: https://www.gene.com/download/pdf/kadcyla_prescribing.pdf

Genentech, Inc.: POLIVY™ (polatuzumab vedotin-piiq). Highlights of prescribing information (June 2019b) [cited 2020 Jan 22]. Available from: https://www.gene.com/download/pdf/polivy_prescribing.pdf

Gibiansky, L., et al.: Model-based pharmacokinetic analysis of elotuzumab in patients with relapsed/refractory multiple myeloma. J. Pharmacokinet. Pharmacodyn. **43**(3), 243–257 (2016)

Gorovits, B., et al.: Bioanalysis of antibody-drug conjugates: American Association of Pharmaceutical Scientists Antibody-Drug Conjugate Working Group position paper. Bioanalysis. **5**(9), 997–1006 (2013)

Gupta, M., et al.: Effects of trastuzumab emtansine (T-DM1) on QT interval and safety of pertuzumab plus T-DM1 in patients with previously treated human epidermal growth factor receptor 2-positive metastatic breast cancer. Clin. Pharmacol. Drug. Dev. **2**(1), 11–24 (2013)

Han, T.H., et al.: CYP3A-mediated drug-drug interaction potential and excretion of brentuximab vedotin, an antibody-drug conjugate, in patients with CD30-positive hematologic malignancies. J. Clin. Pharmacol. **53**(8), 866–877 (2013a)

Han, T.H., et al.: Brentuximab vedotin does not cause clinically relevant QTc interval prolongation in patients with CD30-positive hematologic malignancies. Cancer Chemother. Pharmacol. **72**(1), 241–249 (2013b)

Hibma, J., Knight, B.: Population pharmacokinetic modeling of gemtuzumab ozogamicin in adult patients with acute myeloid leukemia. Clin. Pharmacokinet. **58**(3), 335–347 (2019)

Iasonos, A., O'Quigley, J.: Design considerations for dose-expansion cohorts in phase I trials. J. Clin. Oncol. **31**(31), 4014–4021 (2013)

ICH Expert Working Group: ICH harmonized tripartite guideline: nonclinical evaluation for anticancer pharmaceuticals (2009) [cited 2019 Dec 14]. Available from: https://database.ich.org/sites/default/files/S9_Guideline.pdf

Kagedal, M., et al.: Platform model describing pharmacokinetic properties of vc-MMAE antibody-drug conjugates. J. Pharmacokinet. Pharmacodyn. **44**(6), 537–548 (2017)

Kraynov, E., et al.: Current approaches for absorption, distribution, metabolism, and excretion characterization of antibody-drug conjugates: an industry white paper. Drug Metab. Dispos. **44**(5), 617–623 (2016)

Krop, I.E., et al.: Phase I study of trastuzumab-DM1, an HER2 antibody-drug conjugate, given every 3 weeks to patients with HER2-positive metastatic breast cancer. J. Clin. Oncol. **28**(16), 2698–2704 (2010)

Kumar, S., et al.: Antibody-drug conjugates nonclinical support: from early to late nonclinical bioanalysis using ligand-binding assays. Bioanalysis. **7**(13), 1605–1617 (2015)

Le Tourneau, C., Lee, J.J., Siu, L.L.: Dose escalation methods in phase I cancer clinical trials. J. Natl. Cancer Inst. **101**(10), 708–720 (2009)

Li, H., et al.: Population pharmacokinetics of brentuximab vedotin in patients with CD30-expressing hematologic malignancies. J. Clin. Pharmacol. **57**(9), 1148–1158 (2017)

Li, D., et al.: Evaluation and use of an anti-cynomolgus monkey CD79b surrogate antibody-drug conjugate to enable clinical development of polatuzumab vedotin. Br. J. Pharmacol. **176**(19), 3805–3818 (2019)

Lu, D., et al.: Strategies to address drug interaction potential for antibody-drug conjugates in clinical development. Bioanalysis. **5**(9), 1115–1130 (2013a)

Lu, D., et al.: An integrated multiple-analyte pharmacokinetic model to characterize trastuzumab emtansine (T-DM1) clearance pathways and to evaluate reduced pharmacokinetic sampling in patients with HER2-positive metastatic breast cancer. Clin. Pharmacokinet. **52**(8), 657–672 (2013b)

Lu, D., et al.: Integrated two-analyte population pharmacokinetic model of polatuzumab vedotin in patients with non-Hodgkin lymphoma. CPT Pharmacometrics Syst. Pharmacol. **9**(1), 48–59 (2020)

Palanca-Wessels, M.C., et al.: Safety and activity of the anti-CD79B antibody-drug conjugate polatuzumab vedotin in relapsed or refractory B-cell non-Hodgkin lymphoma and chronic lymphocytic leukaemia: a phase 1 study. Lancet Oncol. **16**(6), 704–715 (2015)

Roberts, S.A., et al.: Considerations for the nonclinical safety evaluation of antibody drug conjugates for oncology. Regul. Toxicol. Pharmacol. **67**(3), 382–391 (2013)

Rock, E.P., et al.: Assessing proarrhythmic potential of drugs when optimal studies are infeasible. Am. Heart J. **157**(5), 827–836 (2009)., 836.e1

Saber, H., Leighton, J.K.: An FDA oncology analysis of antibody-drug conjugates. Regul. Toxicol. Pharmacol. **71**(3), 444–452 (2015)

Schrieber, S.J., et al.: Considerations for biologic product drug-drug interactions: a regulatory perspective. Clin. Pharmacol. Ther. **105**(6), 1332–1334 (2019)

Seattle Genetics, Inc.: ADCETRIS® (brentuximab vedotin). Highlights of prescribing information (Oct 2019) [cited 2020 Jan 22]. Available from: https://seagendocs.com/adcetris/pi/734371533_80542-10_Adcetris_Full_Ltr_Master.pdf

Shankar, G., et al.: Assessment and reporting of the clinical immunogenicity of therapeutic proteins and peptides-harmonized terminology and tactical recommendations. AAPS J. **16**(4), 658–673 (2014)

Sievers, E.L., et al.: Selective ablation of acute myeloid leukemia using antibody-targeted chemotherapy: a phase I study of an anti-CD33 calicheamicin immunoconjugate. Blood. **93**(11), 3678–3684 (1999)

Simon, R., Norton, L.: The Norton-Simon hypothesis: designing more effective and less toxic chemotherapeutic regimens. Nat. Clin. Pract. Oncol. **3**(8), 406–407 (2006)

Singh, A.P., Shin, Y.G., Shah, D.K.: Application of pharmacokinetic-pharmacodynamic modeling and simulation for antibody-drug conjugate development. Pharm. Res. **32**(11), 3508–3525 (2015)

Wang, D.D., et al.: Fixed dosing versus body size-based dosing of monoclonal antibodies in adult clinical trials. J. Clin. Pharmacol. **49**(9), 1012–1024 (2009)

Watson, R.G., et al.: Implementing a tiered approach to bioanalytical method validation for large-molecule ligand-binding assay methods in pharmacokinetic assessments. Bioanalysis. **9**(18), 1407–1422 (2017)

Wyeth Pharmaceuticals LLC, a subsidiary of Pfizer Inc.: BESPONSA (inotuzumab ozogamicin). Highlights of prescribing information (Mar 2018a) [cited 2020 Jan 22]. Available from: http://labeling.pfizer.com/ShowLabeling.aspx?id=9503&format=PDF

Wyeth Pharmaceuticals LLC, a subsidiary of Pfizer Inc.: MYLOTARG (gemtuzumab ozogamicin). Highlights of prescribing information (Apr 2018b) [cited 2020 Jan 22]. Available from: http://labeling.pfizer.com/ShowLabeling.aspx?id=9548

Younes, A., et al.: Brentuximab vedotin (SGN-35) for relapsed CD30-positive lymphomas. N. Engl. J. Med. **363**(19), 1812–1821 (2010)

Zhang, S., et al.: Fixed dosing versus body size-based dosing of therapeutic peptides and proteins in adults. J. Clin. Pharmacol. **52**(1), 18–28 (2012)

Zhao, B., et al.: Brentuximab vedotin, an antibody-drug conjugate, in patients with CD30-positive haematologic malignancies and hepatic or renal impairment. Br. J. Clin. Pharmacol. **82**(3), 696–705 (2016)

Bioanalytical Methodologies for Pharmacokinetic Characterization of Antibody Drug Conjugates

Y. -J. Xue, Matthew V. Myers, Jian Chen, Mu Chen, Brian Melo, Martha Vallejo, Stephen E. Maxwell, Jim X. Shen, Johanna R. Mora, and Sekhar Surapaneni

Abstract The structure of antibody-drug conjugates (ADCs) consists of a cytotoxic drug conjugated to a monoclonal antibody. Due to its multiple components, complex bioanalytical strategies are needed to characterize an ADC's pharmacokinetic properties. This chapter discusses bioanalytical assays and parameters required for the PK characterization, including: total antibody, total conjugated drug, unconjugated payload and its metabolites, drug-to-antibody ratio, and immunogenicity, using ligand binding and mass spectrometry-based approaches. Comprehensive ADC analysis entails an integrated understanding of antibody stability, linker chemistry, payload characteristics, and potential immunogenicity. Robust bioanalytical packages are imperative to elucidate exposure-response relationships and support further ADC optimization and development. Overall, analytical selectivity, sensitivity, and novelty are paramount to decipher the intricate disposition mechanisms of ADCs.

Keywords Pharmacokinetics · Bioanalysis · Antibody-drug conjugates · Ligand binding assays · Mass spectrometry · Immunocapture · Unconjugated payload and its metabolite analysis · Drug-to-antibody ratio · Immunogenicity · Pharmacodynamics

---

Y. -J. Xue (✉) · J. Chen · B. Melo · M. Vallejo · J. X. Shen · J. R. Mora
Bristol-Myers Squibb, Princeton, NJ, USA
e-mail: Yongjun.Xue@bms.com

M. V. Myers
Gilead Sciences, Foster City, CA, USA

M. Chen
Regeneron Pharmaceuticals Inc., Tarrytown, NY, USA

S. E. Maxwell
Schrodinger, New York, NY, USA

S. Surapaneni
Nimbus Therapeutics, Boston, MA, USA

H. J. Gukasyan, A. Kulkarni (eds.), *Antibody Drug Conjugates*, AAPS Advances in the Pharmaceutical Sciences Series 52, https://doi.org/10.1007/978-3-031-89319-3_6

Introduction

Antibody–drug conjugates (ADCs) offer a unique opportunity for synergistically combining a therapeutic antibody (mAb) for specific target delivery with a highly potent cytotoxic payload for anticancer activity, and more recently other areas such as infectious diseases, immunoscience and neuroscience (Sasso et al. 2023; Hejmady et al. 2023; Tsuchikama et al. 2024). The unique properties of ADCs makes them a valuable and increasingly important tool in the pharmaceutical industry. ADCs have potential to enhance targeted therapy, increase therapeutic index, and address drug resistance. As application of ADCs became normalized in the field of medicine, regulatory thinking evolved as well. Recently, the FDA issued a final guidance on "Clinical Pharmacology Considerations for Antibody-Drug Conjugates" where bioanalytical approaches, dosing and exposure response, intrinsic factors (i.e. organ impairment, and pharmacogenomics), QTc assessment, immunogenicity and DDIs were discussed in detail (FDA 2024). Successful ADC drug development requires a comprehensive understanding of target biology, mAb properties, linker chemistry and payload characteristics. The safety and efficacy of an ADC depends on optimizing each individual component (antibody, linker, and toxin). Likewise, the development of a complete bioanalytical package that can successfully characterize the ADC pharmacokinetic properties relies on the optimization of multiple assays directed toward the individual components. The development of a promising ADC requires a comprehensive pharmacokinetic/pharmacodynamic (PK/PD) package that includes characterization throughout the development process to delineate the safety and efficacy of the compound. Each individual component of the biotherapeutic (mAb, linker, and toxin) contributes to the overall absorption, distribution, metabolism and excretion (ADME) properties of an ADC. To fully characterize the ADME of a novel ADC, the disposition and clearance of the mAb, the deconjugation and release of payload, and the metabolism and excretion of the toxin need to be assessed. Due to this inherent structural and functional complexity, the successful characterization of an ADC's PK properties becomes increasingly difficult and entails additional assay development when compared to either a small molecule or therapeutic antibody alone (Stephan et al. 2011; Kaur et al. 2013; Mu et al. 2022; Beaumont et al. 2022). To suitably characterize an ADC and what could potentially be a variety of molecular species, a distinctive panel of bioanalytical assays are required. These include total antibody, ADC (or total conjugated payload), unconjugated payload and its metabolites, along with the assessment of the drug-to-antibody ratio (DAR) in animal studies. The combined data set from these methods enables a comprehensive understanding of an ADC's PK, safety, and efficacy profiles (Jenkins et al. 2015; Kotapati et al. 2021; Qin and Gong 2022; Yin et al. 2023a; Zhu et al. 2020). Total antibody (tAb), ADC (an antibody or fragment with at least one payload), unconjugated payload (free payload), or pharmacologically active metabolite are typical PK assays derived from individual methodologies routinely used for both small molecules or biologics, while DAR assessment is a new genre of assays indicative of DAR distributions (Stephan et al. 2011; Kaur et al. 2013; Qin and

Gong 2022; Zhu et al. 2020; Gorovits et al. 2013). Per FDA's ADC guidance (FDA 2024), beginning with first-in-human studies, the ADC and its constituent parts, and its pharmacologically active metabolite should be measured to inform exposure-response of the drug. All bioanalytical methods should be validated and reported as outlined in the FDA guidance (FDA 2018) along with ICH M10 Bioanalytical Method Validation (November 2022) (Dere et al. 2013; Huang et al. 2020; ICH 2022). A fit-for-purpose qualification can be developed for the DAR method since it is typically only used for profiling DAR species in preclinical studies. In addition, potential immunogenicity of the ADC resulting from mAb, toxin or the linker should be assessed with appropriate anti-drug antibody assays. Each of these five assays will be discussed in the following sections of this chapter.

Total Antibody Quantitation (TAb)

Due to the bivalent nature of antibodies, they can exist in a variety of forms in the presence of the drug target ranging from free, to partially bound, to fully bound. While a free or partially bound antibody is still available to bind target, a fully bound antibody is completely engaged with either soluble drug target or anti-drug antibodies (ADA). Total antibody refers to all forms of an ADC present including fully bound and "free" antibodies, which refers to free and partially bound antibodies with either both or one arm available to engage target, and all those forms can be conjugated to payload or unconjugated (Mu et al. 2022; Qin and Gong 2022; Yin et al. 2023a; Lee et al. 2011; Ahene 2011). Understanding the distinction between total or free forms of the antibody is imperative for developing the appropriate assay to help characterize the ADC and support drug development. Knowing precisely what the bioanalytical assay is measuring will allow for proper interpretation of the data and characterization of the PK properties. From a bioanalytical perspective of ADCs, total antibody can be defined as the anti-target antibody with a DAR of 0 to *N*. The completely conjugated version of the ADC with a specific and designed DAR is the desired form of the ADC; however, the reality is that a small percentage of un-conjugated or "naked" antibody may be present in the dosing material. Un-conjugated antibody could also arise as a result of linker cleavage where the payload is released from the ADC in systemic circulation (Hejmady et al. 2023; Tsuchikama et al. 2024; Stephan et al. 2011; Gorovits et al. 2013). Non-cleavable linkers can minimize this effect, as the payload is primarily released through antibody endocytosis and lysosomal degradation (Ritchie et al. 2013). In such cases, the quantity of naked antibody returning to circulation should not be significant.

The fully conjugated ADC is the therapeutically active form of the drug, and the cardinal version of the drug for understanding the PK profile of the molecule. However, a naked antibody remains capable of binding target and imparting some therapeutic effects separate from the conjugated form, such as antibody dependent cellular cytotoxicity (ADCC) or complement dependent cytotoxicity (CDC) (Presta 2002). Since the naked antibody retains its ability to bind target, its presence in circulation may alter the PK profile, affect efficacy and safety and complicate interpretation of data by competing for target binding with the ADC. The potential for a shift in the ratios of ADC to naked antibody over time is a legitimate concern that necessitates monitoring of both total antibody and ADC concentrations, especially in preclinical and early clinical studies, when first characterizing the properties of the ADC. Naked antibody concentrations can be inferred from the tAb and the ADC measurements by taking the difference between the two measured concentrations (Kaur et al. 2013). Understanding the linker chemistry of the ADC (cleavable vs. non-cleavable), in addition to closely monitoring multiple ADC variants throughout preclinical and early clinical studies, will demonstrate whether the ratio of tAb to ADC concentrations shift significantly over time. If the ratio between these two forms remains constant or concentration changes are nominal, it could serve as justification to only monitor the ADC moving forward through development and into late phase studies. The following sections will discuss the use of both ligand binding assays (LBA) and liquid chromatography-tandem mass spectrometry (LC-MS/MS) approaches commonly employed to assess total antibody (tAb) levels in circulation (Stephan et al. 2011; Kaur et al. 2013; Mu et al. 2022; Qin and Gong 2022; Yin et al. 2023a; Dere et al. 2013; van den Broek et al. 2013).

Ligand Binding Assay

Ligand binding assays (LBA) are regularly used to determine the concentrations of monoclonal therapeutics in circulation and the tAb form of an ADC is no exception. Although LBAs continued to be utilized for tAb assessments, it is important to note that, depending on the program requirements, more particularly the LLOQ needs, a hybrid LCMS assay may be suitable and allow for efficiency gains when utilized for both tAb and total ADC quantification (see section "LC-MS/MS Assay").

There are many formats that may be implemented for tAb quantitation (see Fig. 1). Different stages of drug development can necessitate different approaches, for preclinical PK/TK studies, a generic assay approach can be developed utilizing anti-human antibodies that bind to the non-variable regions of the antibody as both the capture and detection reagents. There are affinity-purified polyclonal and monoclonal antibodies commercially available that do not cross-react with endogenous IgGs in animal plasma or serum (Wang et al. 2019). Other formats, more commonly utilized to support clinical studies, include the use of an anti-idiotypic (anti-ID) antibody or a target protein as capture reagent with an anti-ID for detection (Stephan et al. 2011; Kaur et al. 2013) (Fig. 1a–c). These formats are able to achieve LLOQs at lower concentrations, hence their implementation in clinical settings where ADCs are not dose as high as in TK studies. Aside from the format, the most typically used platforms are plate-based with colorimetric, chemiluminescent, or electro-chemiluminescent detection, and Gyros. The public domain is teeming with a vast collection of peer-reviewed literature and regulatory guidance documents detailing

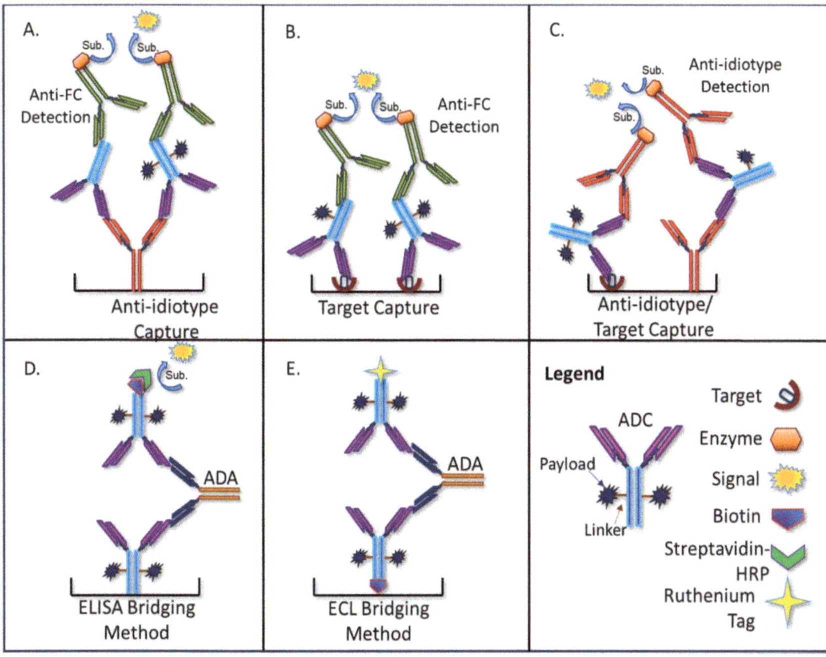

Fig. 1 Typical LBA and ADA method formats. Panels (**a**)–(**c**) represent potential strategies for LBA analysis of total antibody. (**a**) An anti-idiotype antibody used to capture the ADC which is then detected with an anti-Fc detection antibody; (**b**) Target used as capture reagent for ADC and subsequent detection using an anti-Fc antibody. (**c**) Anti-idiotype/target used as capture reagent for ADC, and an anti-idiotype antibody for detection. Panels (**d**) and (**e**) show common bridging assay formats for anti-drug antibody assessments using ELISA and ECL methods respectively

the development and validation of LBAs for therapeutic antibodies (FDA 2018; ICH 2022; Smolec et al. 2005; EMA 2011).

Additional assessments during the early development of the program may be necessary to offer confidence that the assay is indeed measuring the intended species, *i.e.*, all forms of the antibody irrespective of conjugated form or DAR. One crucial consideration when developing a tAb assay is to ensure that the assay used to detect the concentrations of total antibody is insensitive to the conjugated status of the ADC. Meaning that despite the number of payloads conjugated to the ADC (0–*N*), the tAb assay performance should not be affected. Comparing quality control samples with different DAR species, ranging from naked to fully, conjugated to those of reference standards or dosing solution containing intact ADC will provide such information. Like other regulatory validations, applying acceptance criteria for the accuracy and precision to these subsets of quality controls will verify whether the assay is insensitive to DAR. Newer generations of ADCs are being engineered with site-specific conjugation. For these ADC iterations, it may be important to demonstrate that the site of conjugation does not interfere with the detection of the antibody in the assay (Gorovits et al. 2013). Using the naked antibody as a reference and comparing its assay performance to the assay performance of antibodies conjugated at one or more varying sites may be desirable or even necessary to ensure that the selected site of conjugation is not interfering with tAb detection. The principal obstacle in addressing these questions is the availability of the critical reference reagents, such as ADC preparations with different DARs, anti-idiotype antibodies, and target antigens which are necessary for appropriate and comprehensive ADC characterization (Gorovits et al. 2013).

Kadcyla® (Trastuzumab emtansine) [T-DM1] is an ADC with a non-cleavable thioether linker clinically approved for the treatment of patients with HER2+ breast cancer. LBAs were used to measure total-trastuzumab in monkey and human serum with recombinant HER2 extracellular domain (ECD) as the capture and goat anti-human F(ab′)2 IgG Fc for detection. The validated total-antibody assays were capable of quantifying all T-DM1 DAR species immunoreactive to HER2 ECD ranging from fully conjugated T-DM1 to partially and fully unconjugated T-DM1 (Kaur et al. 2013; Dere et al. 2013). Additionally, Adcetris® (Brentuximab vedotin) is an ADC approved for the treatment of patients with relapsed or refractory Hodgkin's lymphoma that has also been supported with a validated ELISA assay for tAb detection. The method employs a bridging format using anti-ID30 as both a capture and detection reagent with a reported assay range of 12.5–400 ng/mL (Zhao et al. 2016; Han et al. 2013).

LC-MS/MS Assay

A myriad of publications have highlighted the utility of using LC-MS/MS for the quantitation of therapeutic proteins including monoclonal antibodies (Dubois et al. 2008; Wang and Heilig 2012; Fernandez Ocana et al. 2012; Neubert et al. 2013; Furlong et al. 2012, 2013; Ouyang et al. 2012; Jiang et al. 2013; Yang et al. 2007; Lassman et al. 2014). Furthermore, recent publications have even reported the development of methods for the simultaneous detection of tAb and antibody-conjugated drug (Xu et al. 2017, 2019; Huang et al. 2021; Suh et al. 2023). A common theme of the currently reported LC-MS/MS approaches involves protease digestion and monitoring the peptides generated as surrogate markers. The MS-based approaches for total antibody analysis can include: (1) direct LC-MS/MS analysis without sample enrichment; (2) solvent-mediated coprecipitation, LC-MS/MS using protein precipitation for sample enrichment; or (3) hybrid methodologies combining LC-MS/MS detection and immunoaffinity sample enrichment. Hybrid approaches have recently gained popularity as promising methods for antibody quantification (Suh et al. 2023; Faria et al. 2019). The combination of immunoaffinity enrichment coupled with the enhanced selectivity of LC-MS/MS works well in concert to detect and quantify the therapeutic protein. However, the limitations for hybrid LC/MS are its relative low throughput and sensitivity, which often requires higher sample volume than LBAs. All of these MS-based approaches share two identical steps: enzymatic digestion and LC-MS/MS analysis; however, the latter two methods employ enrichment techniques to simplify the sample matrix prior to subsequent processing and analysis.

Sample Enrichment

Plasma and serum are two matrices commonly used for total antibody bioanalysis. These matrices are complex, heterogenous mixtures dominated by highly abundant endogenous proteins that present significant challenges for the detection of the total antibody by LC-MS/MS. Sample enrichment (i.e. immunocapture or solvent-mediated precipitation) can ameliorate challenges inherent to LC-MS/MS-based antibody detection by: (1) enriching the sample from a biological matrix to increase mass spectrometry signal of the total antibody to achieve the desired detection limit; (2) removing highly abundant endogenous components to reduce the background noise of mass spec measurements; and/or (3) increasing digestion efficiency and decreasing the consumptions of protease enzyme for digestion. While alternative methods such as immuno-depletion and filtration have been successfully implemented as enrichment steps, these approaches are less frequently used than immunocapture (Dubois et al. 2008; Wang and Heilig 2012; Fernandez Ocana et al. 2012; Neubert et al. 2013) and solvent-mediated coprecipitation (Furlong et al. 2012; Ouyang et al. 2012; Jiang et al. 2013) approaches; therefore, they will not be addressed in this chapter. The implementation of anti-ID or anti-human Fc immunocapture in the sample extraction generally results in very clean samples ready for LC-MS/MS analysis. In contrast, less selective protein A and G pull-downs, solvent-mediated coprecipitation, or direct-digestion methods will likely result in "dirtier" sample preparations with lower signal to noise ratios and higher background due to interfering signals. In these cases, peptide level sample clean-up (e.g., solid-phase extraction or peptide immunocapture) might be necessary if higher sensitivity is required (Yang et al. 2007; Mark Rose et al. 2014).

Immunoaffinity Capture

Similar to LBAs, different reagents can be used for immunocapture of the tAb (of the ADC) from complex matrices prior to detection by LC-MS/MS analysis, ranging from protein A/G, anti-human IgG Fc, to target antigen and anti-idiotypic antibodies (Fig. 2). In terms of the pull-down selectivity, protein A/G generally bind to all mammalian IgGs (Fig. 2a), anti-human IgG Fc reagents only bind to human IgGs (Fig. 2b), target antigen (Fig. 2c) or anti-idiotypic antibody (Fig. 2d) selectively binds to the complement determining regions (CDRs) of the Ab portion of the ADC (free and partially bound forms), and anti-drug antibody binds to the payload (Fig. 2e). For a generic immunocapture procedure, either Protein A/G or anti-human IgG Fc (preferred since not typically cross-reactive with endogenous animal IgGs)

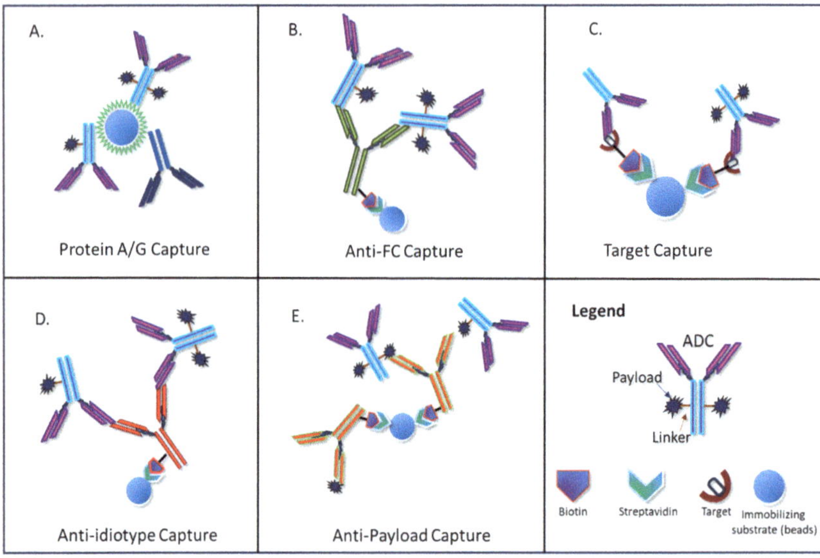

Fig. 2 ADC immunocapture methods. Immunocapture methods for ADC include a variety of means by which to capture the ADC for assessment. The properties of the ADC and the bioanalytical measurement should be considered when selecting the approach: (**a**) Protein A/G capture of IgG Fc regions; (**b**) anti-Fc antibody capture; (**c**) target capture; (**d**) anti-idiotype capture; (**e**) anti-payload capture

can be used to pull-down the tAb. Specifically, in pre-clinical studies when anti-idiotype reagents may not yet be available for use, these generic approaches can be developed and often applied successfully across different animal species to quantitate tAb levels. Anti-ID immunocapture is the preferred capture for support of clinical trials to selectively pull-down the target ADC (free and partially bound forms). However, method development for both preclinical and clinical studies will be heavily dependent on the availability of high-quality reagents directed at the total antibody form of the target ADC.

Solvent-Mediated Coprecipitation

As an alternative to immunocapture, methanol or other organic solvents can be used to coprecipitate the ADC along with other endogenous proteins as the pellet for enzymatic digestion (Furlong et al. 2012; Ouyang et al. 2012; Jiang et al. 2013; Liu et al. 2014). However, this sample enrichment is not selective and contains most endogenous proteins; therefore, it is not well-suited for highly sensitive tAb assays and cannot be used to measure unbound antibody (cannot distinguish free tAb from bound or partially bound tAb).

Enzymatic Digestion

Numerous protease enzymes (e.g., trypsin, Arg-C, Asp-N, Glu-C, Lys-C and papain) are available for the digestion of proteins to peptides with each protease possessing distinct cleavage site specificity. Trypsin can selectively cleave the peptide bond at C-terminal of Lys and Arg (except when N-terminal to a proline) and has been widely used as a protease of choice for many standard protein digestion methods (Dubois et al. 2008; Wang and Heilig 2012; Fernandez Ocana et al. 2012; Neubert et al. 2013; Furlong et al. 2012; Ouyang et al. 2012; Jiang et al. 2013). In addition, Lys and Arg are highly abundant and evenly distributed within antibody sequences. Following reduction of disulfide bonds and alkylation of cysteines, tryptic digestion produces peptides (7–20 amino acids) which are normally suitable for LC-MS/MS analysis. While trypsin is a widely used and robust protease for the digestion of proteins to peptides, protease selection should be individually based on the ADC peptide sequence, ensuring that an appropriate, unique, surrogate peptide can be generated for quantitation. More recent work demonstrates that an approach using nano-surface and molecular-orientation limited (nSMOL) chemistry provides selective proteolysis of the ADC FAb domain, while minimizing interference from additional analyte peptides (Iwamoto et al. 2014, 2016, 2017).

LC-MS/MS Analysis

In the field of mass spectrometry-based bioanalytical quantitation of proteins, it is common practice to rely on the measurement of a representative (surrogate) peptide rather than the entire protein. This approach is driven by sensitivity considerations and has become a widely accepted strategy. In these applications, surrogate peptides are selected based on their ability to serve as representative molecules of interest, enabling efficient and reproducible analysis without direct quantitation of the full protein. The selection of a representative peptide involves careful consideration of factors such as amino acid sequence, physicochemical properties, and enzymatic digestion characteristics. The chosen peptide should accurately reflect the presence and abundance of the target protein in the sample.

The use of *in silico* digestion tools (e.g. Skyline) can accurately predict potential tryptic (or other protease) peptides of a given ADC, and neat ADC digestion can tell which peptides offer suitable sensitivity & selectivity for mass spectrometry. A selected surrogate peptide should have the following characteristics: (1) approximately 7–20 amino acids without significant interference from endogenous components; (2) unique to your ADC sequence and (3) stable during sample preparation from digestion to analysis. For human IgG1 or IgG4 quantitation of the ADC, this common peptide in the Fc region, VVSVLTVLHQDWLNGK, can be used as the surrogate peptide in animal studies (Wang and Heilig 2012; Furlong et al. 2012, 2013; Jiang et al. 2013), while a peptide from the idiotype region should be used as the surrogate in clinical trials to minimize matrix interference and achieve the desired sensitivity (Dubois et al. 2008; Fernandez Ocana et al. 2012; Neubert et al.

2013; Yang et al. 2007). Most advanced triple-quadrupole mass spectrometers (e.g. Sciex 7500, Sciex 6500+, ThermoFisher TSQ, etc.) operating in selected-reaction monitoring (SRM) mode can provide adequate sensitivity (ng/mL) for the surrogate peptide analysis. Long analytical columns (e.g. 2.1 × 100 mm) are normally selected to minimize the potential chromatographic interference from the endogenous components as well as other peptides from the ADC. In some cases, micro or nano-LC can be used if additional sensitivity is required (Dubois et al. 2008; Suh et al. 2023).

In summary, a typical workflow may involve sample enrichment, digestion, and LC-MS/MS analysis for total mAb quantitation. Among sample enrichment methodologies, hybrid immunocapture LC-MS/MS workflows often provide better sensitivity than either solvent-mediated coprecipitation or direct LC-MS/MS approaches. Anti-Fc pull-down with LC-MS/MS is preferred for tAb quantitation (bound, unbound) in preclinical settings, while anti-idiotype with LC-MS/MS is preferred in clinical settings for more sensitive measurement of unbound tAb. Due to the less selective nature of protein A&G pull-down, solvent-mediated coprecipitation, or direct LC-MS/MS methods, they are less frequently used for tAb quantitation and may not provide the desired sensitivity for the assay being developed.

Summary for Total Antibody Quantitation

LBAs offer excellent sensitivity and throughput for tAb quantitation, with inexpensive instrumentation and plenty of validation experience. However, they rely on critical reagents for assay specificity and can be susceptible to matrix interference (Su et al. 2019) and differential binding affinity of DARs. On the other hand, LC-MS/MS, especially hybrid immunocapture LC-MS/MS methodologies, can provide superior specificity (minimal matrix effects), adequate sensitivity, and wide curve ranges with short method development times. Thus, hybrid immunocapture LC-MS/MS methodologies have gained popularity at an increasing rate as promising approaches for tAb quantification. However, they require expensive instrumentation and additional LC-MS operational, technical, and method development familiarities. While either LBA or hybrid immunocapture LC-MS/MS approaches can be used for total antibody quantitation, LBAs are still routinely the method of choice unless specific development issues (e.g. matrix effect) are encountered. Traditional therapeutic protein method validation should be followed along with any potential assessments that may be specific to the total antibody measurement of ADCs (Jenkins et al. 2015; FDA 2018; ICH 2022). Regardless of selecting an LBA or LC-MS/MS based approach for development and validation, it is important to consider whether the total antibody method is measuring the total or free form of the ADC in order to properly interpret the data.

An additional consideration when establishing a validated total antibody assay is the potential for an immunogenic response toward the administered ADC and the possible subsequent formation of anti-drug antibodies (ADA). These ADAs bind to the circulating drug and can form complexes with ADCs that may prevent ADC target binding or are endocytosed by non-target immune cells, resulting in cell death. For this reason, ADA impact on total antibody quantitation should be adequately evaluated during method validation prior to sample analysis and will be discussed in a later section.

Total Conjugated Drug or ADC Quantitation

Ideally, ADCs are intended to be stable during systematic circulation, reach the target for internalization, and release the payload for the maximum desired pharmacological effects. In addition to targeted release of the payload, there is a potential for off-target release of payload, through non-specific cellular uptake (phagocytosis, or pinocytosis), non-specific binding (FcyR, N-glycan receptors) or during circulation. High concentrations of circulating ADC are desirable to maximize efficacy, while low concentrations of circulating payload can minimize non-tumor specific toxicity. Therefore, the total conjugated drug quantitation provides an accurate measurement of the mean DAR concentrations. ADC systemic exposures are the critical PK component needed to establish the overall PK/PD correlation. Among the FDA approved ADCs so far (Mu et al. 2022), LBAs were more prevalent for ADC quantitation because the same technologies have been routinely used for tAb quantitation. More recently, LC-MS/MS (Stephan et al. 2011; Kaur et al. 2013; Dere et al. 2013; Suh et al. 2023) (specifically hybrid LC-MS/MS) has emerged as the leading option for ADC quantitation due to its unique assay selectivity. LC-MS/MS-based approaches can be advantageous when using a hybrid LC-MS/MS method that takes advantage of the efficiencies of both the immunoaffinity capture step (directed at the antibody portion) and the quantitative detection of the released payload by LC-MS/MS. Both approaches will be discussed in this section.

Ligand Binding Assays

For ADC quantitation by LBA, either the capture or detection reagent must bind to the payload while the other antibody binds to the monoclonal antibody. Since the molecular weight of the payload (even including the linker) is so small (1500 Dalton or less), generation of an antibody with adequate affinity and specificity to the payload is challenging. With the same linker chemistry as used in the ADC, the payload is linked to an immunogen/adjuvant (such as Keyhole limpet hemocyanin) for the immunization of animals. Animal sera are collected and purified for antibodies specific to the payload.

There are many factors that influence if a LBA assay for quantitation of ADC is DAR sensitive or not. For it to be sensitive to DAR, the signal would have to increase depending on the number of payloads. Therefore, if an ADC has a very high DAR it could result in steric hindrance impacting anti-payload antibodies from binding, so even if the same format is used for two different ADCs one assay could end up not being DAR sensitive. The affinity of the anti-payload antibody is another important factor, as the antibodies must remain bound for the signal to increase. Finally, the format of the assay would determine DAR sensitivity because if the capture antibody is an anti-payload, the signal would not increase according to the number of payloads.

DAR Sensitive Total Conjugated Drug Assay

A DAR sensitive method typically employs a recombinant target-protein or an antibody directed toward the mAb portion of the ADC as the capture reagent. The affinity between the target protein or anti-idiotype antibody and the mAb portion of the ADC is generally high and more than strong enough to serve as an efficient and specific capture reagent to provide acceptable assay sensitivity. During early development, it is imperative to demonstrate that the capture step is not affected by steric hindrance from varying conjugated forms of the ADC. In this format, a labeled anti-payload antibody is used for detection. The binding affinity between anti-payload antibodies and the captured ADC should be strong enough so that a bound anti-payload antibody would not be displaced upon binding of a second Ab. It is important to understand that the relationship between the number of conjugated payloads and assay response may not have a linear correlation. This will have to be characterized during method development and validation to ensure the consistent binding efficiency or binding ratio between the anti-cytotoxic antibody and the payload. A prime example in literature uses human CD22-ECD or goat antihuman IgG antibody (Fc or F(ab′)2 fragment specific) for capture of anti-CD22-MCC–DM1 or MC–MMAF ADCs, respectively, followed by incubation with a biotinylated anti-DM1 or anti-MMAF antibody for subsequent detection (Stephan et al. 2008).

DAR Insensitive Total Conjugated Antibody Assay

For DAR insensitive method development, an anti-toxin-antibody can be used to capture the ADC, and labeled antibody directed towards the antibody portion of the ADC or labeled target protein is used for detection. Detecting with the recombinant target or an Ab against the monoclonal portion of the ADC should render the assay insensitive to DAR. Ensuring the anti-cytotoxin capture antibody has sufficient affinity toward the payload is essential, since this interaction determines the efficiency of the capture step. A lower affinity antibody may not be as effective at capturing ADCs with low DARs (Xu et al. 2013), as it is at capturing ADCs with higher DARs due to the increased avidity multiple cytotoxic payloads can offer. During method development and validation, DAR assay sensitivity should be evaluated using ADCs with specific DARs or ADCs enriched for unique DARs species. This assay format measures conjugated antibody concentrations, and arguably does not provide a direct measurement of the efficacious entity (conjugated drug). Since in this format, an antibody with a single conjugated payload is expected to produce the same assay response as an antibody with multiple sites of conjugated payload. In the bioanalytical analysis of trastuzumab–maytansinoid conjugates supporting monkey and human studies, capture of the ADC was carried out using an anti-DM1 antibody followed by subsequent incubation with biotinylated HER2-ECD and streptavidin–HRP for detection. With this conjugated antibody being insensitive to the number of payloads, standards/QCs with DAR of 3.5 could be used to accurately quantify T-DM1 with DARs ranging from 2.58 to 4.10 (Kaur et al. 2013; Dere et al. 2013; Xu et al. 2013; Lewis Phillips et al. 2008). Similarly, when a DAR insensitive assay was used to quantitate the anti-STEAP1 ADC by using an anti-toxin mAb for capture of the ADC and a biotinylated anti-idiotype mAb for detection, it resulted in significant under quantitation of DAR1 and DAR6 (11% and 64% recovery), but acceptable quantitation of DAR2 and DAR4 (102% and 99% recovery) (Kaur et al. 2013).

LC-MS/MS

In contrast to LBAs for ADC quantitation, LC-MS/MS assays can measure the cleaved payload from the ADC directly. LC-MS/MS assays typically involve three key steps: sample enrichment, digestion, and LC-MS/MS analysis (Stephan et al. 2011; Kaur et al. 2013; Dere et al. 2013; Suh et al. 2023; Buckwalter et al. 2004).

Sample Enrichment

Comparable to the hybrid LC-MS/MS methods for total antibody quantitation discussed previously, enrichment prior to sample analysis is still recommended to decrease sample complexity and eliminate interference. Similar capture reagents can be used for ADC enrichment and, when possible, should be used to allow for more accurate comparison of mAb and ADC data within a study. Sample enrichment methods should be selected according to study designs: total ADC enrichment for preclinical studies or free form of ADC for clinical trials. While ADC concentrations are normally at μg/mL levels or higher, sample enrichment may still be necessary to ensure selectivity and assay performance throughout all phases of development.

Digestion

As previously mentioned, using LC-MS/MS to determine ADC concentration involves detection of the cleaved payload from the antibody. This is generally accomplished by removal of the payload by cleaving linker between ADC and payload. Early generations of cleavable ADCs were comprised of:

1. Acid-labile (hydrazone) linkers (Stephan et al. 2011; Kaur et al. 2013; Gorovits et al. 2013; Bross et al. 2001; Hamann et al. 2002)—titration of plasma or serum pH from 7.4 to 4.5 releases the conjugated drug (drug + linker) for ADC analysis.
2. Disulfide cleavable linkers (Stephan et al. 2011; Kaur et al. 2013; Gorovits et al. 2013; Widdison et al. 2006; Chari 2008; Lee et al. 2020a)—disulfide bond reduction (e.g. dithiothreitol) followed by alkylation (e.g. iodoacetamide) release the conjugated drug with thiol group and form a stable thioether bond.
3. Peptide cleavable linkers (e.g. Val-Cit, Phe-Lys) (Kaur et al. 2013; Gorovits et al. 2013; Suh et al. 2023; Carter and Senter 2008; Doronina et al. 2003)—lysosomal protease digestion (e.g. cathepsin B, Papain) releases the conjugated drug.
4. Beta-glucuronide cleavable linkers (Jeffrey et al. 2006; Gianolio et al. 2012; Lee et al. 2020b)—beta-glucuronidase hydrolysis releases the conjugated drug.

For example, the quantitation of anti-STEAP1 ADC (with Val-Cit linker) was accomplished by immunocapture/enzymatic digestion/LC-MS/MS, since LBA method could not accurately measure DAR1 (Kaur et al. 2013).

Later generations of ADCs were designed with noncleavable linkers including:

1. Non-site specific conjugation (e.g. cysteine)—protease digestion results multiple species of the conjugated drug; as the digested specie may or may not contain payload, this type of the noncleavable ADC is not well suited for LC-MS/MS quantitation. A good example was that Kadcyla® ADC quantitation which was conducted by LBA instead of immunocapture/digestion-LC-MS/MS (Stephan et al. 2011; Kaur et al. 2013; Dere et al. 2013) because of the random thiol noncleavable conjugations.
2. Site specific conjugation—protease digestions yield various forms of the payload that are still attached to the linker, as well as sub-species of the antibody-drug conjugate (ADC). Although quantifying the payload through LC-MS is highly challenging in this case, it is possible to develop a method using the aforementioned analytical approach to measure the subunit of the ADC.

Understanding these linker factors highlights the importance of comprehensive characterization of the ADC linker chemistry and stability early on in development to design an appropriate method for accurate quantitation.

LC-MS/MS Analysis

Through immunocapture pull-down and enzymatic or chemical digestion, the ADC sample extracts should be relatively free from interference and ready for LC-MS/MS detection without further sample cleanup, unless assay ranges dictated by study requirements necessitate additional sample preparation steps. Since the released payloads (as a surrogate of ADC) behave very much like small molecule drugs, reverse-phase chromatography should provide adequate separation. A stable-labeled ADC internal standard is preferred to track the released payload from immunoaffinity capture and digestion to LC-MS/MS analysis. A stable-labeled drug (or drug + linker) can serve as an alternative, but it will not track immunocapture pull-down and digestion steps. Highly sensitive mass spectrometry in SRM mode typically accommodates the required assay sensitivity and selectivity. A good example was the use of a hybrid immunocapture, LC-MS/MS assay for the quantitation of anti-STEAPI ADC for a clinical study (Kaur et al. 2013).

In summary, by taking advantages of two orthogonal platforms (LBA and LC-MS/MS), hybrid immunocapture LC-MS/MS provides unique assay selectivity and sensitivity for the detection of the total conjugated ADC unrealized by either platform independently. As a generic ADC sample enrichment, anti-human Fc or Protein A/G are normally used to measure total ADC (bound and unbound) in animal species. For clinical trials, anti-idiotype methods are typically used to selectively enrich and detect only unbound ADC.

Summary of Total Conjugated Drug or ADC Quantitation

Both LBA and hybrid LC-MS/MS assays can be used to quantify ADCs; the platform of choice depends on specific ADC linker chemistry and sensitivity requirements (Jenkins et al. 2015; Su et al. 2019). Similar to mAb quantitation, Ligand binding ADC assays offer excellent assay sensitivity, sample throughput with relatively inexpensive instrumentation. The drawbacks for LBAs are poor assay selectivity or differential recovery (of DARs) due to low and variable conjugated drug binding of individual DARs. Additionally, it is not a direct measurement of the conjugated drug with complex relationship between detection signal to payload. While hybrid assays require more expensive instrumentation and may only be suitable for the certain ADC linker chemistry, they offer excellent assay selectivity for a direct and sensitive measurement of the conjugated drug, with relatively short assay development time. In addition, the circulating target protein and ligand can impact the assay performance of either platform; therefore, it is important to evaluate their impact during method development or validation.

Measurement of DAR Distribution

Stability of an ADC therapeutic form is a paramount concern with the composition of an ADC designed with the intent to deliver a potent agent to targeted tissue while minimizing its toxicity to normal tissues. ADCs are required to be stable during systemic circulation to be viable drug candidates and confer desired efficacy and safety. However, first generation ADC linkers lacked robust stability and were more prone to releasing the toxic payload into the blood resulting in toxicity. Although new generations of ADCs employ non-cleavable or enzymatically cleavable linkers in an attempt to overcome such a shortcoming (Junutula et al. 2010; Xu et al. 2011; Todoroki et al. 2020); the loss of payload still occurs *in vivo* from linker cleavage or other degradation pathways. Significant loss of the payload in systemic circulation not only compromises the desired therapeutic effect, but also raises a safety concern regarding the administered ADC. Understanding these factors highlights the need to monitor DAR shift *in vivo* and understand the stability and activity of ADC drugs (Mu et al. 2022; Huang et al. 2021; Xu et al. 2013).

LBAs may be able to distinguish ADC from drug-free antibody using payload specific antibodies; however, they cannot discern various DAR species due to indirect detection. High-resolution mass spectrometry (HRMS), with the help of advanced deconvolution software, has proven to be an indispensable approach for investigating the DAR distribution of ADCs. Not only can this LC-MS approach differentiate between conjugated antibody and drug-free antibody, but it can also detect an array of DAR species generated by the manufacturing process, deconjugation or degradation *in vivo*. Efforts to accurately characterize DAR distribution in ADC molecules by analyzing fully intact or partially digested protein (e.g. light chain and heavy chains) on HRMS with electrospray ionization (ESI) are highly valuable. One caution is that the partial digestion can lead the loss of some structural information necessary to understand the DAR and drug loss, although its smaller ADC components can make DAR measurement simpler (Wagner-Rousset et al. 2014). Using an HRMS platform to assess DAR distributions typically involves three steps: immunocapture, deglycosylation (for antibodies generated in cell cultures), and high-resolution mass spectrometry analysis.

Immunocapture of ADCs

Extensive sample cleanup and enrichment for DAR samples are critical because chromatographic separation cannot be achieved using standard reverse phase chromatography for detecting different DAR species among a host of plasma or serum proteins. Therefore, immunocapture has become a standard practice for enriching the target ADC and removing endogenous proteins for DAR measurement.

Similar to immunocapture for total antibody or conjugated drug, polyclonal anti-human IgG Fc antibody, target antigen/anti-idiotype antibody or anti-toxin antibody have been employed as capture reagents to enrich ADCs from plasma or serum samples. To decide which capture reagent is more appropriate, one should consider the type of sample being analyzed, the sites of drug conjugation, and the availability of the reagents in order to choose the optimal enrichment method for sample cleanup (Dere et al. 2013; Dubois et al. 2008).

It is important to note that the first approach, anti-Fc capture, is an ideal option for the capture and measurement of all DAR species in preclinical studies but not in human samples because ADCs with humanized or fully human antibodies have almost identical sequences as endogenous IgGs. The anti-Fc capture approach can be used for multiple structural variants of conjugated antibody or across multiple species. This generic and robust capture method was used for immunocapture and subsequent LC-ESI-MS quantification of the human IgG1 mAb, αPCSK9 in cynomolgus monkey serum (Dubois et al. 2008; Stubenrauch et al. 2010).

The second approach uses target antigen or anti-idiotype antibodies to capture the ADC for DAR (free or partially bound from 0 to *N*) measurement in both animals and humans, in some cases, with commercial kits available. For example, the biotinylated ligand (MUC16) was used to pull down a cysteine engineered anti-MUC16 ADC, followed by LC-MS/MS for DAR distribution analysis (Xu et al. 2011). The same immunocapture methodology was subsequently used to investigate other ADCs (Xu et al. 2013; Shen et al. 2012a; Kaur 2013a).

The last approach is to capture the ADC using an anti-toxin antibody specific to the conjugated drug. This method requires that antibodies be raised against the drug portion of the ADC, not mAb portion of the ADC or endogenous antibodies. This approach applies to both animals and human samples. However, the anti-toxin antibody can have different binding affinities to various DAR species and also will not pull down DAR0. One specific method used an anti-drug antibody against MMAE of the ADC to measure the DAR values (Sanderson et al. 2005).

Regardless of the different immunocapture approaches, sample cleanup and enrichment are required for the accurate measurements of the DAR distribution in *in vivo* samples. Researchers should note that the three capture approaches may exhibit different binding affinities to various DAR forms, and it will be helpful to examine immunocapture with individual DARs for more consistent pull-down of all DAR forms.

Deglycosylation for Captured ADCs

Deglycosylation of captured ADCs from plasma or serum samples is needed for antibodies produced by cell-based systems. There are two N-linked glycosylation sites (ASN-297 of each Fc heavy chain) (Arnold et al. 2005) with considerable heterogeneity due to the various combinations of the glycans. Glycan modifications lead to complex MS spectra of the ADCs and inaccurate measurements of DAR values. N-Glycanase (PNGase F) is commonly used to remove N-linked oligosaccharides from glycosylated proteins (Tarentino and Plummer Jr. 1994), and a variety of deglycosylation kits or reagents are available from commercial sources. Following incubation with the enzyme in either nondenaturing or denaturing conditions (Prozyme, *Glycopro Enzymatic Deglycosylation Kit.* 2013), the treated ADC samples can be analyzed by HRMS without further preparation.

High Resolution Mass Spectrometry Analysis

Measuring DAR distributions of ADCs in *in vivo* samples is a critical part of ADC drug development. It provides important information such as *in vivo* linker stability and biotransformation regarding different DAR species. Some early mass spectrometry-based methods developed for DAR measurement employed matrix-assisted laser desorption for ionization mass spectrometry (MALDI/MS) of the ADC samples (Junutula et al. 2010; Xu et al. 2011; Todoroki et al. 2020). For example, MALDI-time-of-flight mass spectrometry (TOF-MS) was used to measure DAR values of multiple conjugated species to prostate specific membrane antigen (PSA)-specific monoclonal antibodies. These PSA-mAbs were conjugated to either a radio-labeled form of 1,4,7,10-tetraazacyclododecane-N,N0,N00,N000-tetraaceticacid or the cytotoxic drug maytansine (Wagner-Rousset et al. 2014). Although these early DAR methods were straightforward to use, they had difficulty differentiating between DAR variants due to overlapping spectra from multiple isotopic species (Stubenrauch et al. 2010).

Currently, high-resolution mass spectrometers in combination with electro-spray ionization (LC-ESI-HRMS) are widely used for the quantitation of DAR distribution for ADCs. Total ion current (TIC) spectra consisting of a multiply charged envelope of ADC species are acquired using quadrupole time-of-flight mass spectrometers (Q-TOF) in positive ESI mode or orbitrap mass spectrometers such as Thermo Q-exactive. Specialized software such as BioPharmaView™ (AB Sciex) and Protein Deconvolution Software (Thermo Scientific) are used to deconvolute the full mass spectra and generate the *in vivo* DAR profile of circulating ADCs that include a variety of conjugated antibody forms with varying drug loads. DAR distribution can be estimated by comparing the deconvoluted peak area for each DAR species present, with the assumption that each ADC species has similar ionization efficiency. A prime example of this was the use of reverse-phase (RP) and size exclusion HPLC coupled with Q-TOF to measure the DAR distribution for the recombinant human C242 antibody and its maytansinoid ADCs (Shen et al. 2012a).

While DAR analysis for lysine-linked and engineered cysteine-linked ADCs with RP-LC-Q-TOF has been well established (Hamann et al. 2002; Stubenrauch et al. 2010; Shen et al. 2012a), DAR analysis for labile ADCs (e.g. reduced interchain cysteine residues for drug conjugation sites) held together by non-covalent interaction between the subunits have been challenging with the similar RP-LC-TOF approach due to instability of the denatured ADCs (Kaur 2013a). Instead, the DAR analysis of these unstable ADCs has been accomplished by hydrophobic interaction chromatography (HIC) and UV detection (HIC-UV) (Arnold et al. 2005) or with a combination of HIC-UV and SEC-MS (size exclusion chromatography mass spectrometry) (Tarentino and Plummer Jr. 1994). More recently, a nano-ESI-MS platform has been reported for the determination of average DAR in the native or denatured stage (Sanderson et al. 2005; Arnold et al. 2005). Ion mobility mass spectrometry (IM-MS) has also been demonstrated as a useful tool when combine with high-resolution MS to measure DAR in complex ADC samples (Debaene et al. 2014; Huang et al. 2015). A recently published "middle-up" approach used the IgG-degrading enzyme of streptococcus pyogenes (IdeS) and DTT reduction to produce light chain (LC) and Fd' fragments for DAR analysis (Davis et al. 2017).

A monkey study with Kadcyla® highlights the use of MS methods to measure DAR changes in early ADC development. In this approach, biotinylated extracellular domain of recombinant HER2 or anti-DM1 antibody was used to capture T-DM1 from plasma. After on-bead deglycosylation and elution of DAR species, capillary liquid chromatography (RP-LC) coupled with Q-TOF MS was used for

DAR measurement. A gradual DAR shift to lower values over 28 days was observed in the study samples, providing evidence that fresh ADCs used to prepare standards/QCs may not fully represent the analytes measured at later PK time points (Kaur et al. 2013; Dere et al. 2013). The downward DAR shift indicates the extent of drug deconjugation and change in clearance rates among DAR species (FDA 2018).

Summary of DAR Assessment

While current DAR analysis methods are notoriously not sensitive and only semi-quantitative in nature, they can provide important supplementary information to help support ADC drug development. DAR distribution assessments in the discovery setting guide the engineering of better ADC candidates with optimal linker chemistry/stability and other desirable "drug-able" properties; DAR distribution assessments in the development setting allow for the generation of robust quantitative PK assays for more accurate measurements of ADC species *in vivo* (FDA 2018). New generations of high-resolution mass spectrometry with better sensitivity and increased resolution have enabled more accurate and sensitive DAR monitoring *in vivo* or even total antibody/ADC quantitation (Huang et al. 2021). Therefore, DAR assessment should be considered in a preclinical study, especially for labile ADCs.

Unconjugated Payload Quantitation by LC-MS/MS

The catabolism (*in vivo* breakdown) of an ADC can occur either on the mAb portion or the linker and conjugated drug. The mAb portion is catabolized into intact proteins (e.g. different DAR species), protein sub-units, and eventually peptides and amino acids (Saad et al. 2015). The fate of mAb catabolites is well understood to pose little safety concerns (Hall 2014; Han and Zhao 2014; Ezan et al. 2014). Conversely, unconjugated payload and its metabolites derived from the conjugated drug (e.g., drug, drug-linker, drug-linker-peptide), need to be thoroughly evaluated to understand potential off-target toxicity effects and to help inform on the safety and efficacy of the ADC (Erickson and Lambert 2012; Prueksaritanont and Tang 2012).

Unconjugated payloads are often identified through *in vitro* studies including metabolism studies in target-expressing cell lines and plasma stability studies across species. These early development studies can help identify ADC free payloads, define disposition mechanisms and highlight relevant preclinical species. One such example is the identification of three free payloads (MCC-DM1, Lys-MCC-DM1, and DM1) for Kadcyla®. Monitoring of these free payloads was beneficial in elucidating the dynamic and temporal free payload/metabolite's characteristics of the ADC during this process (Han and Zhao 2014; Shen et al. 2012b; Lin and Tibbitts 2012; Li et al. 2024; Wu et al. 2024; Yin et al. 2023b).

For *in vivo* studies, the quantitation of unconjugated payload and its pharmacologically active metabolite, is required by FDA (2024) and typically performed using LC-MS/MS methods. The inherent advantage of improved selectivity (Stephan et al. 2011; Kaur et al. 2013) compared to ligand binding assays (e.g. ELISA methods), makes MS-based approaches the primary choice for small molecule unconjugated payload quantitation in complex biological matrices such as plasma, serum, and tissue samples (Zhang et al. 2016; Lanshoeft et al. 2016; Jeon et al. 2023). It is worth noting, however, that LBA methods have been reportedly used for such assessments as well (Salomon and Singh 2015).

LLOQ Challenges

Developing bioanalytical methods with sufficient sensitivity for unconjugated payload and any pharmacologically active metabolites poses a significant challenge, particularly when working with plasma or serum matrices where the levels of circulating unconjugated payload and its metabolites are typically very low to minimize off-target toxicity. To address this concern, the FDA recommends that the assay for unconjugated payload should possess a high level of sensitivity to detect very low levels of unconjugated payload and its pharmacologically active metabolite in circulation.

Fortunately, unconjugated payload and its metabolites usually exist in a single charge state (and occasionally in a double charge-state for specific cases), simplifying mass spectrometer tuning. Additionally, these compounds are more hydrophobic than most endogenous components, allowing them to be retained well on reverse-phase LC columns, facilitating chromatographic separation and efficient electrospray ionization. Due to these characteristics, typical sample preparation procedures used for small molecule bioanalysis can be applied to quantify unconjugated payload and its pharmacologically active metabolite, often achieving the desired sensitivity. Even straightforward protein precipitation (PPT) methods have been reported to provide sufficient sensitivity, with lower limits of quantification (LLOQs) as low as 0.1 ng/mL (Nagai et al. 2019; Yang et al. 2019) using modern LC-MS/MS platforms like Sciex 7500, Sciex 6500+. For more challenging LLOQs, liquid/liquid extraction or solid phase extraction can be employed, as described in the subsequent section.

Sample Preparation Considerations

Depending on the structure and chemistry of the linker, addition of stabilizers (e.g. protease inhibitors) might be needed to prevent the payload release during sample storage and processing (Wei et al. 2018). This is because even a minor payload release can have a large impact on quantitation due to the low level of free payload.

Based on the specific requirement of the study (e.g. low dose level clinical study) or nature of the ADC, the unconjugated payload and its metabolites, more complicated sample treatments and LC separations may be necessary. Compared to PPT, liquid-liquid extraction (LLE) and solid phase extraction (SPE) methods clean up and concentrate the samples more efficiently when higher sensitivity and selectivity are required. Novel liquid extraction surface analysis and microcapillary LC techniques have also been reported for high sensitivity quantitation of a lys-MMC-DM1 free payload in rat liver and tumor tissue samples (Lanshoeft et al. 2016).

In addition to measuring circulating unconjugated payload and metabolite concentrations in plasma or serum, it has been shown that identification and quantitation in target tissue samples can provide important efficacy information (Zhang et al. 2016; Lanshoeft et al. 2016). For sample preparation, homogenization of the tissue samples in blank plasma is usually applied and followed by typical small molecule sample preparation procedures.

Summary of the Unconjugated Payload and Its Metabolites Analysis

Although the unconjugated payload and its metabolite assays are just a few of the many assays required for comprehensive ADC characterization, they play a crucial role in assessing ADC toxicity and should be carefully considered during development. As previously mentioned, maintaining low levels of unconjugated payload and its metabolites in circulation is advantageous for minimizing systemic toxicity associated with ADCs. Therefore, it is essential to employ sensitive LC-MS methodologies to monitor these species effectively.

Immunogenicity Analysis

Immunogenicity is the ability of a biotherapeutic, such as an ADC, to induce an immune response. The consequences of an immune response to a therapeutic protein can range from no clinical significance to life-threatening (Casadevall et al. 2002); like other biotherapeutics, consequences of ADA may also include impact on pharmacokinetics, pharmacodynamics, safety, and efficacy (Grimm et al. 2019; Bartelds et al. 2011). The FDA guidance "Clinical Pharmacology Considerations for Antibody-Drug Conjugates Guidance for Industry" (FDA 2024) provides details for the assessment of ADC immunogenicity, it references Immunogenicity Assessment for Therapeutic Protein Products (August 2014) (FDA 2014) and Immunogenicity Testing of Therapeutic Protein Products—Developing and Validating Assays for Anti-Drug Antibody Detection (January 2019) (FDA 2019), and it indicates that it could be appropriate to measure the immune responses to the constituent parts of the ADC. Current industry guidelines require the generation of an appropriate approach including sensitive and validated methods for the assessment of anti-drug antibodies. Myler and colleagues (Myler et al. 2022) provide industry-wide recommendations on the method validation and reporting of immunogenicity assessments. Additionally, it is recommended by the FDA to implement a risk-assessment strategy to identify and mitigate potential risks related to immune responses that may occur as a result of treatment with the therapeutic protein (FDA 2019; Mora and Richards 2023; EMA 2017). When establishing the risk-based assessment and developing methods to detect immunogenicity, the goal should be to investigate the presence of an immune response to the therapeutic protein and understand its clinical impact (Baldo 2013). This involves a multi-tiered approach including a screening assay for identifying ADA positive samples/patients, a competition assay for confirming the presence and specificity of ADA and functional assays for assessing the neutralizing capability of the ADA (FDA 2019; EMA 2017). Thus, the comprehensive evaluation of ADC immunogenicity should be based on integrated analysis of immunological, pharmacokinetic, and pharmacodynamic, as well as clinical efficacy and safety data. Hock et al. (2015) provide guidance on immunogenicity monitoring strategy specific to ADCs and elaborates on how to perform the risk assessment that would drive the strategy (https://pmc.ncbi.nlm.nih.gov/articles/PMC4287284/).

Immunogenicity and Cut Points

Immunogenicity assays are qualitative and employ statistically derived cut points (CP) to assess sample positivity. The CP serves as a threshold level of response in an immunogenicity assay at or above which a sample is denoted as positive; responses below the cut point are defined as negative (Carrasco-Triguero et al. 2013). Establishing an appropriate CP is crucial for developing assays that would accurately report incidence, and is performed for each individual tier of assays in an immunogenicity assessment (e.g. screening CP, confirmatory CP, titration CP, etc.) (Myler et al. 2022; Shankar et al. 2008). Cut points not only serve as a positive/negative threshold for bioanalytical assays, but also as means of normalizing data across runs. Cut point calculation is not a trivial task; it requires statistical analysis tools and samples from treatment-naïve subjects to generate robust cut point values and appropriately account for assay variability (Shankar et al. 2008, 2014). Current guidance suggests a minimum of 50 samples run on at least three different days to capture enough information for proper statistical analysis. Depending on the study, this may be required for disease-specific matrices as well since different patient populations may have matrix components that impact signal and subsequently the cut point (FDA 2019). Several literature sources have already thoroughly addressed the importance of identifying the appropriate statistical methods and how to calculate the CPs within regulatory considerations (FDA 2019; EMA 2017; Carrasco-Triguero et al. 2013, 2019; Shankar et al. 2008, 2014; Shen et al. 2014; Boehncke and Brembilla 2018); a detailed discussion of CP statistics is out of scope for this chapter.

Screening, Confirmatory and Titer

Detecting and characterizing ADAs generated toward an ADC follows the same multi-tiered strategy used for other therapeutic antibodies. First, a screening assay is performed capable of binding low- and high-affinity ADAs. Because the cut point in the screening tier is established to report 5% false positives, those samples that test positive for the presence of ADA will need to be further verified using a confirmatory assay. Generally, confirmatory assays involve addition of the drug (ADC) in the assay to result in signal inhibition and confirmation that ADAs are binding specifically to the therapeutic protein. Once specificity of the ADA for the ADC has been confirmed, titration assessments are performed to understand the magnitude of the immune response to the conjugated drug (Shankar et al. 2008, 2014). Development and validation of the screening, confirmatory and titration assays for ADC immunogenicity assessment should follow standard regulatory recommendations applied to other therapeutic proteins (FDA 2019; Myler et al. 2022; EMA 2017). This inclusion of several levels of immunogenicity testing effectively reduces false positive identifications, thus ensuring the finally "positive" result is not an artifact of non-specific binding between detection reagents and matrix components.

Screening Assays

Screening assays, also known as binding antibody assays, are used to detect antibodies that bind to the therapeutic protein product (in this case the ADC). Screening assays are generally regarded as semi-quantitative assays because positive control reagents are not a reference standard, and the immune response is different for every patient. Suitable formats for determining the specificity between an anti-drug antibody and therapeutic protein or ADC commonly include direct/indirect ELISA, bridging ELISA and electrochemiluminescence (ECL) methods (Carrasco-Triguero et al. 2013; Dere et al. 2023; Myler et al. 2015). Figure 1d, e show the format typical bridging assays. Although less common, radioimmunoprecipitation assays (RIPA) and surface plasmon resonance (SPR) techniques have also been reported for panitumumab (Bartelds et al. 2011; Mire-Sluis et al. 2004; Lofgren et al. 2007). Requirements for clinically validated assays can differ from those of preclinical assay where at the early stages of development, multiple variants of the ADC (varying DARs, linker chemistry or payloads) are included in studies across different species. To simplify ADA analysis at this early stage, generic approaches have been reported to save development time and allow for flexibility when there is limited reagent availability; the use of murine anti-human Fc or FAb domain captures and an anti-IgG antibody directed at the test species (mouse or monkey) for detection of

ADAs has been reported (Stubenrauch et al. 2010, 2012; Carrasco-Triguero et al. 2016). In some cases, the generic approaches performed equal or better than the therapeutic-specific bridging assay with better drug tolerance and adequate sensitivity. However, the caveats of the generic formats is that they do not permit domain mapping of ADA binding sites on the therapeutic protein and are usually isotype specific (Carrasco-Triguero et al. 2016; Partridge et al. 2016).

To improve the drug tolerance of ADA assays, sample pre-treatment approaches such as acid dissociation and solid-phase extraction with acid dissociation (SPEAD) have been implemented; these pre-treatments are needed when a therapeutic protein, such as an ADC, is present at high concentrations in systemic circulation (Patton et al. 2005; Song et al. 2016; Smith et al. 2007; Dai et al. 2014). It is worth noting, that before proceeding with either of these pre-treatment processes, it is imperative to have a good understanding of the ADC properties and linker chemistry; acid used in a pretreatment sample preparation step should be adequately neutralized before incubation with labeled ADCs to ensure integrity of critical reagents (Hock et al. 2015).

Confirmatory Assays

Confirmatory assays are performed as a follow up assessment on the samples that "screen" positive for ADAs, to ensure specificity of the ADAs for the biotherapeutic and reduce false-positives identified in the screening assay. While a variety of method formats are available for the confirmatory assay including immunoprecipitation and depletion, the typical format employed for the confirmatory assay uses competitive inhibition with the addition of excess of unlabeled drug (Driscoll et al. 2013; Jani et al. 2015). Specific signals caused by the presence of ADA in the sample should be suppressed by addition of the exogenous drug competing for binding to the ADA with the labeled drugs used to capture (direct assay) or capture and detection (bridging assay), while the nonspecific signals should be not be suppressed (Hock et al. 2015). The degree of competition is determined by comparing drug-spiked samples to unspiked samples and calculating the percent inhibition of the spiked drug. Preference is placed on developing a confirmatory assay based on the screening assay to increase the likelihood of comparable assay sensitivities and verifiable false-positives (Swanson and Chirmule 2009). Similar to screening assays, confirming a putative positive sample as ADA positive involves establishing a confirmatory cut point above which all samples are denoted as confirmed positive (Kubiak et al. 2020).

Like screening assays, an acid dissociation step may be necessary when a therapeutic drug, such as an ADC, is in high concentrations in circulation. This step carries the same pros (increased drug tolerance) and cons (potential to release ADC from the antibody) as discussed above for screening assays. For a competitive assay, it is recommended to add exogenous drug after sample neutralization and before capture to allow the exogenous drug a chance to bind the ADA (Jani et al. 2015).

Titration Assays

ADAs present in samples are further characterized using titration assays, which measure the magnitude of the ADA response. The assessment is important, as it may be the magnitude of the ADA response, and not the incidence that has an impact on pharmacokinetics, pharmacodynamics, safety, and/or efficacy (Cohen and Rivera 2010). A titration assay is a semi-quantitative method to determine relative levels of ADA in a confirmed positive sample. Samples are typically analyzed via the same assay format used for screening with cut point set to give 0.1% false positives (FDA 2019); alternative statistical methods have been proposed (Wakshull et al. 2011) to set the cut point based on only the negative-control values and their variability. This of course has the advantage of having the same CP across disease populations and avoids the need of patient specific sample pools for sample dilutions.

Titer values are commonly reported as the reciprocal of the highest dilution that gives a value above the assay cut point but may also be determined by extrapolating the dilution to the cut point using the linear portion of the dose response curve. When reporting titer values, all sample dilutions including acid dissociation and MRD (Matrix Required Dilution) should be included in the calculations (FDA 2019). Measuring titer values and understanding changes that may occur in titer levels can help inform on the immunogenicity of the drug and supplement the bioanalytical data package for the drug in development. For example, titer levels become increasingly useful in the clinical setting since comparing post-dose ADA levels to pre-dose samples can inform if a subject is experiencing treatment-induced ADA formation, treatment-boosted ADA formation (pre-existing ADA that increased post-dose) or just a transient ADA response that does not persist post dose (Carrasco-Triguero et al. 2019). Recent publications (Starcevic Manning et al. 2017, 2022) have proposed the use of signal to noise instead of titer to report the magnitude of the ADA response. As of the writing of this chapter, it is recommenced that prior to implementation of this approach, Sponsors seek feedback from Health Authorities (Mora et al. 2024).

Neutralizing Antibody Assays

Neutralizing antibodies (NAbs) are those with the ability to interfere with interactions between a therapeutic protein product and its target. Anti-drug antibodies can neutralize ADC activity in a variety of ways. ADAs that interfere with ADC target binding will generally be neutralizing; however, the concentrations of drug, drug target and neutralizing antibodies present will ultimately control whether the effect on ADC activity is completely or only partially neutralized; for example, there could be NAbs present, but the ADC still achieves efficacy. For ADCs that also rely on inducing cell receptor cross-linking engaging effector functions (i.e. CDC or ADCC), antibodies that bind to the drug but do not directly obstruct target binding may still interfere with the functionality and reduce efficacy (Nejadmoghaddam et al. 2019; van Schie et al. 2015).

Developing a robust assay capable of detecting NAbs formed against an ADC is imperative to provide a comprehensive picture of the clinical impact of an ADA response to the ADC being administered (FDA 2019; Wu et al. 2016). The health authority guidance documents (FDA 2019; EMA 2017) provide some high-level recommendations regarding method development and validation of NAb assay, and a recent white paper (Myler et al. 2023) provides clear advice on the validation testing and reporting strategies. Cell-based and non-cell-based assays are the two approaches highlighted in the 2019 FDA guidance as acceptable approaches to measure NAb activity (FDA 2019; Wu et al. 2016). Although cell-based neutralization assays are complex, they are preferred because they represent a physiological system for characterizing the neutralizing capacity of any ADA. The use of non-cell-based assays is considered appropriate if the therapeutic MoA targets the blocking of a soluble ligand, but Sponsors may also justify their implementation if efforts to develop a cell-based assay demonstrate to be futile.

The format of cell-based assays can be direct or indirect. Direct assays are generated when biotherapeutics exert their effect directly on a cell by binding to a ligand-specific receptor on the cell surface and stimulating a cellular response. In the presence of NAbs, the response is either diminished or deactivated. Indirect assays are used when the drug product (typically a monoclonal antibody fusion protein, or a soluble receptor) works by blocking the binding of a ligand to a specific cell-surface receptor (Wu et al. 2011, 2016; Jolicoeur and Tacey 2012). The desired result is a reduction or absence of a cellular response. When NAbs are present, they bind to the drug product, preventing it from attaching to the cellular receptor. As a result, the ligand can bind to its intended receptor and trigger a specific cellular response (Wu et al. 2016; Gupta et al. 2007). For both formats, considerable thought and planning needs to be invested prior to development to ensure selection of the appropriate stable cell lines, identification of meaningful biological readouts or attainable levels of sensitivity can be achieved (Jolicoeur and Tacey 2012).

In the event an ADC has an inhibitory mechanism of action or suitable cell-lines are not available for NAb assay development with sufficient specificity or sensitivity requirements, non-cell-based assays may be sufficient to characterize the NAb. Non-cell binding assays typically employ competitive ligand binding or enzyme activity for the NAb assessment. Competitive ligand binding assays are designed around competition between drug target and NAbs for binding to drug. When neutralizing antibodies are present, drug target binding is inhibited. In contrast to cell-based assays, non-cell-based assays are easier to set up, and boast better sensitivity, higher accuracy and may be more tolerant of matrix effects (Wu et al. 2011, 2016).

While NAb assays are normally not required prior to late stage (Phase 3) development, potential assay formats should at least be considered early on and, may be required prior to pivotal studies if the immunogenicity risk for the ADC is deemed high. Overall, NAb assays are important in assessing ADAs clinical impact. Selecting the appropriate format for the NAb assay should be based on regulatory recommendations and include three fundamental considerations including ADC mechanism of action, desired assay requirements and risk of immunogenicity (Wu et al. 2016).

Immunogenicity Summary

As with any biotherapeutic protein, since ADAs have the potential to impact efficacy and safety, immunogenicity of ADCs remains a critical assessment of a drug development program. It is recommended to perform an immunogenicity risk assessment that would inform the bioanalytical strategy. Adopting a risk-assessment strategy encompasses more than just developing an assay for the detection of ADA, it involves a myriad of considerations including but not limited to mechanism of action, statistical analysis for cut point determination, critical reagent generation, and relevant disease states. From the development of sensitive assays to detect and confirm ADA specificity for the biotherapeutic to measuring the degree of immunogenic response and identifying neutralizing antibodies decreasing PK exposure, an inclusive understanding of the immunogenicity associated with the ADC is important to help support the bioanalytical package for drug development. Novel modalities such as ADC may be perceived to be at a higher risk of eliciting an immune response compared to traditional mAb therapeutics due to their multidomain nature; however, incidence also depends on the MoA, for example an ADC that targets antigen presenting cells may have a lower ADA incidence than the monoclonal alone, but an ADC that boots NK effector function may have higher ADA incidence than the monoclonal alone. Current regulatory guidance for conventional biotherapeutics provides a framework to initiate a conservative approach early in development with an immunogenicity risk that may be lowered as the program moves to later stages of development and more is known about the ADC. Common bridging assay formats for anti-drug antibody assessments are shown in Fig. 1d, e.

Application of Bioanalytical Methodologies to Study Support

Consisting of both an antibody and a cytotoxic payload, full pharmacokinetic characterization of an ADC presents more complexities than the standard small molecule or therapeutic protein bioanalytical packages. To understand ADC pharmacokinetics and possible exposure-response relationships, bioanalysis typically requires a multi-pronged analytical approach, minimally measuring total antibody, total conjugated drug and unconjugated drug, as well as ADA. Given the nature of longer half-lives associated with antibodies compared to small molecules, the PK profile of an ADC is likely predominantly dictated by the antibody portion of the ADC; however, other inherent structures of the ADC such as linker chemistry (stability, polarity) and the payload can also influence the characteristic PK properties and even trigger harmful immune responses. In addition to the multiple intrinsic factors which can influence bioanalysis and data interpretation, other factors including target-mediated drug disposition (sink), target turnover rate, neonatal Fc receptor (FCRn)-dependent recycling and Fc effector functions will affect antibody clearance and, therefore, the PK of an ADC. A total antibody assay of the ADC can inform on PK parameters related to the antibody, while a total conjugated drug assay provides information on the conjugation status of the antibody (Hejmady et al. 2023; FDA 2024; Stephan et al. 2011; Kaur et al. 2013; Qin and Gong 2022; Yin et al. 2023a; Suh et al. 2023; Lin and Tibbitts 2012; Sapra et al. 2013; Kaur 2013b). While circulating concentrations of the total antibody or total conjugated drug are anticipated to decline over time, a rapid decline of the total conjugated drug relative to the total antibody measurements represents ADC deconjugation (payload loss) and the generation of lower DAR species (e.g. DAR_{n-1} to DAR0). These deconjugated variants potentially have a fraction of the desired biotherapeutic effect of the intended ADC species. In contrast, low circulating free payloads along with similar total antibody and conjugated drug PK profiles suggest a relatively stable linker *in vivo*. Measuring DAR values will help elucidate the relationship between DAR distribution and the mean DAR value. Representative concentration vs. time profiles for total antibody, total conjugated drug, and free payload are presented in Fig. 3.

Performing the panel of assays described above including DAR distribution, total antibody, total conjugated drug, free payload and immunogenicity assessment is likely essential to characterize the multifaceted bioanalytical package of an ADC especially in the preclinical setting when little is known about the drug *in vivo*. Understanding the properties intrinsic to the antibody, linker, and payload early on in development can help guide the bioanalytical approaches as the ADC program moves through development into later stage trials (Hejmady et al. 2023; FDA 2024; Qin and Gong 2022; Yin et al. 2023a; Junutula et al. 2010; Erickson et al. 2012). Because the human antibody of an ADC is generally cross-reactive to the monkey target protein, preclinical toxicological and toxicokinetic evaluations of ADC drug candidates in cynomolgus monkeys represent critical assessments that can provide important information regarding target or non-target mediated clearance mechanisms in human. The same assays can be adapted for clinical trials to assess maximum tolerable dose, characterize a therapeutic window, and understand PK variability and the PK/PD relationship. Direct comparison of ADC PK and PK/PD data between different species and studies should be made only with careful consideration of the differences in analytical methods, assay formats, and reference materials used. Although each of these ADC assays has their own challenges, an integrated analysis of all relevant ADC PK data is critical in fully understanding each ADC.

As an example, Kadcyla® (T-DM1) is an ADC comprising a emtansine (DM1) payload covalently bound via a stable thioether linker to trastuzumab a humanized IgG1 anti-HER2 mAb (Krop et al. 2010; Burris 3rd et al. 2011; Poon et al. 2013). The total- and conjugated-trastuzumab concentration-time profile share similar trends in cynomolgus monkeys and humans. The difference between the total- and conjugated-trastuzumab concentration-time profiles indicates slow deconjugation

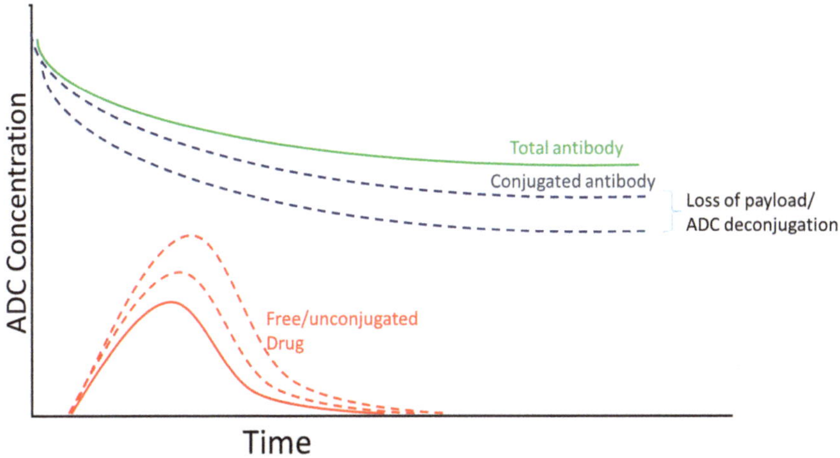

Fig. 3 Representative ADC concentration vs time profiles. Concentration vs. time plots represent three ADC components: red lines for free or unconjugated drug; green lines for total antibody and blue for conjugated drug or conjugated antibody

of DM1 from trastuzumab *in vivo* to release low levels of the free payload. This data is consistent with an orthogonal measurement of the DAR distribution data (Dere et al. 2013; Xu et al. 2013). Despite the time-dependent deconjugation, conjugated T-DM1 was selected as the bioanalytically relevant species for determining PK parameters of the ADC with a volume of distribution of 3.3 L, clearance of 0.7 L/day, and a half-life of 4.5 days in human. Kadcyla immunogenicity assessment was supported using an electrochemilumiecent (ECL)/ELISA bridging assay and domain specificity competition. With a low incidence of ADA across multiple studies (~4–6%), conclusions cannot yet be determined regarding the impact of anti-drug antibodies on the pharmacokinetics, safety, and efficacy of Kadcyla®. However, in a clinical study of 466 patients, 24 tested positive for ADA and 13 of those 24 were found to have neutralizing antibodies reiterating the need to have comprehensive bioanalytical packages for characterization. For Adcetris® (brentuximab vedotin) with a protease cleavable linker joining a chimeric IgG1 anti-CD30 antibody and a monomethyl auristatin E (MMAE) payload. This relatively stable protease cleavable linker results in a favorable PK profile with payload deconjugation occurring primarily in lysosomes (Sapra et al. 2013). With the total antibody and conjugated MMAE concentrations correlating well, conjugated MMAE was selected as the principal analyte for PK assessment displaying a volume of distribution of 8.2 L, clearance of 1.7 L/day, and a half-life of 4–6 days in human. Immunogenicity assessments for Adcetris (which demonstrated infusion related reactions), were monitored with a domain specificity ECL assay. In patients with Hodgkin's lymphoma (cHL) and systemic anaplastic large cell lymphoma (sALCL), approximately 7% (11/156) tested positive for persistent Anti-Adcetris antibodies while 30% (47/156) were positive for transient anti-Adcetris antibodies. Of these 58 ADA positive patients, 36 (62%) tested positive for neutralizing antibodies. In all patients with either persistent or transient ADA, the anti-Adcetris antibodies were directed at the antibody portion of the ADC, highlighting the importance of assays to distinguish ADA binding locations. Mylotarg® (Gemtuzumab ozogamicin) is another example of an ADC with a cleavable linker. This drug comprises a humanized IgG4 mAb against the CD33 antigen and a derivative of calicheamicin (*N*-acetyl-γ calicheamicin 1,2-dimethyl hydrazine dichloride) joined with an acid-labile cleavable linker. Due to the increased lability of the linker, this ADC demonstrates a less desirable PK profile than either T-DM1 or brentuximab vedotin (Sapra et al. 2013) with a volume of distribution of 10–18 L, clearance of 6.4 L/day, and a half-life of 3 days in human.

It is worth taking a closer look at Mylotarg® approval. Created in 1991, it was initially approved in 2000 under an accelerated-approval process by the FDA. Within the first year of approval, the FDA required the issuance of a black box warning (noted to increase the risk of veno-occlusive disease (VOD) in the absence of bone marrow transplantation) (Giles et al. 2001). Later (2003), it was shown that even following bone marrow transplants VOD onset occurred at an increased rate (Wadleigh et al. 2003). In fact, Mylotarg® was removed from the market in 2010 at the request of the FDA. In 2017, Pfizer reapplied for US an EU approval based on data from AML study (ALFA-0701) and a meta-analysis of prior trials. In 2018, the drug was again approved in both EU and US. Mylotarg® serves as a unique example for ADC drug development and highlights the need to ensure ADC bioanalytical data packages are comprehensive.

Conclusions

A combination of cytotoxic potency and antibody target specificity, ADCs represent a significant advance in targeted cancer therapy as demonstrated by the recent regulatory approval of more than 11 ADCs (e.g. Kadcyla®, Adcetris®, Besponsa™ and Mylotarg™) for treatment of various oncology indications (Tsuchikama et al. 2024). The structural complexity of ADCs poses a unique bioanalytical challenge compared to standard small molecule or therapeutic protein assay development. To fully characterize the PK properties of each ADC and establish PK/PD, five distinct bioanalytical methods are essential: total antibody, total conjugated drug, unconjugated payload and its metabolites, DAR distribution and immunogenicity assessment. In cases where reference standards are not exactly representative of the ADC species present *in vivo*, additional assessments may be necessary during method development and validation to accurately measure concentrations and understand PK *in vivo*. In addition to routine assays for small molecules and biotherapeutics, specialized assays are needed to understand the fate of ADCs and facilitate PK/PD characterization. For instance, total conjugated drug methods demand the full implementation of multiple workflows and technologies from immunocapture and digestion to LC-MS/MS analysis. Moreover, a DAR distribution assay requires a powerful high-resolution mass spectrometer for assessing the intact form of the

ADC. These are important points to consider when developing an ADC bioanalytical package.

A significant increase in regulatory approvals can be expected within the next few years, as there are currently more than 100 ADCs at various stages of clinical trials (Tsuchikama et al. 2024). This highly promising field will have a fresh landscape as new antibody and antibody targets, novel conjugation chemistries, and unique payloads are selected for the next generation of ADCs. Additionally, as we increase our understanding of ADC mechanisms of action, it will lead to improved specificity and safety allowing for expansion into additional disease states and patient populations. The continued synergy between LBA and mass spectrometry-based approaches will drive the development of more robust bioanalytical strategies to better track all ADC species *in vivo*.

References

Ahene, A.B.: Application and interpretation of free and total drug measurements in the development of biologics. Bioanalysis. **3**(11), 1287–1295 (2011)

Arnold, J.N., et al.: Human immunoglobulin glycosylation and the lectin pathway of complement activation. Adv. Exp. Med. Biol. **564**, 27–43 (2005)

Baldo, B.A.: Adverse events to monoclonal antibodies used for cancer therapy: focus on hypersensitivity responses. Oncoimmunology. **2**(10), e26333 (2013)

Bartelds, G.M., et al.: Development of antidrug antibodies against adalimumab and association with disease activity and treatment failure during long-term follow-up. JAMA. **305**(14), 1460–1468 (2011)

Beaumont, K., et al.: ADME and DMPK considerations for the discovery and development of antibody drug conjugates (ADCs). Xenobiotica. **52**(8), 770–785 (2022)

Boehncke, W.-H., Brembilla, N.C.: Immunogenicity of biologic therapies: causes and consequences. Expert Rev. Clin. Immunol. **14**(6), 513–523 (2018)

Bross, P.F., et al.: Approval summary: gemtuzumab ozogamicin in relapsed acute myeloid leukemia. Clin. Cancer Res. **7**(6), 1490–1496 (2001)

Buckwalter, M., et al.: Pharmacokinetics of gemtuzumab ozogamicin as a single-agent treatment of pediatric patients with refractory or relapsed acute myeloid leukemia. J. Clin. Pharmacol. **44**(8), 873–880 (2004)

Burris 3rd, H.A., et al.: Phase II study of the antibody drug conjugate trastuzumab-DM1 for the treatment of human epidermal growth factor receptor 2 (HER2)-positive breast cancer after prior HER2-directed therapy. J. Clin. Oncol. **29**(4), 398–405 (2011)

Carrasco-Triguero, M., et al.: Immunogenicity assays for antibody–drug conjugates: case study with ado-trastuzumab emtansine. Bioanalysis. **5**(9), 1007–1023 (2013)

Carrasco-Triguero, M., et al.: Application of a plug-and-play immunogenicity assay in cynomolgus monkey serum for ADCs at early stages of drug development. J. Immunol. Res. **2016**, 2618575 (2016)

Carrasco-Triguero, M., et al.: Immunogenicity of antibody–drug conjugates: observations across 8 molecules in 11 clinical trials. Bioanalysis. **11**(17), 1555–1568 (2019)

Carter, P.J., Senter, P.D.: Antibody-drug conjugates for cancer therapy. Cancer J. **14**(3), 154–169 (2008)

Casadevall, N., et al.: Pure red-cell aplasia and antierythropoietin antibodies in patients treated with recombinant erythropoietin. N. Engl. J. Med. **346**(7), 469–475 (2002)

Chari, R.V.: Targeted cancer therapy: conferring specificity to cytotoxic drugs. Acc. Chem. Res. **41**(1), 98–107 (2008)

Cohen, B.A., Rivera, V.M.: PRISMS: the story of a pivotal clinical trial series in multiple sclerosis. Curr. Med. Res. Opin. **26**(4), 827–838 (2010)

Dai, S., et al.: Development of a method that eliminates false-positive results due to nerve growth factor interference in the assessment of fulranumab immunogenicity. AAPS J. **16**(3), 464–477 (2014)

Davis, J.A., et al.: Immunoprecipitation middle-up LC-MS for in vivo drug-to-antibody ratio determination for antibody-drug conjugates. Bioanalysis. **9**(20), 1535–1549 (2017)

Debaene, F., et al.: Innovative native MS methodologies for antibody drug conjugate characterization: high resolution native MS and IM-MS for average DAR and DAR distribution assessment. Anal. Chem. **86**(21), 10674–10683 (2014)

Dere, R., et al.: PK assays for antibody-drug conjugates: case study with ado-trastuzumab emtansine. Bioanalysis. **5**(9), 1025–1040 (2013)

Dere, R.C., et al.: Integrated summary of immunogenicity of polatuzumab vedotin in patients with relapsed or refractory B-cell non-Hodgkin's lymphoma. Front. Immunol. **14**, 1119510 (2023)

Doronina, S.O., et al.: Development of potent monoclonal antibody auristatin conjugates for cancer therapy. Nat. Biotechnol. **21**(7), 778–784 (2003)

Driscoll, R.O., et al.: Statistical and bioanalytical considerations for establishing a depletion criterion for specificity testing during immunogenicity assessment of a biotherapeutic. AAPS J. **15**(4), 1160–1167 (2013)

Dubois, M., et al.: Immunopurification and mass spectrometric quantification of the active form of a chimeric therapeutic antibody in human serum. Anal. Chem. **80**(5), 1737–1745 (2008)

EMA: Guideline on Bioanalytical Method Validation. EMA, London (2011)

EMA: Guideline on Immunogenicity Assessment of Biotechnology-Derived Therapeutic Proteins. EMA, London (2017)

Erickson, H.K., Lambert, J.M.: ADME of antibody-maytansinoid conjugates. AAPS J. **14**(4), 799–805 (2012)

Erickson, H.K., et al.: The effect of different linkers on target cell catabolism and pharmacokinetics/pharmacodynamics of trastuzumab maytansinoid conjugates. Mol. Cancer Ther. **11**(5), 1133–1142 (2012)

Ezan, E., Becher, F., Fenaille, F.: Assessment of the metabolism of therapeutic proteins and antibodies. Expert Opin. Drug Metab. Toxicol. **10**(8), 1079–1091 (2014)

Faria, M., et al.: Multiplex LC-MS/MS assays for clinical bioanalysis of MEDI4276, an antibody-drug conjugate of tubulysin analogue attached via cleavable linker to a biparatopic humanized antibody against HER-2. Antibodies. **8**(1), 11 (2019)

FDA: Guidance for Industry: Immunogenicity Assessment for Therapeutic Protein Products. FDA, Maryland (2014)

FDA: Bioanalytical Method Validation Guidance for Industry. FDA, Maryland (2018)

FDA: Guidance for Industry: Immunogenicity Testing of Therapeutic Protein Products—Developing and Validating Assays for Anti-Drug Antibody Detection. FDA, Maryland (2019)

FDA: Guidance for Industry: Clinical Pharmacology Considerations for Antibody-Drug Conjugates. FDA, Maryland (2024)

Fernandez Ocana, M., et al.: Clinical pharmacokinetic assessment of an anti-MAdCAM monoclonal antibody therapeutic by LC-MS/MS. Anal. Chem. **84**(14), 5959–5967 (2012)

Furlong, M.T., et al.: A universal surrogate peptide to enable LC-MS/MS bioanalysis of a diversity of human monoclonal antibody and human Fc-fusion protein drug candidates in pre-clinical animal studies. Biomed. Chromatogr. **26**(8), 1024–1032 (2012)

Furlong, M.T., et al.: Dual universal peptide approach to bioanalysis of human monoclonal antibody protein drug candidates in animal studies. Bioanalysis. **5**(11), 1363–1376 (2013)

Gianolio, D.A., et al.: Targeting HER2-positive cancer with dolastatin 15 derivatives conjugated to trastuzumab, novel antibody-drug conjugates. Cancer Chemother. Pharmacol. **70**(3), 439–449 (2012)

Giles, F.J., et al.: Mylotarg™ (gemtuzumab ozogamicin) therapy is associated with hepatic veno-occlusive disease in patients who have not received stem cell transplantation. Cancer. **92**(2), 406–413 (2001)
Gorovits, B., et al.: Bioanalysis of antibody-drug conjugates: American Association of Pharmaceutical Scientists Antibody-Drug Conjugate Working Group position paper. Bioanalysis. **5**(9), 997–1006 (2013)
Grimm, H.P., et al.: PKPD assessment of the anti-CD20 antibody obinutuzumab in cynomolgus monkey is feasible despite marked anti-drug antibody response in this species. J. Pharm. Sci. **108**, 3729–3736 (2019)
Gupta, S., et al.: Recommendations for the design, optimization, and qualification of cell-based assays used for the detection of neutralizing antibody responses elicited to biological therapeutics. J. Immunol. Methods. **321**(1–2), 1–18 (2007)
Hall, M.P.: Biotransformation and in vivo stability of protein biotherapeutics: impact on candidate selection and pharmacokinetic profiling. Drug Metab. Dispos. **42**(11), 1873–1880 (2014)
Hamann, P.R., et al.: Gemtuzumab ozogamicin, a potent and selective anti-CD33 antibody-calicheamicin conjugate for treatment of acute myeloid leukemia. Bioconjug. Chem. **13**(1), 47–58 (2002)
Han, T.H., Zhao, B.: Absorption, distribution, metabolism, and excretion considerations for the development of antibody-drug conjugates. Drug Metab. Dispos. **42**(11), 1914–1920 (2014)
Han, T.H., et al.: CYP3A-mediated drug-drug interaction potential and excretion of brentuximab vedotin, an antibody-drug conjugate, in patients with CD30-positive hematologic malignancies. J. Clin. Pharmacol. **53**(8), 866–877 (2013)
Hejmady, S., et al.: Pharmacokinetics and toxicity considerations for antibody-drug conjugates: an overview. Bioanalysis. **15**(19), 1193–1202 (2023)
Hock, M.B., et al.: Immunogenicity of antibody drug conjugates: bioanalytical methods and monitoring strategy for a novel therapeutic modality. AAPS J. **17**(1), 35–43 (2015)
Huang, R.Y., et al.: Utility of ion mobility mass spectrometry for drug-to-antibody ratio measurements in antibody-drug conjugates. J. Am. Soc. Mass Spectrom. **26**(10), 1791–1794 (2015)
Huang, Y., et al.: Multifaceted bioanalytical methods for the comprehensive pharmacokinetic and catabolic assessment of MEDI3726, an anti-prostate-specific membrane antigen pyrrolobenzodiazepine antibody-drug conjugate. Anal. Chem. **92**(16), 11135–11144 (2020)
Huang, Y., et al.: Characterization of antibody-drug conjugate pharmacokinetics and in vivo biotransformation using quantitative intact LC-HRMS and surrogate analyte LC-MRM. Anal. Chem. **93**(15), 6135–6144 (2021)
ICH: ICH Guideline: Bioanalytical Method Validation and Study Sample Analysis. ICH (2022)
Iwamoto, N., et al.: Selective detection of complementarity-determining regions of monoclonal antibody by limiting protease access to the substrate: nano-surface and molecular-orientation limited proteolysis. Analyst. **139**(3), 576–580 (2014)
Iwamoto, N., et al.: Multiplex LCMS bioanalysis of brentuximab vedotin, rituximab and cetuximab towards therapeutic drug monitoring application by combined calibration curve using fab-selective limited proteolysis nSMOL. Clin. Pharmacol. Biopharm. **5**, 164 (2016)
Iwamoto, N., et al.: LC–MS bioanalysis of trastuzumab and released emtansine using nano-surface and molecular-orientation limited (nSMOL) proteolysis and liquid–liquid partition in plasma of trastuzumab emtansine-treated breast cancer patients. J. Pharm. Biomed. Anal. **145**, 33–39 (2017)
Jani, D., et al.: Recommendations for the development and validation of confirmatory anti-drug antibody assays. Bioanalysis. **7**(13), 1619–1631 (2015)
Jeffrey, S.C., et al.: Development and properties of beta-glucuronide linkers for monoclonal antibody-drug conjugates. Bioconjug. Chem. **17**(3), 831–840 (2006)
Jenkins, R., et al.: Recommendations for validation of LC-MS/MS bioanalytical methods for protein biotherapeutics. AAPS J. **17**(1), 1–16 (2015)

Jeon, E.J., et al.: Implementation of systematic bioanalysis of antibody-drug conjugates for preclinical pharmacokinetic study of ado-trastuzumab emtansine (T-DM1) in rats. Pharmaceutics. **15**(3), 756 (2023)

Jiang, H., et al.: Fully validated LC-MS/MS assay for the simultaneous quantitation of coadministered therapeutic antibodies in cynomolgus monkey serum. Anal. Chem. **85**(20), 9859–9867 (2013)

Jolicoeur, P., Tacey, R.L.: Development and validation of cell-based assays for the detection of neutralizing antibodies to drug products: a practical approach. Bioanalysis. **4**(24), 2959–2970 (2012)

Junutula, J.R., et al.: Engineered thio-trastuzumab-DM1 conjugate with an improved therapeutic index to target human epidermal growth factor receptor 2-positive breast cancer. Clin. Cancer Res. **16**(19), 4769–4778 (2010)

Kaur, S.: Analysis of antibody drug conjugates by bead-based affinity capture and mass spectrometry (2013a)

Kaur, S.: Bioanalysis special focus issue on antibody-drug conjugates. Bioanalysis. **5**(9), 981–983 (2013b)

Kaur, S., et al.: Bioanalytical assay strategies for the development of antibody-drug conjugate biotherapeutics. Bioanalysis. **5**(2), 201–226 (2013)

Kotapati, S., et al.: The role of ligand-binding assay and LC-MS in the bioanalysis of complex protein and oligonucleotide therapeutics. Bioanalysis. **13**(11), 931–954 (2021)

Krop, I.E., et al.: Phase I study of trastuzumab-DM1, an HER2 antibody-drug conjugate, given every 3 weeks to patients with HER2-positive metastatic breast cancer. J. Clin. Oncol. **28**(16), 2698–2704 (2010)

Kubiak, R.J., et al.: Confirmatory cut point has limited ability to make accurate classifications in immunogenicity assays. Bioanalysis. **12**(4), 245–256 (2020)

Lanshoeft, C., et al.: Analysis of small molecule antibody-drug conjugate catabolites in rat liver and tumor tissue by liquid extraction surface analysis micro-capillary liquid chromatography/tandem mass spectrometry. Rapid Commun. Mass Spectrom. **30**(7), 823–832 (2016)

Lassman, M.E., et al.: Practical immunoaffinity-enrichment LC-MS for measuring protein kinetics of low-abundance proteins. Clin. Chem. **60**(9), 1217–1224 (2014)

Lee, J.W., et al.: Bioanalytical approaches to quantify "total" and "free" therapeutic antibodies and their targets: technical challenges and PK/PD applications over the course of drug development. AAPS J. **13**(1), 99–110 (2011)

Lee, M.V., Kaur, S., Saad, O.M.: Conjugation site influences antibody-conjugated drug PK assays: case studies for disulfide-linked, self-immolating next-generation antibody drug conjugates. Anal. Chem. **92**(18), 12168–12175 (2020a)

Lee, B.I., et al.: Quantification of an antibody-conjugated drug in fat plasma by an affinity capture LC-MS/MS method for a novel prenyl transferase-mediated site-specific antibody-drug conjugate. Molecules. **25**(7), 1515 (2020b)

Lewis Phillips, G.D., et al.: Targeting HER2-positive breast cancer with trastuzumab-DM1, an antibody-cytotoxic drug conjugate. Cancer Res. **68**(22), 9280–9290 (2008)

Li, X., et al.: Development and validation of pharmacokinetics assays for a novel HER2-targeting antibody-drug conjugate (SHR-A1201): application to its dose-escalation pharmacokinetic study. J. Pharm. Biomed. Anal. **240**, 115964 (2024)

Lin, K., Tibbitts, J.: Pharmacokinetic considerations for antibody drug conjugates. Pharm. Res. **29**(9), 2354–2366 (2012)

Liu, G., et al.: A novel and cost effective method of removing excess albumin from plasma/serum samples and its impacts on LC-MS/MS bioanalysis of therapeutic proteins. Anal. Chem. **86**(16), 8336–8343 (2014)

Lofgren, J.A., et al.: Comparing ELISA and surface plasmon resonance for assessing clinical immunogenicity of panitumumab. J. Immunol. **178**(11), 7467–7472 (2007)

Mark Rose, L.T., Sharifi, A., Li, H.: Validation of LC-MS/MS Methods for the Determination of Large Molecules: Promises and Challenges. AAPS NBC (2014)

Mire-Sluis, A.R., et al.: Recommendations for the design and optimization of immunoassays used in the detection of host antibodies against biotechnology products. J. Immunol. Methods. **289**(1–2), 1–16 (2004)

Mora, J.R., Richards, S.M.: The AAPS journal theme issue: compendium of immunogenicity risk assessments: an industry guidance built on experience and published work. AAPS J. **25**(3), 43 (2023)

Mora, J., et al.: 2023 White paper on recent issues in bioanalysis: ISR for ADA assays, the rise of dPCR vs qPCR, international reference standards for vaccine assays, anti-AAV TAb post-dose assessment, NanoString validation, ELISpot as gold standard (Part 3 – Recommendations on gene therapy, cell therapy, vaccines immunogenicity & technologies; biotherapeutics immunogenicity & risk assessment; ADA/NAb assay/reporting harmonization). Bioanalysis. **16**(7), 77–119 (2024)

Mu, R., et al.: Bioanalytical methods and strategic perspectives addressing the rising complexity of novel bioconjugates and delivery routes for biotherapeutics. BioDrugs. **36**(2), 181–196 (2022)

Myler, H., et al.: An integrated multiplatform bioanalytical strategy for antibody-drug conjugates: a novel case study. Bioanalysis. **7**(13), 1569–1582 (2015)

Myler, H., et al.: Anti-drug antibody validation testing and reporting harmonization. AAPS J. **24**(1), 4 (2022)

Myler, H., et al.: Neutralizing antibody validation testing and reporting harmonization. AAPS J. **25**(4), 69 (2023)

Nagai, Y., et al.: Comprehensive preclinical pharmacokinetic evaluations of trastuzumab deruxtecan (DS-8201a), a HER2-targeting antibody-drug conjugate, in cynomolgus monkeys. Xenobiotica. **49**(9), 1086–1096 (2019)

Nejadmoghaddam, M.-R., et al.: Antibody-drug conjugates: possibilities and challenges. Avicenna J. Med. Biotechnol. **11**(1), 3–23 (2019)

Neubert, H., et al.: Sequential protein and peptide immunoaffinity capture for mass spectrometry-based quantification of total human beta-nerve growth factor. Anal. Chem. **85**(3), 1719–1726 (2013)

Ouyang, Z., et al.: Pellet digestion: a simple and efficient sample preparation technique for LC-MS/MS quantification of large therapeutic proteins in plasma. Bioanalysis. **4**(1), 17–28 (2012)

Partridge, M.A., et al.: Emerging technologies and generic assays for the detection of anti-drug antibodies. J. Immunol. Res. **2016**, 6262383 (2016)

Patton, A., et al.: An acid dissociation bridging ELISA for detection of antibodies directed against therapeutic proteins in the presence of antigen. J. Immunol. Methods. **304**(1–2), 189–195 (2005)

Poon, K.A., et al.: Preclinical safety profile of trastuzumab emtansine (T-DM1): mechanism of action of its cytotoxic component retained with improved tolerability. Toxicol. Appl. Pharmacol. **273**(2), 298–313 (2013)

Presta, L.G.: Engineering antibodies for therapy. Curr. Pharm. Biotechnol. **3**(3), 237–256 (2002)

Prueksaritanont, T., Tang, C.: ADME of biologics-what have we learned from small molecules? AAPS J. **14**(3), 410–419 (2012)

Qin, Q., Gong, L.: Current analytical strategies for antibody-drug conjugates in biomatrices. Molecules. **27**(19), 6299 (2022)

Ritchie, M., Tchistiakova, L., Scott, N.: Implications of receptor-mediated endocytosis and intracellular trafficking dynamics in the development of antibody drug conjugates. MAbs. **5**(1), 13–21 (2013)

Saad, O.M., et al.: Bioanalytical approaches for characterizing catabolism of antibody-drug conjugates. Bioanalysis. **7**(13), 1583–1604 (2015)

Salomon, P.L., Singh, R.: Sensitive ELISA method for the measurement of catabolites of antibody-drug conjugates (ADCs) in target cancer cells. Mol. Pharm. **12**(6), 1752–1761 (2015)

Sanderson, R.J., et al.: In vivo drug-linker stability of an anti-CD30 dipeptide-linked auristatin immunoconjugate. Clin. Cancer Res. **11**(2 Pt 1), 843–852 (2005)

Sapra, P., Betts, A., Boni, J.: Preclinical and clinical pharmacokinetic/pharmacodynamic considerations for antibody-drug conjugates. Expert Rev. Clin. Pharmacol. **6**(5), 541–555 (2013)

Sasso, J.M., et al.: The evolving landscape of antibody-drug conjugates: in depth analysis of recent research progress. Bioconjug. Chem. **34**(11), 1951–2000 (2023)

Shankar, G., et al.: Recommendations for the validation of immunoassays used for detection of host antibodies against biotechnology products. J. Pharm. Biomed. Anal. **48**(5), 1267–1281 (2008)

Shankar, G., et al.: Assessment and reporting of the clinical immunogenicity of therapeutic proteins and peptides-harmonized terminology and tactical recommendations. AAPS J. **16**(4), 658–673 (2014)

Shen, B.Q., et al.: Conjugation site modulates the in vivo stability and therapeutic activity of antibody-drug conjugates. Nat. Biotechnol. **30**(2), 184–189 (2012a)

Shen, B.Q., et al.: Catabolic fate and pharmacokinetic characterization of trastuzumab emtansine (T-DM1): an emphasis on preclinical and clinical catabolism. Curr. Drug Metab. **13**(7), 901–910 (2012b)

Shen, M., Dong, X., Tsong, Y.: Statistical evaluation of several methods for cut-point determination of immunogenicity screening assay. J. Biopharm. Stat. **25**, 269–279 (2014)

Smith, H.W., Butterfield, A., Sun, D.: Detection of antibodies against therapeutic proteins in the presence of residual therapeutic protein using a solid-phase extraction with acid dissociation (SPEAD) sample treatment prior to ELISA. Regul. Toxicol. Pharmacol. **49**(3), 230–237 (2007)

Smolec, J., et al.: Bioanalytical method validation for macromolecules in support of pharmacokinetic studies. Pharm. Res. **22**(9), 1425–1431 (2005)

Song, S., et al.: Understanding the supersensitive anti-drug antibody assay: unexpected high anti-drug antibody incidence and its clinical relevance. J. Immunol. Res. **2016**, 3072586 (2016)

Starcevic Manning, M., et al.: Assay signal as an alternative to titer for assessment of magnitude of an antidrug antibody response. Bioanalysis. **9**(23), 1849–1858 (2017)

Starcevic Manning, M., et al.: Comparison of titer and signal to noise (S/N) for determination of anti-drug antibody magnitude using clinical data from an industry consortium. AAPS J. **24**(4), 81 (2022)

Stephan, J.P., et al.: Anti-CD22-MCC-DM1 and MC-MMAF conjugates: impact of assay format on pharmacokinetic parameters determination. Bioconjug. Chem. **19**(8), 1673–1683 (2008)

Stephan, J.P., Kozak, K.R., Wong, W.L.: Challenges in developing bioanalytical assays for characterization of antibody-drug conjugates. Bioanalysis. **3**(6), 677–700 (2011)

Stubenrauch, K., et al.: Evaluation of a generic immunoassay with drug tolerance to detect immune complexes in serum samples from cynomolgus monkeys after administration of human antibodies. J. Pharm. Biomed. Anal. **52**(2), 249–254 (2010)

Stubenrauch, K., et al.: Generic anti-drug antibody assay with drug tolerance in serum samples from mice exposed to human antibodies. Anal. Biochem. **430**(2), 193–199 (2012)

Su, D., et al.: Antibody–drug conjugates derived from cytotoxic seco-CBI-dimer payloads are highly efficacious in xenograft models and form protein adducts in vivo. Bioconjug. Chem. **30**(5), 1356–1370 (2019)

Suh, M.J., et al.: Enhanced pharmacokinetic bioanalysis of antibody-drug conjugates using hybrid immunoaffinity capture and microflow LC-MS/MS. AAPS J. **25**(4), 68 (2023)

Swanson, S.J., Chirmule, N.: Assessing specificity for immunogenicity assays. Bioanalysis. **1**(3), 611–617 (2009)

Tarentino, A.L., Plummer Jr., T.H.: Enzymatic deglycosylation of asparagine-linked glycans: purification, properties, and specificity of oligosaccharide-cleaving enzymes from Flavobacterium meningosepticum. Methods Enzymol. **230**, 44–57 (1994)

Todoroki, K., et al.: Bioanalytical methods for therapeutic monoclonal antibodies and antibody-drug conjugates: a review of recent advances and future perspectives. J. Pharm. Biomed. Anal. **179**, 112991 (2020)

Tsuchikama, K., et al.: Exploring the next generation of antibody-drug conjugates. Nat. Rev. Clin. Oncol. **21**(3), 203–223 (2024)

van den Broek, I., Niessen, W.M., van Dongen, W.D.: Bioanalytical LC-MS/MS of protein-based biopharmaceuticals. J. Chromatogr. B Analyt. Technol. Biomed. Life Sci. **929**, 161–179 (2013) van Schie, K.A., Wolbink, G.J., Rispens, T.: Cross-reactive and pre-existing antibodies to therapeutic antibodies--effects on treatment and immunogenicity. MAbs. **7**(4), 662–671 (2015)

Wadleigh, M., et al.: Prior gemtuzumab ozogamicin exposure significantly increases the risk of veno-occlusive disease in patients who undergo myeloablative allogeneic stem cell transplantation. Blood. **102**(5), 1578–1582 (2003)

Wagner-Rousset, E., et al.: Antibody-drug conjugate model fast characterization by LC-MS following IdeS proteolytic digestion. MAbs. **6**(1), 273–285 (2014)

Wakshull, E., et al.: Proposal for a new protein therapeutic immunogenicity titer assay cutpoint. Bioanalysis. **3**(23), 2627–2636 (2011)

Wang, Y., Heilig, J.S.: Differentiation and quantification of endogenous and recombinant-methionyl human leptin in clinical plasma samples by immunocapture/mass spectrometry. J. Pharm. Biomed. Anal. **70**, 440–446 (2012)

Wang, W., et al.: Development and validation of indirect and generic immunoassays to quantify free and total evolocumab in rat serum. Bioanalysis. **11**(8), 679–687 (2019)

Wei, B., et al.: Discovery of peptidomimetic antibody-drug conjugate linkers with enhanced protease specificity. J. Med. Chem. **61**(3), 989–1000 (2018)

Widdison, W.C., et al.: Semisynthetic maytansine analogues for the targeted treatment of cancer. J. Med. Chem. **49**(14), 4392–4408 (2006)

Wu, B.W., Gunn III, G.R., Shankar, G.: Competitive ligand-binding assays for the detection of neutralizing antibodies. In: Detection and Quantification of Antibodies to Biopharmaceuticals, pp. 175–192. Wiley, Chicester (2011)

Wu, B., et al.: Strategies to determine assay format for the assessment of neutralizing antibody responses to biotherapeutics. AAPS J. **18**(6), 1335–1350 (2016)

Wu, B., et al.: Development and validation of bioanalytical methods to support clinical study of disitamab vedotin. Bioanalysis. **16**(9), 385–400 (2024)

Xu, K., et al.: Characterization of intact antibody-drug conjugates from plasma/serum in vivo by affinity capture capillary liquid chromatography-mass spectrometry. Anal. Biochem. **412**(1), 56–66 (2011)

Xu, K., et al.: Characterization of the drug-to-antibody ratio distribution for antibody-drug conjugates in plasma/serum. Bioanalysis. **5**(9), 1057–1071 (2013)

Xu, L., et al.: A generic approach for simultaneous measurements of total antibody and cleavable antibody-conjugated drug by LC/MS/MS. Anal. Biochem. **537**, 33–36 (2017)

Xu, L., et al.: Simultaneous quantification of total antibody and antibody-conjugated drug for XMT-1522 in human plasma using immunocapture-liquid chromatography/mass spectrometry. J. Pharm. Biomed. Anal. **174**, 441–449 (2019)

Yang, Z., et al.: LC-MS/MS approach for quantification of therapeutic proteins in plasma using a protein internal standard and 2D-solid-phase extraction cleanup. Anal. Chem. **79**(24), 9294–9301 (2007)

Yang, C., et al.: Preclinical pharmacokinetics of a novel anti-c-Met antibody-drug conjugate, SHR-A1403, in rodents and non-human primates. Xenobiotica. **49**(9), 1097–1105 (2019)

Yin, L., et al.: Bioanalytical assays for pharmacokinetic and biodistribution study of antibody-drug conjugates. Drug Metab. Dispos. **51**(10), 1324–1331 (2023a)

Yin, F., et al.: A sensitive LC-MS/MS assay to quantitate free payload Aur0101 from ADC PYX-201 in rat and monkey plasma. Bioanalysis. **15**(14), 833–843 (2023b)

Zhang, D., et al.: Chemical structure and concentration of intratumor catabolites determine efficacy of antibody drug conjugates. Drug Metab. Dispos. **44**(9), 1517–1523 (2016)

Zhao, B., et al.: Brentuximab vedotin, an antibody-drug conjugate, in patients with CD30-positive haematologic malignancies and hepatic or renal impairment. Br. J. Clin. Pharmacol. **82**(3), 696–705 (2016)

Zhu, L., Glick, J., Flarakos, J.: Bioanalytical challenges in support of complex modalities of antibody-based therapeutics. AAPS J. **22**(6), 130 (2020)

Considerations for Antibody-Drug Conjugate Design: Stability and Formulation

Isis Janilkarn-Urena, Nicole A. Lynn, and N. Kalapatapu

Abstract Antibody drug conjugates (ADCs) are specific tumor targeting complexes that are typically composed of monoclonal antibodies (mAbs) attached covalently to a cytotoxic payload via a chemical linker. The precision-based targeting ability of ADCs has revolutionized the world of cancer therapeutics. Currently, 13 ADCs have received FDA-approval, with the majority receiving breakthrough or accelerated approvals for various cancers. At present, an excess of 400 drug candidates are being investigated, tested, or are undergoing clinical trials worldwide. Designing an 'ideal' ADC presents unique challenges, as production and conjugation of the individual elements (antibody, linker, and payload) requires extremely high precision and control. Of the ADCs that have been approved as well as those under investigation, formulation is strictly guided to maintain effective antibody-antigen targeting and linker-payload compatibility with the goal of preserving stability in circulation; allowing specific release of the payload once it reaches it's target and as such, upholding optimal therapeutic output. These complexities extend to antibody: structure and post-translational modifications, linker: location, size, length and hydrophobicity, and payload: cytotoxicity, immunogenicity and solubility. Herein, we review design and formulation strategies used in the approved ADCs since the first FDA-approved ADC through today, discussing biologic engineering modifications that have shown to improve clinical efficacy and the therapeutic index, while maintaining ADC stability.

Authors Janilkarn-Urena and Lynn contributed equally to the manuscript.

I. Janilkarn-Urena (✉) · N. Kalapatapu
USC Alfred E. Mann School of Pharmacy and Pharmaceutical Sciences,
Los Angeles, CA, USA
e-mail: janilkar@usc.edu

N. A. Lynn
ASBMB Today, Rockville, MD, USA

H. J. Gukasyan, A. Kulkarni (eds.), *Antibody Drug Conjugates*, AAPS Advances in the Pharmaceutical Sciences Series 52, https://doi.org/10.1007/978-3-031-89319-3_7

Keywords Antibody-drug conjugates · Targeted therapies · Chemical biology · Monoclonal antibodies · Linker · Payload · Combination therapies · Immunotherapy · Antibody engineering

Introduction

Antibody drug conjugates (ADCs) combine the object specificity of monoclonal antibodies (mAbs) and small molecules with the lethality of cytotoxic drugs to target and eliminate cancer cells (De Cecco et al. 2021; Mahmood 2021). Introduced in clinical trials over two decades ago, ADCs quickly rose in popularity as a promising alternative to traditional methods like aggressive surgical tissue excision, radiotherapy, and toxic chemotherapies. The concept of targeted disease therapy, however, is not entirely novel. Almost a century before the first ADC-focused clinical trial, German immunologist Paul Ehrlich would establish the foundations for what is now known as precision medicine. Ehrlich coined the term 'magic bullet', as a more enticing name for his 'Zauberkugel-Theorie', where he describes finely-tuned medicine capable of eradicating disease while leaving healthy tissues unharmed (Valent et al. 2016).

Monoclonal antibody research of the mid-twentieth century established a proof-of-concept for the combination and re-association of antigen binding fragments (Fabs) from polyclonal serum (Brinkmann and Kontermann 2017). Hybridoma technology, established by Köhler and Milstein, is the well-established method wherein antibody-producing B-cells are fused to immortal myeloma cells to produce large supplies of monoclonal antibody; this technology heavily fueled antibody production, discovery, and is still widely-used in antibody development (Köhler and Milstein 1975). The first described antibody-toxin conjugates, called immunotoxins, merged the variable domains of the interleukin 2 (IL-2) mouse antibody with a truncated pseudomonas endotoxin (Chaudhary et al. 1989). In 1986, muromonab-CD3 (Orthoclone OKT3™), an antibody developed to mitigate organ transplant rejections, became the first FDA-approved mAb for clinical use (Cosimi 1987). At present, over 150 FDA-approved mAbs are available on the market for varying therapeutic indications.

The year 2000 marked a great success for ADCs, with the anti-CD33-targeting gemtuzumab ozogamicin (Mylotarg®), designated for adults with acute myeloid leukemia (AML), garnering accelerated approval by the FDA (Norsworthy et al. 2018). In 2013, ado-trastuzumab emtansine (Kadcyla®/T-DM1) was the first ADC approved for the treatment of solid tumors, specifically targeting human epithelial growth factor receptor 2 (HER2) positive metastatic breast cancer (Amiri-Kordestani et al. 2014). At present, 13 ADCs have received FDA approval since 2000, with 11 currently available in market (Table 1) (Liu et al. 2024). While first generation ADCs laid the foundation for targeted therapies, each successive generation has improved on the weaknesses of their predecessors. First-generation ADCs encountered obstacles with hepatotoxicity (Rajvanshi et al. 2002) and low efficacy

Table 1 List of FDA-Approved ADCs

ADC	Sponsor	Approved Indication(s)	FDA Approval Date	FDA Designation(s)
Gemtuzumab ozogamicin (Mylotarg®)	Pfizer	Relapsed or refractory (R/R) CD33-positive acute myeloid leukemia (AML) in patients 60 years of age or older	May 17, 2000 (AA) **WITHDRAWN**	Orphan Drug, Priority Review
		Relapsed or refractory (R/R) CD33-positive acute myeloid leukemia (AML) in adults and pediatric patients 2 years or older	September 2, 2017 **REAPPROVED**	
		Newly diagnosed CD33-positive AML to include pediatric patients 1 month and older	June 16, 2020	
Brentuximab vedotin (Adcetris®)	Pfizer and Takeda	Classical Hodgkin lymphoma (cHL) after failure of ASCT or after failure of at least two prior multi-agent chemotherapy regimens in patients who are not ASCT candidates	August 19, 2011 (AA)	Breakthrough Therapy, Orphan Drug, Priority Review
		Relapsed or refractory systemic anaplastic large-cell lymphoma	August 19, 2011 (AA)	
		cHL at high risk of relapse or progression as post-autologous hematopoietic stem cell transplantation (auto-HSCT) consolidation	August 17, 2015	
		MF or pcALCL who had previously received one prior systemic therapy	November 9, 2017	
		cHL previously untreated stage III or IV in combination with chemotherapy	March 20, 2018	
		Previously untreated systemic anaplastic large cell lymphoma (sALCL)	November 16, 2018	
		Previously untreated other CD30-expressing peripheral T-cell lymphomas (PTCL), including angioimmunoblastic T-cell lymphoma, and PTCL not otherwise specified	November 16, 2018	
		Pediatric patients with previously untreated high-risk classical Hodgkin lymphoma	November 10, 2022	
Ado-trastuzumab emtansine (Kadcyla®)	Genentech	HER2-positive metastatic breast cancer who have received prior therapy	February 22, 2013	Breakthrough Therapy, Priority Review
		HER2-positive early breast cancer as a single agent, for the adjuvant treatment of patients with residual invasive disease after neoadjuvant taxane and trastuzumab-based treatment	May 3, 2019	
Moxetumomab pasudotox (Lumoxiti®)	AstraZeneca	Relapsed or refractory hairy cell leukemia with a history of at least two prior systemic therapies, including treatment with a purine nucleoside analog	September 13, 2018 **WITHDRAWN**	Fast Track, Orphan Drug, Priority Review
Belantamab mafodotin (Blenrep®)	GlaxoSmithKline	Adult patients with either relapsed or refractory multiple myeloma who have received at least four prior treatments which must include an anti-CD38 monoclonal antibody	August 5, 2020 (AA) **WITHDRAWN**	Breakthrough Therapy, Orphan Drug, Priority Review

(continued)

Table 1 (continued)

ADC	Sponsor	Approved Indication(s)	FDA Approval Date	FDA Designation(s)
Fam-trastuzumab deruxtecan (Enhertu®)	AstraZeneca and Daiichi Sankyo	Unresectable or metastatic HER2-positive breast cancer who have received two or more prior anti-HER2-based regimens in the metastatic setting	December 20, 2019 (AA)	Breakthrough Therapy, Fast Track
		Locally advanced or metastatic HER2-positive gastric or gastroesophageal junction adenocarcinoma who have received a prior trastuzumab-based regimen	January 15, 2021	
		Unresectable or metastatic HER2-low (IHC 1+ or IHC 2+/ISH-) breast cancer who have received a prior chemotherapy in the metastatic setting or developed disease recurrence during or within 6 months of completing adjuvant chemotherapy	May 4, 2022	
		Unresectable or metastatic HER2-positive breast cancer who have received a prior anti-HER2-based regimen either in the metastatic setting, or in the neoadjuvant or adjuvant setting and have developed disease recurrence during or within 6 months of completing therapy	August 5, 2022	
		Unresectable or metastatic non-small cell lung cancer (NSCLC) whose tumors have activating HER2 (ERBB2) mutations and who have received a prior systemic therapy	August 11, 2022 (AA)	
		Adult patients with unresectable or metastatic HER2-positive solid tumors who have received prior treatment or for whom no satisfactory treatment alternatives exist	April 5, 2024 (AA)	
Sacituzumab govitecan (Trodelvy®)	Gilead	Adult patients who have received at least two prior therapies with triple-negative breast cancer that has spread to other parts of the body	April 22, 2020 (AA)	Breakthrough Therapy, Fast Track, Priority Review
		Adult patients with unresectable locally advanced or metastatic triple-negative breast cancer who have previously received two or more systemic therapies, with at least one therapy for metastatic disease	April 7, 2021	
		Adult patients with locally advanced or metastatic urothelial cancer who have previously received a PD-1 or PD-L1 inhibitor	April 13, 2021 (AA)	
		Adult patients with unresectable, locally advanced or metastatic, hormone receptor (HR)-positive, HER2-negative breast cancer who have received endocrine-based therapy and at least two additional systemic therapies in the metastatic setting	February 3, 2023	
Loncastuximab tesirine (Zynlonta®)	ADC Therapeutics	Adult patients with relapsed or refractory large B-cell lymphoma after at least two lines of systemic therapy, including diffuse large B-cell lymphoma not otherwise specified, diffuse large B-cell lymphoma arising from low grade lymphoma, and high-grade B-cell lymphoma	April 23, 2021 (AA)	Orphan Drug, Priority Review

Enfortumab vedotin (Padcev®)	Seagen and Astellas	Adult patients with locally advanced or metastatic urothelial cancer who have previously received a programmed death receptor-1 (PD-1) or programmed death ligand 1 (PD-L1) inhibitor and a platinum-containing chemotherapy	December 18, 2019 (AA)	Breakthrough Therapy, Priority Review
		Advanced or metastatic urothelial cancer in patients who have previously received a PD-1 or PD-L1 inhibitor and a platinum-containing chemotherapy	July 9, 2021	
		Advanced or metastatic urothelial cancer in patients who are ineligible for cisplatin-containing therapy and have previously received one or more prior lines of therapy	July 9, 2021	
		Locally advanced or metastatic urothelial carcinoma who are ineligible for cisplatin-containing chemotherapy in combination with pembrolizumab	April 3, 2023 (AA)	
		Locally advanced or metastatic urothelial carcinoma who are not eligible for cisplatin-containing chemotherapy as a first line in combination with pembrolizumab	December 15, 2023	
Polatuzumab vedotin (Polivy®)	Genentech	Adult patients with diffuse large B-cell lymphoma (DLBCL) that has progressed or returned after at least two prior therapies in combination with the chemotherapy bendamustine and a rituximab product	June 10, 2019 (AA)	Breakthrough Therapy, Orphan Drug, Priority Review
		Adult patients in combination with a rituximab product, cyclophosphamide, doxorubicin, and prednisone (R-CHP) for the treatment of adult patients who have previously untreated diffuse large B-cell lymphoma (DLBCL), not otherwise specified (NOS) or high-grade B-cell lymphoma (HGBL) and who have an International Prognostic Index score of 2 or greater	April 19, 2023	
Inotuzumab ozogamicin (Besponsa®)	Pfizer	Adults with relapsed or refractory B-cell precursor acute lymphoblastic leukemia (ALL)	August 17, 2017	Breakthrough Therapy, Orphan Drug, Priority Review
		Pediatric patients 1 year and older with relapsed or refractory CD22-positive B-cell precursor acute lymphoblastic leukemia (ALL)	March 6, 2024	
Mirvetuximab soravtansine (Elahere®)	AbbVie	Adult patients with folate receptor alpha (FRα) positive, platinum-resistant epithelial ovarian, fallopian tube, or primary peritoneal cancer, who have received one to three prior systemic treatment regimens	November 14, 2022 (AA)	Priority Review
		Folate receptor α positive, platinum-resistant epithelial ovarian, fallopian tube, or primary peritoneal cancer for patients who have received 1 to 3 previous systemic treatment regimens	March 22, 2024	
Tistotumab vedotin (Tivdak®)	Pfizer	Adult patients with recurrent or metastatic cervical cancer with disease progression on or after chemotherapy	September 20, 2021 (AA)	Priority Review
		Adult patients with recurrent or metastatic cervical cancer with disease progression on or after chemotherapy	April 29, 2024	
Cetuximab sarotalocan (Akalux®)	Rakuten	Unresectable locally advanced or recurrent head and neck cancer	PENDING	NA
Disitamab vedotin (Aidixi®)	Seagen	HER2-positive locally advanced or metastatic urothelial carcinoma following treatment with platinum-based chemotherapy	PENDING	Breakthrough Therapy

(Egan and Reagan 2018). Second-generation ADCs addressed these drawbacks by refining approaches to reduce toxicity and improve plasma half-life (changing IgG subtype, increasing stability of *in-situ* cleavable linkers and early site-specific conjugation strategies) (Chen et al. 2016; Lamb 2017; van de Donk and Dhimolea 2012), but still suffered from narrow therapeutic window as well as off-target toxicity due to increased heterogeneity and a wide range of drug-antibody-ratio, or DAR (Vollmar et al. 2021). Third-generation ADCs improved upon concerns of DAR and species heterogeneity with more controlled site-specific conjugation. Currently, 7/8 third-generation ADCs implement a cysteine-based conjugation technique, yielding reproducible and homogeneous ADCs, with predictable/controllable DAR (Tzogani et al. 2021; Yip et al. 2021; Heo 2023). New linker chemistries and antibody engineering techniques are constantly emerging for use in the genesis of next-generation ADCs. In this chapter, we will evaluate the innovations and developments in composition of ADCs of past and present and provide considerations for design and formulation while highlighting milestones from the advent of ADCs.

The Required Elements of an Antibody Drug Conjugate

Three features consistent of a typical ADC are (1) A solid or hematological tumor protein-targeting monoclonal antibody (mAb) (2) A cytotoxic small molecule or drug (payload) and (3) A linker that bridges the two (Fig. 1). After recognition and binding to its target antigen (often a surface receptor), the ADC becomes internalized and chemically or enzymatically separated from the antibody shuttle in the endosomal or lysosomal compartments of the cell; the released drug then

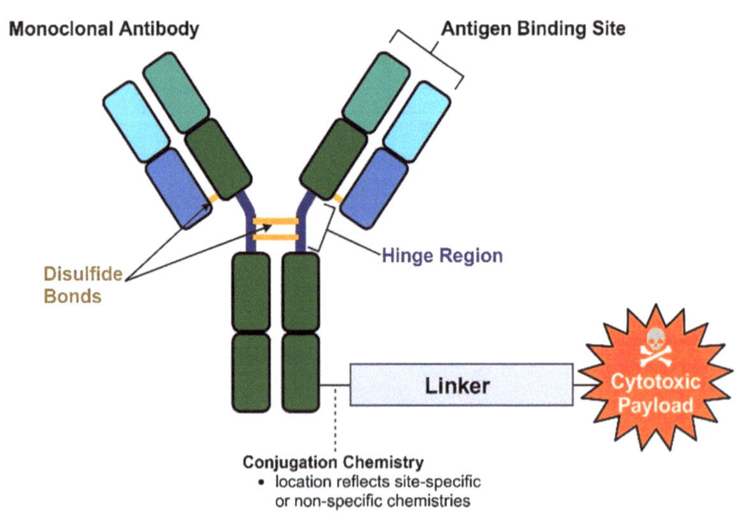

Fig. 1 Components of an antibody drug conjugate: monoclonal antibody + Linker + cytotoxic payload. (Image made with Biorender.com)

promotes cellular destruction. Antibodies are sizably large, approximately 150 kDa, they form up to 90% of the mass of the ADC as a whole. In comparison, many of the commonly used cytotoxic payloads are simple structures with very low molecular weight (Wang et al. 2023). Each individual component of the ADC (linker, mAb, and payload) can influence stability, efficacy, safety, and toxicity (Gogia et al. 2023). As such, a careful holistic approach focused on evaluating each module within the ADC is required when considering design and formulation.

Choosing an Antibody

Choosing the 'right' antibody to shuttle your cytotoxic payload to its desired destination requires the consideration of many factors: antigen specificity, antigen abundance in cancerous cells but negligible expression in healthy cells, and low host immunogenicity (De Cecco et al. 2021). Selected antibodies should also exhibit efficient cellular internalization, and long plasma half-life ($t_{½}$). Antibody bioavailability and $t_{1/2}$ in plasma is critical for ensuring the dose concentration which correlates to maintained therapeutic efficacy (Di et al. 2005). The formulation design of the mAb portion of the ADC should attempt to follow its original unconjugated mAb formulation with minimal modifications, preserving its naïve characteristics.

Antibodies, or immunoglobulins (Igs), are canonically described as Y-shaped proteins composed of two light and two heavy chain dimers (Alberts et al. 2002). The heavy chain of an antibody, or fragment crystallizable (Fc) region, constitutes the base of the Y-shaped protein, and is largely responsible for specific interactions with proteins of the complement system (Fig. 2a) (Chiu et al. 2019; Natsume et al. 2009). The light chain, or fragment-antigen binding (Fab) regions, harbor constant (C_H1 and C_H2) and variable domains (V_H and V_L), which serve to recruit the host-immune system, and recognize foreign antigens in the body, respectively (Surowka et al. 2021). The variable domains of the Fab region contain the antigen-binding motifs and facilitate cellular entry.

Early approved immunotherapies used mouse-derived (murine) monoclonal antibodies (mAbs) that were characterized by enhanced tumor specificity and clearance (Courtenay-Luck et al. 1986; Sharma et al. 2023). The non-human species element, however, drew concern. In clinical trials, repeated dosing with mouse-derived antibodies over time elicited rapid primary and secondary human-anti-mouse Ig immune responses; while no clinical toxicity was observed, therapeutic efficacies were reportedly reduced (Courtenay-Luck et al. 1986; Hedrich et al. 2018; Schroff et al. 1985). One example is the Phase 1 clinical trial assessing the safety and efficacy of murine-derived mAb KS1/4, produced from lung adenocarcinomas. KS1/4 was assessed as a stand-alone antibody and conjugated therapeutic, using the antimetabolite methotrexate as the payload, for treatment of Stage IIIB and IV non-small cell lung cancer (NSCLC). While both conjugated and unconjugated species were well-tolerated throughout the trial, the KS1/4-methotrexate conjugate exhibited cross-reactivity in healthy tissues, raising concern for future clinical trials (Elias et al. 1990).

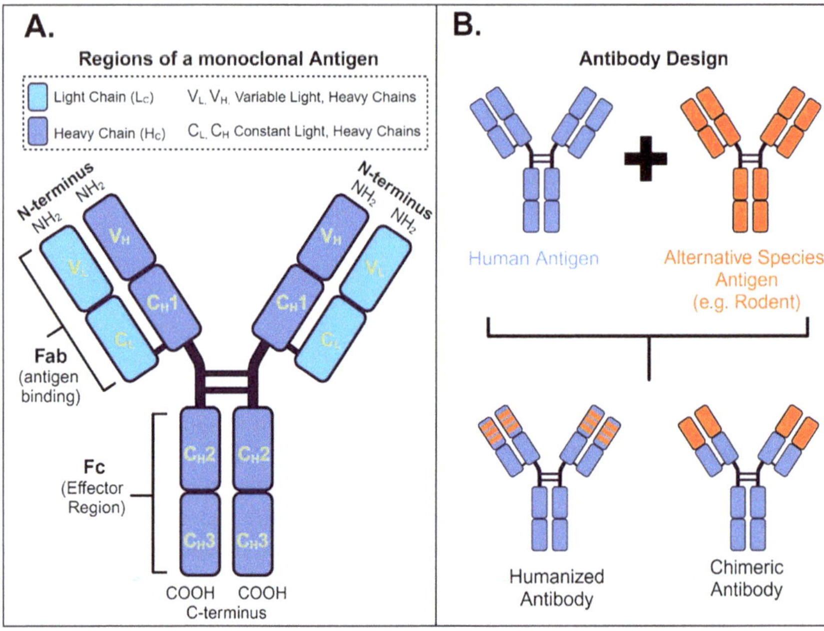

Fig. 2 Antibody structure and design. (**A**) Regions of a monoclonal antibody. (**B**) Chimeric and humanized antibody design (Image made with Biorender.com)

Antibody Engineering

Compared to first-generation therapeutic antibody design, subsequent iterations of antibody-based therapeutics expanded to engineered chimeric and humanized antibodies (Sharma et al. 2023). First described in 1984, the chimeric design merges the Fab variable region of one species with the constant Fc region of another; in a majority of cases, using mouse and human, respectively (Fig. 2b) (Morrison et al. 1984). Combining the two regions resolves early concerns of anti-human antibody immune responses, such as those observed in early murine-based ADCs (Schroff et al. 1985; Elias et al. 1990). The first chimeric mAb with oncologic indication was Rituximab®, an anti-CD20 targeting antibody that has been used for the treatment of non-Hodgkin's lymphoma since 1997 (Maloney et al. 1997).

Humanized antibodies are created by grafting the complimentary determining (CD) region of a non-human antibody (rodent, e.g. mouse) to the frame of a human antibody (Fig. 2b) (Lu et al. 2020; Jones et al. 1986). In combining these regions, the humanized mAb gains enhanced tumor-targeting from the mouse antibodies, while maintaining low immunogenicity with the human frame. In this regard, humanization greatly, if not completely reduces the potential for the human anti-human antibody (HAHA) response (Holgate and Baker 2009). A recent assessment of ADC immunogenicity compared to clinically approved mAb therapeutics surveyed 8 molecules across 11 clinical trials, and found the immunogenic effect presented by ADCs is no greater than known effects of current mAbs (Carrasco-Triguero et al. 2019). Antibody humanizing grafting technologies largely influenced future accelerated mAb approvals (Lu et al. 2020).

The Immunoglobulin (Ig) subtype largely informs plasma stability and mechanism of action for biologics (Di et al. 2005; Yu et al. 2020). Ko et al., assessed Fc-domain mutations in the cynomolgus macaque, identified the alterations at Q311R and M428L greatly improved antibody serum $t_{1/2}$, and effector functions (Ko et al. 2022). Of the many human Igs, immunoglobulin G (IgG) is the most abundant, comprising up to 75% of Ig protein in circulation and; 10–20% total plasma protein, making it desirable for therapeutic use (Yu et al. 2020; Bournazos and Ravetch 2017). Notably, IgG is the only Ig capable of crossing the placental barrier (Vidarsson et al. 2014). Within IgG exist four subclasses: IgG1, IgG2, IgG3 and IgG4; while nearly identical in sequence identity, they differ largely in their structure, function, and antigen binding properties (de Taeye et al. 2019). Of the four, IgG1 is the most abundant in serum and is the more commonly used subtype for the antibody arm of next-generation ADCs. IgG1 is also known to be a strong inducer of Fc-mediated antibody-dependent cellular cytotoxicity (ADCC), complement-dependent cytotoxicity (CDC), antibody-dependent cellular phagocytosis (ADCP), and can bind with high affinity to all Fc gamma receptors (FcγRs) (Fig. 3) (Yu et al. 2020; de Taeye et al. 2019). IgG4 is uniquely unstable in the Fc domain, and susceptible to reduction of the disulfide bonds of its hinge region; destabilization in this manner produces two half molecules, making IgG4 well suited for bispecific production (de Taeye et al. 2019). Due to its high affinity for FcγRI, IgG4 presents well for targeting pathways specific to programmed death-ligand 1 (PDL-1), a protein

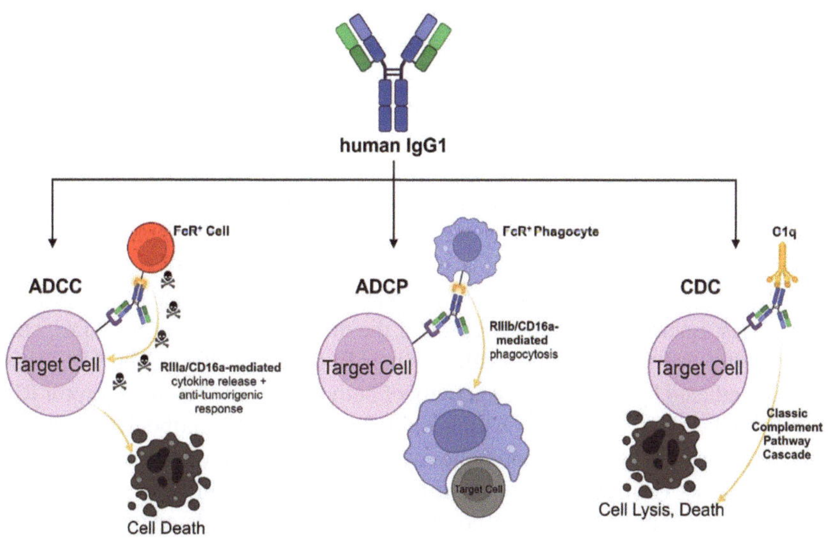

Fig. 3 IgG1 Roles in antibody-dependent cellular cytotoxicity (ADCC), antibody-dependent cellular phagocytosis (ADCP), and complement-dependent cytotoxicity (CDC) pathways for IgG1 mediated targeted tumor destroying mechanism of action. (Image made with Biorender.com)

commonly overexpressed on the surface of various cancer cells. The FDA-approved ADC, gemtuzumab ozogamicin (Mylotarg®) which is used to treat acute myeloid leukemia (AML), as well as FDA-approved lintuzumab ozogamicin (Besponsa®), used for treatment of relapsed or refractory B-cell precursor acute lymphoblastic leukemia, both incorporated IgG4 in their design (Mahmood 2021; Abdollahpour-Alitappeh et al. 2019; Fu et al. 2022).

The target antigen motif plays a large role in ADC directed delivery. Cell surface expression explicit to, if not heavily increased in cancer cells, with little- or minimal presence in normal tissue is critical to reducing off-target toxicity (Fu et al. 2022). Antigen-specific endocytosis, as described by Ritchie et al. is the predominant mechanism for first- through third-generation ADC cellular entry and warhead delivery within the cell (Ritchie et al. 2013). In the last 10 years, established receptor-mediated mechanisms of ADC delivery have become increasingly characterized, with predictable lysosomal pathways. Clathrin, and caveolin-mediated, endocytosis are both well-characterized cellular internalization pathways that can occur at the plasma membrane, and have been used for basic drug delivery platforms, however, in the context of ADCs much remains unknown (Ritchie et al. 2013; Hammood et al. 2021). Of the approved ADCs used for tumorigenic and hematological delivery, many if not all, employ receptor-driven internalization, often in the form of clathrin-mediated endocytosis (Gogia et al. 2023).

Engineering the Fc domain to optimize target moiety binding, increasing plasma stability, and induce an immunogenic response is a method of increasing popularity for next-generation ADCs. One such approach is by way of fine tuning of Fc effector functions via glycosylation engineering. Glycosylation is a post-translational modification on the mAb that results in improved pharmacokinetic parameters, including increased bioavailability, circulation $t_{1/2}$, and the duration of ADC therapeutic effect. Structural studies, for example, have suggested that conserved glycosylation in the constant region of IgG1 can maintain the conformation of the Fc region for receptor mediated binding (Jennewein and Alter 2017). As mentioned, the main cytotoxic effector activities that contribute to the anti-tumor efficacy of ADCs are the CDC, ADCC, and the ADCP pathways. Wherein CDC is activated upon the binding of C1q (the initiating molecule of the classical complement pathway) to the IgG-Fc region of antibodies triggers the classical complement cascade, resulting in cell lysis; ADCC is a mechanism in which IgG-Fc receptor RIIIa (FcγIIIa/CD16a), which is expressed in abundance on the surface of NK cells and monocytes, interacts with the Fc region of the mAb component of the ADC, effectively destroying targeted cancer cells (Gómez Román et al. 2014). The ADCP pathway is mediated by the combined activity of both CD16a and IgG-Fc receptor RIIIb (FcγIIIb/CD16b). Regarding pharmacokinetic parameters, antibodies alone have relatively long $t_{1/2}$; typically, antibodies can be recycled and found in circulation for up to 4 weeks (Liu 2018; Roopenian and Akilesh 2007). In fact, several studies conducted using mAb technology confirmed that non-glycosylated IgGs had no significant effect on antibody clearance and can still have a normal $t_{1/2}$, however, N-glycosylation in the Fc region has been shown to enhance stability and have protective effects against degradation, thermal denaturation and aggregation, thereby producing mAbs that express various glycoforms for improved therapeutic efficacy during ADC design has its advantages (Higel et al. 2016; Chen et al. 2009; Batra and Rathore 2016; Liu et al. 2011).

IgG molecules contain a variety of glycan structures attached to their Fc domains, with all four subclasses harboring N-glycosylation sites in the constant region of the heavy chain, but vary in glycosylation patterns; the major N-glycans present in this region are biantennary, largely core-fucosylated, partially truncated, and can carry sialic acid or bisecting N-acetylglucosamine (GlcNAc) species (Wuhrer et al. 2007). Glycoengineering of the Fc region of antibodies by way of afucosylation (removal of Fc N-glycans) have been shown to enhance the ADCC effect by intensifying the binding to and activation of FcγRIIIa, resulting in higher and stronger ADCC activity compared to highly fucosylated antibodies; this strategy was previously used by monoclonal cancer therapies obinutuzumab (Gazyva®) and mogamulizumab (Poteligeo®) leading to superior *in vivo* efficacy (Tai et al. 2014; Pereira et al. 2018; Tai and Anderson 2015; Giddens et al. 2018). Recent developments from *in vitro* and *in vivo* models have revealed that the degree of fucosylation in the heavy chain of human IgG at the highly glycosylated and conserved N-linked glycosylation residue, Asn297, plays a pivotal role in regulating both complement and FcγR-driven effector functions (Arnold et al. 2007; Golay et al. 2022). In fact, the affinity of antibodies for antigens can be further influenced by changes to the IgG-Fc glycan at the Asn297 site, which is composed of an GlcNAc and a mannose biantennary core structure (Jennewein and Alter 2017). To date, Belantamab mafodotin-blmf (Blenrep®), a monotherapy designed to treat adult patients with refractory multiple myeloma (RRMM) is the only FDA-approved ADC that is composed of an afucosylated humanized IgG1 mAb conjugated with a non-cleavable maleimidocaproyl linker to monomethyl auristatin F (MMAF) (Tai et al. 2014; U.S. Food and Drug Administration 2020). The afucosylation of Blenrep's monoclonal antibody was shown to increase binding to FcγRIIIa receptors, which increased recruitment and activation of immune effector cells (Tzogani et al. 2021). However, in 2022, Blenrep® was withdrawn from the market after failing to meet the primary endpoint of improving the progression free survival (PFS) during a phase III confirmatory trial (DREAMM-3), (Table 2) (GSK 2022).

Furthermore, the pattern of glycosylation in the Fc regions are known to affect immunogenicity, the bioavailability, and pharmacokinetics of ADCs. Changes to these regions may elicit immunological responses to ADCs that not only include the activity of effector functions, but they also include the generation of antidrug antibodies (ADAs). ADAs can be either neutralizing or non-neutralizing antibodies, in which they bind to the idiotope of the antibody component of ADCs, resulting in reduced drug target binding and increased clearance of ADCs, respectively. The two main causes for glycosylation-mediated immunogenicity can be distinguished as (i) non-human glycosylation and (ii) glycan heterogeneity (Wolf et al. 2022). Non-human glycosylation is of concern as most antibodies are derived from non-human cell lines such as Chinese hamster ovary (CHO) or murine myeloma cells (NSO; SP2/0).

Table 2 FDA-Approved ADC Chemistry Profiles

	ADC	IgG	Target	Conjugation Method	Linker motif	Linker Chemistry	Payload	DAR	Average DAR	Half-Life ($t_{1/2}$)
First Generation	Gemtuzumab ozogamicin (Mylotarg®)	Humanized IgG4k	CD33	Non-specific, amide	Acid-labile 4-(4′-acetylphenoxy) butanoic acid (acetyl butyrate)	Cleavable: chemical	Calicheamicin	2–3	2.5	First dose: 62–64 h, second dose: 90 h
Second Generation	Brentuximab vedotin (Adcetris®)	Recombinant chimeric IgG1	CD30	Site-specific, reduced cysteine	Protease-cleavable mc-vc-PABC	Cleavable: enzymatic	Monomethyl auristatin E (MMAE)	4	4	Four to 6 days
	Ado-trastuzumab emtansine (Kadcyla®)	Humanized IgG1	HER2	Non-specific, amide	Non-reducible thioether linker MCC (4-[N-maleimidomethyl] cyclohexane-1-carboxylate)	Non-cleavable	T-DM1 (maytansinoid derivative)	3–4	3.5	Four days
	Moxetumomab pasudotox-tdfk (Lumoxiti®)	Recombinant murine	CD22	Immunotoxin*	ASGGPE peptide bond to VH	Cleavable: enzymatic	PE38	NA	NA	1.4 h

Third Generation	Belantamab mafodotin (Blenrep®)	Afucosylated humanized IgG1k	BCMA	Site-specific, cysteine	Non-cleavable Maleimidocaproyl (mc) cysteine linker	Non-cleavable maleimidocaptoyl (MC)	Auristatin F	4	4	12 days after first dose, 14 days after steady state
	Trastuzumab deruxtecan (Enhertu®)	Humanized IgG1	HER2	Site-specific, cysteine	Protease cleavable tetrapeptide (GFLG) based linker	Cleavable: enzymatic	DXd (Topoisomerase I inhibitor)	7–8	7.5	5.8 days
	Sacituzumab govitecan (Trodelvy®)	Humanized hRS7 IgG1κ	Trop-2	Site-specific, cysteine	Hydrolyzable linker (CL2A)	Cleavable: hydrazone, acid-labile	SN-38, Toposiomerase I inhibitor	7.6	7.6	28 days
	Loncastuximab tesirine-lpy (Zynlonta®)	Fully humanized IgG1k	CD19	Site-specific, cysteine	Valine-alanine linker	Cleavable: enzymatic	SG3199, a cytotoxic pyrrolobenzodiazepine (PBD) dimer alkylating agent	2.3	2.3	7–12.5
	Enfortumab vedotin-ejfv (Padcev®)	Fully humanized IgG1k	Nectin-4	Site-specific, cysteine	Protease-cleavable maleimidocaproyl valine-citrulline (vc) linker	Cleavable: enzymatic	Monomethyl auristatin E (MMAE)	3.8	3.8	3.4 days
	Polatuzumab vedotin (Polivy®)	Fully humanized IgG1k	CD79b	Site-specific, cysteine	Cleavable maleimidocaproyl-valyl-citrullinyl-p-aminobenzyloxycarbonyl (mc-val-cit-PABC)	Cleavable: enzymatic	Monomethyl auristatin E (MMAE)	3.5	3.5	12 days at cycle 6
	Inotuzumab ozogamicin (Besponsa®)	Recombinant humanized monoclonal IgG4 antibody	CD22	Non-specific, amide	Acid-labile 4-(4′-acetylphenoxy) butanoic acid (acetyl butyrate)	Cleavable: hydrazone, acid-labile	Calicheamicin	5–7	6	12.3 days
	Mirvetuximab Soravtansine-Gynx (Elahere®)	IgG1k	Folate factor alpha (FRα)	Non-specific, amide	N-Succinimidyl 4-(2-pyridyldithio)-2-sulfobutanoate linker (Sulfo-SPDB)	Cleavable: Reducible Disulfide bond	DM4 (maytansinoid derivative)	2.5	2.5	4.8
	Tisotumab vedotin-tftv (Tivdak®)	Hman IgGk	Tissue factor (TF)	Site-specific, cysteine	Protease-cleavable mc-vc-PABC	Cleavable: enzymatic	Monomethyl auristatin E (MMAE)	4	4	2 to 7.5

Common non-human glycans that are typically produced by the aforementioned cell lines are the non-human version of sialic acid: N-glycolylneuraminic acid (Neu5GC) and alpha 1–3 gal (Galα1–3Galβ1–4GlcNac extension) (Reusch and Tejada 2015). An example where hypersensitivity to these glycan containing antibodies that resulted in severe anaphylactic shock, is with cetuximab (Erbitux®), a chimeric mouse-human IgG1 mAb produced from murine SP2/0 cells, in which the presence of anti-glycan IgE antibodies in human serum were detected and confirmed to be specific for alpha 1–3 gal, was present on the Fc portion of the heavy chain (Chung et al. 2008; Ghaderi et al. 2010). In contrast, the version of cetuximab where the glycosylation with galactose-α-1,3-galactose on the Fab portion that was missing resulted in lower immunogenicity (Chung et al. 2008). Another source of immunogenicity can stem from mannan glycosylation therefore, the removal of mannose content at the Asp297 position of the Fab region of IgG, may potentially reduce the immunogenic events (Boune et al. 2020). In an effort to circumvent these occurrences, mAbs are developed using expression patterns and systems that closely reflect those found in endogenous human proteins (Almagro et al. 2017; Hötzel et al. 2012).

Linker and Conjugation Strategies

The linker is a multifaceted and critical component to the ADC. In addition to covalently bridging the target-specific antibody to its cytotoxic payload, the linker directly influences efficacy, stability, and can impact the overall safety of an ADC, as instability can trigger premature release (Su et al. 2021). A model linker demonstrates high stability in plasma with a circulation $t_{1/2}$ that is ten-fold greater than that of the antibody arm of the ADC; factors such as linker composition, length, and hydrophobicity can increase or decrease stability of the ADC by contributing to, or reducing bioavailability, or in the case of hydrophobic linkers, decrease overall solubility and promoting aggregation of the ADC (Su et al. 2021; Sheyi et al. 2022). For example, longer linker length may promote flexibility of the ADC but promote steric hinderance; both linker chemistry and hydrophobicity have effect on the $t_{1/2}$. Premature drug release through linker instability is one of the greatest barriers to ADC construction; as such, optimal linker design should both maintain ADC stability in transit without losing the sensitivity required for cargo release in specified conditions (Sheyi et al. 2022; Alas et al. 2021). Designing a linker with a long $t_{1/2}$ guarantees a stable connection between antibody and payload, therefore ensuring stability of the ADC. Similarly, controlled conjugation and release chemistries (cleavable and non-cleavable) have a large influence on homogeneous DAR and targeted delivery. Current ADC linker chemistries in clinic and in development take advantage of both non-cleavable and cleavable linker chemistry types; however, cleavable chemistries are increasingly favored, with nine of the thirteen approved ADCs using cleavable linkers (Sheyi et al. 2022; Alas et al. 2021).

Non-cleavable Linkers

Non-cleavable linkers rely on the complete lysosomal degradation of the ADC after cellular internalization for payload release. Following receptor-driven endocytosis, ADCs are taken into the lysosome, where complete catabolic or hydrolytic breakdown promotes payload release in the target cell (Sheyi et al. 2022; Tsuchikama and An 2018). Kadcyla®, a non-cleavable second-generation ADC, works via receptor-mediated endocytosis, followed by lysosomal breakdown for drug release in target cells (Amiri-Kordestani et al. 2014). Kadcyla® uses a thioether linker, which is more stable in circulation and only separated via complete breakdown of the ADC within the lysosome (Lu et al. 2016). Because of this stability, thioether and maleimidocaproyl (MC) based linkers are the more common choice for non-cleavable linker chemistries among next-generation ADCs (Sheyi et al. 2022). Other second-generation linkers, illustrated with ADCs like brentuximab vedotin (Adcetris®), and loncastuximab tesirine-lpyl (Zynlonta®), employ a mixture of non-cleavable and cleavable linker types.

Cleavable linkers, in contrast, take advantage of known cellular proteins or the cellular microenvironment to trigger payload release (Nolting 2013). There are two classes of cleavable linkers: chemical and enzymatic. Chemically liable linkers exploit intracellular changes to trigger drug release, for example, acid-based linkers are triggered by the lowered pH of the lysosome after cellular internalization. First-generation ADCs like gemtuzumab ozogamicin (Myelotarg®) were designed with a hydrazone linker, which is an acid-liable but relatively unstable in circulation. Myelotarg's hydrazone linker suffered from premature cleavage in circulation, releasing 50% of drug within the first 48 h; this unintended release caused off-target toxicity in the form of severe hepatic injury (Tai and Anderson 2015). Comparatively, glutathione-sensitive or reducible disulfide linkers maintain an intact conjugate in circulation and deliver cargo only in reducing environments (Liu et al. 2016; Ponzianiet al. 2020). Most cancer cells exhibit elevated levels of the reducing agent glutathione, which often contributes to enhanced tumor progression and can create resistance to specific anti-cancer drugs (Kalinina 2024). In this manner, reducible disulfide linkers leverage otherwise difficult-to-target cellular conditions of cancers to deliver payload.

Cleavable Linkers

The second type of cleavable linker leverages the enzymatic reactions that occur within the cell to promote payload release, specifically proteases. Adcetris® and Zynlonta® were the first among second-generation ADCs to implement proteolytic cleavable linker technologies (van de Donk and Dhimolea 2012; Lee 2021). Dipeptide linker strategies are designed to take advantage of proteases predominantly found in cancerous lysosomes; here, linkers contain sequences only recognizable to cancer-specific proteases that subsequently cleave the linker at these regions. One of the most common dipeptide linker strategies takes advantage of the hydrolytic lysosomal β-glucuronidase, Cathepsin B, a protease that cleaves the dipeptide and/or tripeptide sequences of hydrophobic residues between the payload and linker of the ADC (Balamkundu and Liu 2023). Another example of the cleavable dipeptide strategy is the glycosidase-sensitive linker. This linker strategy uses the specificity of *β*-galactosidase, which is overabundant within the lysosome of tumor cells, to cleave the *β*-galactoside moiety and trigger payload release (Su et al. 2021; Kolodych et al. 2017). Linkers using Cathepsin B require hydrolysable residues in the first position of the dipeptide motif, with the most common being Phenylalanine and Valine, however the addition of a lipophilic residue in the second peptide position can enhance stability in human plasma (Kostova et al. 2021).

The dipeptide Phenylalanine-Arginine (Phe-Arg) is the most used in the Cathepsin B cleavage strategy, however, Phenylalanine and Lysine (Phe-Lys), and Valine-Citrulline (Val-Cit) are also used. While stable in human plasma, valine-citruline linkers show instability in mouse plasma because of extracellular carboxylesterases, specifically Carboxylesterase 1C (Dorywalska et al. 2016). This concern was soon resolved with the addition of glutamic acid, creating glutamic acid-valine-citrulline linkers which are resistant to carboxylesterase cleavage, preventing premature degradation in mouse plasma (Dorywalska et al. 2016). Cathepsin B presents as a good target for dipeptide cleavable linkers; it is produced in excess in the lysosomes of multiple cancers and is known to contribute to tumor invasiveness and progression (Aggarwal and Sloane 2014; Egorova et al. 2023). One drawback of Cathepsin B is its non-specific expression profile in normal cells and is known to contribute to off-target toxicity (Balamkundu and Liu 2023; Aggarwal and Sloane 2014). Often, dipeptide linkers require spacer motifs to help with ADC stabilization due to steric hinderance of the drug (Su and Zhang 2021). In this case, spacers can guarantee access to the peptide sequence for efficient in vivo cleavage. One example is through *Para*-aminobenzyl carbamate (PABC) linkages, which allow for clean drug delivery via direct immolation after enzymatic cleavage of the linkage and drug (Dorywalska et al. 2016).

Third generation ADCs, which are represented by trastuzumab deruxtecan (Enhertu®), inotuzumab ozogamicin (Besponsa®), and enfortumab vedotin (Padcev®), largely differ from the first- and second-generation design in conjugation strategy over linker type. Instead of linker-specific changes, the primary differences between third-generation ADCs and their predecessors is the site-specific conjugation strategies that have evolved to allow the production of a homogeneous population of ADCs. Site-specific conjugation not only allows for a known number of payloads to antibody, it is also location-specific with regard to the location of the payload conjugation to the antibody, which is advantageous in that conjugation can be designed for optimal binding that allows optimal delivery, without blocking the antigen binding motif.

Conjugation Approaches and DAR

Ideally, after conjugation with the linker, the conjugation approach must retain the naïve characteristics of the antibody for optimal performance and targeting activity. The most common chemical conjugation methods are targeted towards surface lysine residues, cysteine residues with a functional group, site specific glycosylation, and non-natural amino acid substitutions such as p-acetylphenylalanine. Modification of surface lysine residues may serve to neutralize the positive charges on these residues, particularly in regions with high flexibility and access to the solvent (Sasso et al. 2023). Although modifying lysine residues are more heterogeneous due to stochastic conjugation chemistries, in effect, they retain more naïve mAb traits compared to cysteine-linked ADCs (Singh et al. 2015).

The way by which the antibody is attached to the linker, and subsequently the payload, is a critical component to ADC design, as conjugation chemistry strongly influences the (DAR), which represents the average number of payload molecules attached to the antibody (Mahmood 2021). The optimal DAR for ADCs is between 2 and 4; low DAR is known to reduce drug efficacy, and while high DARs can increase potency, it also negatively impacts antigen structure/function (binding), stability (aggregation), promotes fast clearance, and reduces overall clinical activity and efficacy (Gogia et al. 2023; Hamblett et al. 2004). Conjugation strategies largely inform DAR heterogeneity within the ADC (Mahmood 2021; Kostova et al. 2021). Non-specific conjugation methods can produce a wide range of DAR species due to the wide distribution of reactive residues on the antibody (Chen et al. 2016). In contrast, site-specific chemistries restrict linker and payload attachment to specific sites of the antibody, and have proven more effective in achieving homogeneous populations with clear and reproducible DAR (Gogia et al. 2023).

While the DAR represents the average number of payload molecules attached to the antibody, the type of linker plays a pivotal role in both the binding of the drug to the mAb during conjugation processes, and in the effective release of the payloads at the tumor site. When the ADC complex is in circulation, the linker must be stable enough to maintain the entire 3-prong complex intact and inactive, preventing early release of the cytotoxic payload and reducing any "bystander effects". The above-mentioned conditions are ensured through the two types of linkers as previously described, non-cleavable and cleavable linkers (i.e., SMCC, maleimidocaproyl, peptide-based, ß-glucoronide, acid-sensitive, and disulfide linkers).

Amide coupling is a stochastic conjugation technique by which the available lysine on the surface of the antibody are targeted by carboxylic acid esters attached to the linker (Fig. 4) (Tsuchikama and An 2018). Antibodies have a large distribution of surface lysine (~80) as such, amide conjugation chemistry largely increases species heterogeneity (population of conjugates with variable DAR) (Chari 2008). Lysine-based conjugation requires downstream purification processes to obtain homogeneous populations, which can be rate-limiting to production (Abdollahpour-Alitappeh et al. 2019). Additionally, distribution of payload across a variety of lysine sites creates an additional layer of heterogeneity for amide coupled ADCs.

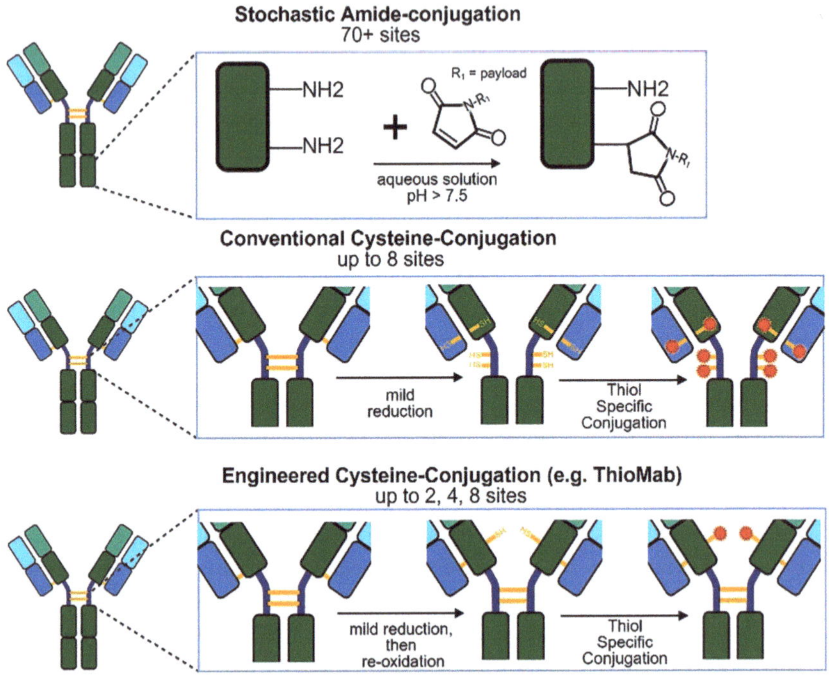

Fig. 4 Conjugation Types: Stochastic amide conjugation, conventional cysteine conjugation, and engineered cysteine conjugation (i.e., ThioMab). (Image made using Biorender.com)

Kadcyla® implements a two-step mechanism to attach linker and payload, first using amide-coupling to attach the linker, then maleimide coupling to the payload (Chen et al. 2016; Kim et al. 2014). This sequential reaction produces an ADC mixture with high heterogeneity, specifically distributing between 0 and 10 payloads per antibody, and obtaining an average DAR of 3.5 (Jackson 2016; Wakankar et al. 2011). The broad amide coupling approach can also alter the lysine residues on the antibody critical to receptor driven endocytosis, reducing ADC internalization (Chiang et al. 2016). Second-generation ADCs showed vast improvements in overall stability, however, concerns with low efficacy and narrow therapeutic window, remained; this was largely due to a lack of homogeneity in ADC species population with regard to DAR); when poorly controlled DAR can reduce ADC efficacy (Gogia et al. 2023).

Chemistries that employ site-specific conjugation are advantageous in that the DAR is controllable and reproducible, yielding highly homogeneous populations with specific DAR species (Indini et al. 2021; Rubinstein and O'Brien 2023; Tang et al. 2024). Site-specific chemistry uses both native and engineered residues to attach linker and payload. In contrast to amide-based conjugation, cysteine-coupling presents an advantage as the number of cysteines present in an antibody are lower, with only sixteen total sites, with four inter- and twelve intra-chain cysteines (Nadkarni 2020). Cysteine conjugation strategies will target native or engineered cysteine residues on the antibody surface, relying on the reaction between cysteine and a thiol reactive functional group attached to the linker, yielding a DAR between 0 and 8 (Tsuchikama and An 2018).

Accounting for Payload Hydrophobicity and Stability

The cytotoxic payload is the defining aspect that makes an ADC an efficient therapeutic once it reaches its targeted destination. ADC payloads utilized currently are mostly small molecules derived from proteins or natural products, with the two most common drug classes being DNA-damaging agents [calicheamicins (e.g., CalichDMH) and pyrrolobenzodiazepines (PBD) dimers (e.g., SG3199-alkylating agent)] and microtubule inhibitors [maytansinoids (e.g., DM1, DM4) and auristatins (e.g., monomethyl auristatin E (MMAE); monomethyl auristatin F (MMAF))]. Due to the low percentage of ADCs that actually reach the target, it is imperative that the payload remain inert in the presence of endogenous molecules as it reaches its target while simultaneously having a high potency for killing the tumor cell ($>10^6$ molecules/cell) (Duerr and Friess 2019). Next-generation ADCs are expanding the cytotoxic activity to include topoisomerase I/II inhibitors (i.e., camptothecin derivates and exatecan extracts such as DXd and SN-38) with increased DAR values such as those seen in Sacituzumab govitecan (Trodelvy®) and Fam-trastuzumab deruxtecan (Enhertu®®), with a DAR of 7.6 and 8, respectively (Tong et al. 2021).

The hallmark of a desirable payload is high potency at low concentrations, and a sub-nanomolar range for the inhibitory concentration (IC_{50}) (Conilh et al. 2023). Many early and current payloads used in ADCs are very hydrophobic, and, as such, prone to aggregation and provoke instability of the ADC (Wang et al. 2023; Anderl et al. 2013). When combined with hydrophobic monoclonal antibodies, which make up the majority of antibodies in clinical use, the likelihood of aggregation and instability is substantially increased (Jain et al. 2015; Mills et al. 2020). Buecheler et al., used dynamic light scattering (DLS) and differential fluorometry to analyze multiple ADCs with comparable DARs, and directly correlated the destabilization of native monoclonal antibodies with increased hydrophobicity of the payload (Buecheler et al. 2018). Hydrophobicity, however, is markedly important to the anti-cancer activity of many well-established and developing cytotoxic drugs, for example, the addition of hydrophobic residues to emerging polymer-based chemotherapeutics enhances the anti-cancer effects of cationic species, who alone, show no anti-cancer activity (Kumar et al. 2023).

The enediyne DNA damaging drug calicheamicin, is very hydrophobic, and used as a payload for both Myelotarg® and Bensposa® (Lamb 2017; Bross et al. 2001). An engineered cysteine approach to produce homogeneous calicheamicin-based ADCs, which would improve upon the stochastic strategy of Myelotarg® and Bensposa®, was designed and evaluated in 2021. In this study, anti-CD22 and anti-Ly6E antibodies were directly conjugated (mAb-drug) to calicheamicin at K149C of the light chain, yielding an average DAR of 1.7 (Vollmar et al. 2021). The linker-less calicheamicin approach not only produced remarkably less aggregation compared to the Myelotarg® and Bensposa®, but also led increased *in-vivo* stability with lower clearance and longer $t_{½}$, performing more similarly to the unconjugated antibody in vivo, with improved tolerability in rats compared to Myelotarg® (Vollmar et al. 2021).

The inherent charge of the chosen payload for ADC construction can also interfere with the physiochemical stability of these conjugates. One example is with the auristatin family of payloads, which include differentially charged species: monomethyl auristatin E (MMAE) is uncharged while monomethyl auristatin F (MMAF) carries the negative charge through its C-terminal carboxylic acid. Johann et al., evaluated the charge heterogeneity, and aggregation of these species under thermal stress, as well as changes to hydrophobicity of two ADCs with identical antibody conjugation chemistry (interchain cysteine), and DAR. The results of this study demonstrated that conjugation to the negatively charged MMAF caused significant increase in aggregation compared to the neutral species, in addition to decrease in ADC net charge and overall charge heterogeneity (Johann et al. 2024). The results of this study show that charge differences of cytotoxic payload should also be evaluated among factors affecting ADC stability. Novel approaches where the linker is used to shield the cytotoxic payload via increased hydrophilic design are currently being used in next-generation ADC construction in order to overcome the challenges encountered with charged payloads (Douez et al. 2024).

The advent of site-specific conjugation strategies allowed for the use of less extreme cytotoxic agents for the payload, which is observed with the TOP1 inhibitors of Trodlevy® and Enhertu®, and the microtubule inhibitors of Adcetris®, Kadcyla®, Tidivac®, Blenrep®, Padcev®, and Polivy®. In addition to changes in payload availability for next generation ADCs, site-specific conjugation chemistry provided a means for constructing homogeneous DAR species which is increasingly important regarding biological clearance of the ADC. Sun et al., analyzed the pharmacokinetics of cleavable and non-cleavable maytansinoid-conjugated ADCs, showing increased potency with conjugates of average DAR below 6 (Sun et al. 2017). Maytansinoid conjugates with an average DAR species of 9–10, however, exhibited increased clearance and rapid accumulation in the liver, which led to decrease in efficacy. Similarly, linker diversification has also expanded ADC design, for example, the development of hydrophilic polyethylene glycol (PEG)-based linkers provided a means to mask they hydrophobicity of cytotoxic drugs (Simmons et al. 2020; Douez et al. 2024; Tedeschini et al. 2021), improve $t_{1/2}$ and reduced cytotoxicity (Li et al. 2021), and reduced occurrence of aggregation during conjugation (Shao et al. 2018). Optimization of PEG linkers, for example increasing the number of PEGylated linkers (up to 24) paired with the self-immolating glucuronide linker resulted in slower drug clearance, providing a wider therapeutic window for monomethyl auristatin (MMAE)-conjugated ADCs (Burke et al. 2017).

Site-specific cysteine chemistry was a primary choice for the drug Adcetris®, along with many other ADCs undergoing clinical trials. Currently greater than 8 of the 13 approved ADCs employ site-specific cysteine chemistry (You et al. 2021). The first description of cysteine engineering for selective antibody attachment was reported in 2008, with increased efficacy and therapeutic window across species (Junutula et al. 2008). The chemistry created in this study, now known as THIOMAB (licensed by Genentech), uses site directed mutagenesis to replace endogenous residues with unpaired cysteines that can be reduced and later attached to the linker and payload through disulfide bonds (Fig. 4). ThioBridge® technology, developed by PolyTherics Research group, now Abenza, works by reducing interchain cysteines to produce stable covalent bonds with the linker-payload, resulting in DAR 4-specific species (Fu et al. 2022; Tsuchikama and An 2018). In a similar manner, antibodies can be engineered with unnatural amino acids for site-specific conjugation of ADCs. One example of this is the optimized unnatural amino acid, *p*-azidomethyl-l-phenylalanine (pAMF), which was incorporated in a site-specific manner to the HER2 antigen, resulting in increased drug potency and efficacy (Zimmerman et al. 2014).

Another method for site-specific ADCs employs engineering or remodeling of the natural glycan residues present within the antibody. Here, a reactive moiety is added to the N-glycan of the antigen, which is followed by conjugation to the linker-payload with the corresponding reactive group (Jaramillo et al. 2023). Short peptide tags have also been used for site-specific strategies for ADC conjugation. More recently, advancements in ADC conjugation chemistry allowed for the implementation of dual-payloads, resolving concerns with varying sensitivity to chemotherapeutic drugs due to tumor cell heterogeneity and drug resistance, by increasing the variety of payload per antigen (Wang et al. 2023; Yamazaki et al. 2021).

Drug Delivery and Stability

In addition to site-specific free sulfhydryl groups in the Fc regions, another method for antibody drug delivery is through the implementation of lipid-based nanoparticles (LNPs) through conjugation to the LNPs via maleimide-activated amino groups. To tether antibodies, LNP-ADC carriers require surface functionalization with their carboxyl groups through known chemical reactions that require 1-ethyl-3-(3-dimethylaminopropyl) carbodiimide (EDC) and N-hydroxy succinimide (NHS) couplings with primary amines of antibodies for effective delivery of the conjugate complex. At present, these delivery approaches are still undergoing clinical evaluation and have not been incorporated into newly approved ADCs. Attaching polyethylene glycol (PEG) polymers to mAbs can shield surface epitopes from immune responses. This shielding by way of PEGylation leads to increased aqueous solubility, improved $t_{1/2}$, and reduced immunogenicity (Dozier and Distefano 2015). However, there are drawbacks to PEGylation of mAbs, which include PEG accumulation in organs, creation or exacerbation of preexisting anti-PEG antibodies, and altered or decreased biological activity. To surmount these potential obstacles, alternatives to PEGylation exist. PASylation technology comprises of polypeptide chains of small biosynthetic residues: Pro, Ala, and Ser (PAS). PAS technology are uncharged biological polymers with biophysical traits that are similar to PEG and demonstrate no immunogenicity and do not interfere with the pharmacological activity of the drug component (Binder and Skerra 2017). PAS polypeptides offer extended pharmacokinetic benefits that include high stability in plasma, decreased kidney filtration, and are quickly metabolized and degraded upon cellular uptake.

To date, most if not all, approved ADCs are formulated and developed in lyophilized form, minimizing the degradation and instability of the linker during storage and shipment before reaching the patient (Cho et al. 2021). All approved ADCs are administered as injectable aqueous solutions containing pharmaceutically inactive excipients and ingredients that allow the facilitation for intravenous delivery (Table 3). Major challenges exist in formulating stable ADC complexes due to their unique antibody-linker-payload design, commonly used standard mAb platform formulations may not apply. The potential for aggregation is every present, where the specificity of the targeted mAb portion allows for lower therapeutic concentrations; the linker-payload portion may increase the hydrophobicity of the ADC complex, which reduces the risk of aggregate formation, providing challenges that need to be circumvented, and advantages that need to be emphasized. Overall, the development of a successful ADC requires defined, underlying biology as well as understanding of (bio)chemistry that is in play. The analytical complexity is significantly magnified relative to that for a mAb alone, when it comes to ADCs. Addressing manufacturing and regulatory issues therefore will involve considerations of biotherapeutics as well as small molecule biopharmaceutics juxtaposed perspectives. The design and development aspects of ADCs with a focus on biophysical and chemical stability of the drug product are equally critical in this context. Basics of the design of ADCs, requirements for each element of the ADC construct, must be taken into account in parallel. The ADC construct has significant impact on its physicochemical stability and therefore the formulation and analytical challenges encountered are interrelated. In this context, considerations of quality and some key regulatory aspects as they relate to ADCs are becoming critical from a quality perspective.

Withdrawals, Failures, and Pending Approvals

Since 2014, a number of ADCs have received either conditional or full approval by the FDA for the treatment of solid and circulating tumors (Table 1). However, several have also faced withdrawal from the US market, underscoring the complexities associated with ADC development. A notable example is Mylotarg®, which, despite its approval from the FDA in 2000 for the treatment of relapsed CD33-positive acute myeloid leukemia (AML) in older patients, was voluntarily withdrawn in 2010 due to safety and efficacy concerns. During its post-marketing period, a confirmatory study failed to show clinical benefit at the current dosing regimen (9 mg/m^2 on days 1 and 15 of a 4-week cycle) and fatal hepatotoxicity and veno-occlusive disease were added as boxed warnings (Norsworthy et al. 2018). Upon revision of the dosing regimen to address the aforementioned concerns (3 mg/m^2 on days 1, 4, and 7 of a 4-week cycle), Mylotarg® was re-approved by the FDA in 2017 for the treatment of relapsed or refractory (R/R) CD33-positive AML in patients 2 years

Table 3 Formulation and Administration Considerations of FDA -Approved ADCs

Trade Name	Delivery	Pharmaceutical Form/ Storage	Formulation Concentration	Dilution Media	Diluted Concentration	Excipients
Mylotarg®	Intravenous	Lyophilized 2–8 °C	1 mg/mL	NaCl	–	Dextran 40, Sucrose, NaCl, Monobasic and dibasic sodium phosphate
Adcetris®	Intravenous	Lyophilized 2–8 °C	5 mg/mL	NaCl, Dextrose, Ringer-Lactate	0.4–1.8 mg/mL	Citric acid monohydrate, Sodium citrate dihydrate, α, α -Trehalose dihydrate, Polysorbate 80
Kadcyla®	Intravenous	Lyophilized 2–8 °C	20 mg/mL	NaCl	–	Succinic acid, Sodium hydroxide, Sucrose, Polysorbate 20
Lumoxiti®	Intravenous	Lyophilized 2–8 °C	1 mg/mL	NaCl	–	Glycine, Polysorbate 80, Monobasic sodium phosphate, Sucrose, Sodium hydroxide
Blenrep®	Intravenous	Lyophilized 2–8 °C	50 mg/mL	NaCl	0.2–2 mg/mL	Sodium citrate, Citric acid, Trehalose dihydrate, Disodium edetate, Polysorbate 80
Enhertu®	Intravenous	Lyophilized 2–8 °C	20 mg/mL	Dextrose	–	L-histidine, L-histidine hydrochloride monohydrate, Sucrose, Polysorbate 80
Trodelvy®	Intravenous	Lyophilized 2–8 °C	10 mg/mL	NaCl	1.1–3.4 mg/mL	2-(N-morpholino)ethane sulfonic acid, Polysorbate 80, Trehalose dihydrate

(continued)

Table 3 (continued)

Trade Name	Delivery	Pharmaceutical Form/ Storage	Formulation Concentration	Dilution Media	Diluted Concentration	Excipients
Zynlonta®	Intravenous	Lyophilized 2–8 °C	5 mg/mL	Dextrose	–	L-histidine, L-histidine monohydrochloride, Polysorbate 20, Sucrose
Padcev®	Intravenous	Lyophilized 2–8 °C	10 mg/mL	NaCl, Dextrose, Ringer-Lactate	0.3–4 mg/mL	Histidine, Histidine hydrochloride monohydrate, Trehalose dihydrate, Polysorbate 20
Polivy®	Intravenous	Lyophilized 2–8 °C	20 mg/mL	NaCl, Dextrose	0.72–2.7 mg/mL	Succinic acid, Sodium hydroxide, Sucrose, Polysorbate 20
Besponsa®	Intravenous	Lyophilized 2–8 °C	0.25 mg/mL	NaCl	0.01–0.1 mg/mL (Infusion bag) 0.025–0.1 mg/mL (Syringe)	Sucrose, Polysorbate 80, NaCl, Tromethamine
Elahere®	Intravenous	Lyophilized 2–8 °C	5 mg/mL	Dextrose	1–2 mg/mL	Glacial acetic acid, polysorbate 20, sodium acetate, sucrose
Tivdak®	Intravenous	Lyophilized 2–8 °C	10 mg/mL	NaCl, Dextrose, Ringer-Lactate	0.7–2.4 mg/mL	D-mannitol, L-histidine, L-histidine monohydrochloride, and Sucrose

and older, and in 2020 for the treatment of newly diagnosed CD33-positive AML to include pediatric patients 1 month and older (Norsworthy et al. 2018; FDA 2020). In 2022, the FDA requested that Blenrep® be withdrawn from market after the required confirmatory study did not meet the primary endpoint of progression free survival (PFS). The US license to manufacture Blenrep® was then withdrawn by the FDA in early 2023. Furthermore, Lumoxiti® was approved in 2018 by the FDA but voluntarily withdrawn from the US market by AstraZeneca in 2023 due to low clinical uptake (AstraZeneca 2022). Lumoxitil® serves as an example to the external factors that can influence ADC market status, outside of the more common concerns with efficacy and toxicity (AstraZeneca 2022).

More than 200 ADCs have undergone evaluation in clinical trials, many of which have encountered clinical failure (Ruan et al. 2024). In June 2017, Seattle Genetics discontinued its phase III trial of vadastuximab talirine (SGN-CD33A) for the treatment of newly diagnosed AML in older patients, after findings revealed that the combination of SGN-CD33A and hypomethylating agents was associated with a higher mortality rate compared to treatment with hypomethylating agents alone (Walter 2021). This followed a clinical hold that was placed on phase I trials of the drug in 2016, after four patient fatalities. In July 2018, a partial clinical hold was placed on a phase I trial of XMT-1522, an ADC manufactured by Mersana Therapeutics for the treatment of HER2-expressing tumors, following a patient death (Therapeutics, M 2018a). This hold was lifted in September 2018 following protocol modifications such as increased monitoring and exclusion of patients with advanced hepatic impairment (Therapeutics, M 2018b). In December 2018, AbbVie terminated research and development of rovalpituzumab tesirine (Rova-T) for the treatment of advanced small-cell lung cancer (SCLC) after a phase III trial failed to demonstrate survival benefit (AbbVie 2019).

As of 2024, among the 15 ADCs that have received market approval on a global scale, two have yet to garner FDA approval. Aidixi® was approved by China's National Medical Products Administration (NMPA) in 2021 for the treatment of HER2-overexpressing locally advanced or metastatic gastric cancer (including gastroesophageal junction adenocarcinoma) in patients who have undergone at least two systemic chemotherapies. Notably, Aidixi® received Breakthrough Therapy designation from the FDA in 2020 for the treatment of HER2-positive locally advanced or metastatic urothelial carcinoma following treatment with platinum-based chemotherapy (RemeGen 2020). Similarly, Akalux® has been approved by the Pharmaceuticals and Medical Devices Agency (PMDA) in 2020 for the treatment of unresectable locally advanced or recurrent head and neck cancer (Riccardi et al. 2023).

Several investigational ADCs are nearing regulatory approval for patients with lung cancers, underscoring significant advancements in this therapeutic area. Noteworthy among them are datopotamab deruxtecan (Dato-DXd), developed by AstraZeneca and Daiichi Sankyo; patritumab deruxtecan (HER3-DXd), a collaborative effort from Daiichi Sankyo and Merck; and telisotuzumab vedotin (ABBV-399) from Abbvie (AstraZeneca 2024; Merck 2023; AbbVie 2022). HER3-DXd received Breakthrough Therapy designation in 2021 and Priority Review status in 2023 for the treatment of adult patients with locally advanced or metastatic EGFR-mutated non-small cell lung cancer (NSCLC) previously treated with two or more systemic therapies. ABBV-399 achieved Breakthrough Therapy designation in 2022 for the treatment of advanced or metastatic NSCLC with c-Met overexpressing tumors in individuals whose disease has progressed on or after platinum-based therapy. Furthermore, a handful of ADCs that have already secured FDA approval are being explored for novel indications. Among these, Trodelvy® stands out as it is currently being investigated in phase III clinical trials for NSCLC, colorectal cancer, and endometrial cancer and phase II trials for numerous other conditions. Enhertu® is being evaluated in a phase III clinical trial for biliary tract cancer. Biosimilars play a crucial role in expanding access to ADCs by offering cost-effective alternatives without compromising safety and efficacy. Currently, five FDA-approved biosimilars exist for trastuzumab (Herceptin®), the mAB used in Kadycla®: Ogivri®, Herzuma®, Ontruzant®, Kanjinti®, Trazimera® (Triantafyllidi and Triantafillidis 2022). Ujvira™, a biosimilar of Kadcyla® that was developed by Zydus Cadila, was approved in 2021 by the Drug Controller General of India (DCGI) (Thungappa et al. 2022).

Conclusions

ADCs have revolutionized traditional cancer therapies for many indications with limited treatments. They provide targeted delivery of cytotoxic payloads that are tumor-focused, binding to specific proteins found on the surface of cancer cells. This approach presents opportunities in next-generation design, where novel surface receptor proteins can be targeted to deliver therapies beyond cancer, for example, surface receptors involved in autoimmune disease. Nearly 70% of ADCs currently marketed in the US, have gained FDA-approvals through Breakthrough Therapy and Priority Review designations (Table 1), and many ADC technologies have demonstrated significant potential for providing greater or substantially improved treatments, based on early clinical evidence (U.S. Food and Drug Administration, F 2023). Among the many advantages of ADC directed therapies, ADC-based therapies require lower concentrations (≤20 mg/mL; Table 3) than traditional antibody-based treatments, which can reach upward of 50 mg/mL concentrations, which is largely facilitated by the targeted delivery of the drug component of the ADC (Garidel et al. 2017; Zidar et al. 2017). The ADCs currently marketed are administered intravenously and available in lyophilized form to improve stability, namely the prevention of linker degradation and subsequent free drug release; future advancements to improve hydrophobicity and solubility at increased dosing concentrations can one day allow prospective ADC delivery via subcutaneous administration. However, there still aspects of ADC design that have yet to be optimized, including the fact that many of the cytotoxic payloads are hydrophobic, further adding to the challenges faced when formulating next-generation ADC complexes. As an infusion-based therapy, dilution of the drug with 0.9% NaCl or 5% dextrose solutions may lead to electrostatic charge shielding, potentially enhancing protein-protein interactions- increasing the possibility of aggregation, particularly those ADCs with high DAR values.

These are just a few of the obstacles faced by formulation scientists; issues that are continuously improved with next-generation technologies. Various approaches have been optimized for conventional antibody formulations. ADCs are more than just antibodies, they are intricate complexes where small molecules to an antibody via a linker, a process that necessitates personalized considerations at each step of the design, and continues to inform formulation strategies.

References

AbbVie: AbbVie discontinues rovalpituzumab tesirine (Rova-T) research and development program. FDA approves gemtuzumab ozogamicin for CD33-positive AML in pediatric patients (2019) bit.ly/2Y86KRI

AbbVie: AbbVie announces U.S. FDA granted breakthrough therapy designation (BTD) to telisotuzumab vedotin (Teliso-V) for previously treated non-small cell lung cancer. FDA approves gemtuzumab ozogamicin for CD33-positive AML in pediatric patients (2022) bit.ly/2Y86KRI

Abdollahpour-Alitappeh, M., et al.: Antibody-drug conjugates (ADCs) for cancer therapy: strategies, challenges, and successes. J. Cell. Physiol. **234**(5), 5628–5642 (2019)

Aggarwal, N., Sloane, B.F.: Cathepsin B: multiple roles in cancer. Proteomics Clin. Appl. **8**(5–6), 427–437 (2014)

Alas, M., Saghaeidehkordi, A., Kaur, K.: Peptide-drug conjugates with different linkers for cancer therapy. J. Med. Chem. **64**(1), 216–232 (2021)

Alberts, B., et al.: Molecular Biology of the Cell, 4th edn. Garland Science, New York (2002)

Almagro, J.C., et al.: Progress and challenges in the design and clinical development of antibodies for cancer therapy. Front. Immunol. **8**, 1751 (2017)

Amiri-Kordestani, L., et al.: FDA approval: ado-trastuzumab emtansine for the treatment of patients with HER2-positive metastatic breast cancer. Clin. Cancer Res. **20**(17), 4436–4441 (2014)

Anderl, J., et al.: Antibody-drug conjugate payloads. Methods Mol. Biol. **1045**, 51–70 (2013)

Arnold, J.N., et al.: The impact of glycosylation on the biological function and structure of human immunoglobulins. Annu. Rev. Immunol. **25**, 21–50 (2007)

AstraZeneca: Permanent withdrawal of LUMOXITI® from the US Market. FDA approves gemtuzumab ozogamicin for CD33-positive AML in pediatric patients (2022) bit.ly/2Y86KRI

AstraZeneca: Datopotamab deruxtecan Biologics License Application accepted in the US for patients with previously treated advanced nonsquamous non-small cell lung cancer. FDA approves gemtuzumab ozogamicin for CD33-positive AML in pediatric patients (2024) bit.ly/2Y86KRI

Balamkundu, S., Liu, C.F.: Lysosomal-cleavable peptide linkers in antibody-drug conjugates. Biomedicines. **11**(11), 3080 (2023)

Batra, J., Rathore, A.S.: Glycosylation of monoclonal antibody products: current status and future prospects. Biotechnol. Prog. **32**(5), 1091–1102 (2016)

Binder, U., Skerra, A.: PASylation®: a versatile technology to extend drug delivery. Curr. Opin. Colloid Interface Sci. **31**, 10–17 (2017)

Boune, S., et al.: Principles of N-linked glycosylation variations of IgG-based therapeutics: pharmacokinetic and functional considerations. Antibodies (Basel). **9**(2), 22 (2020)

Bournazos, S., Ravetch, J.V.: Diversification of IgG effector functions. Int. Immunol. **29**(7), 303–310 (2017)

Brinkmann, U., Kontermann, R.E.: The making of bispecific antibodies. MAbs. **9**(2), 182–212 (2017)

Bross, P.F., et al.: Approval summary: gemtuzumab ozogamicin in relapsed acute myeloid leukemia. Clin. Cancer Res. **7**(6), 1490–1496 (2001)

Buecheler, J.W., et al.: Impact of payload hydrophobicity on the stability of antibody-drug conjugates. Mol. Pharm. **15**(7), 2656–2664 (2018)

Burke, P.J., et al.: Optimization of a PEGylated glucuronide-monomethylauristatin E linker for antibody-drug conjugates. Mol. Cancer Ther. **16**(1), 116–123 (2017)

Carrasco-Triguero, M., et al.: Immunogenicity of antibody-drug conjugates: observations across 8 molecules in 11 clinical trials. Bioanalysis. **11**(17), 1555–1568 (2019)

Chari, R.V.: Targeted cancer therapy: conferring specificity to cytotoxic drugs. Acc. Chem. Res. **41**(1), 98–107 (2008)

Chaudhary, V.K., et al.: A recombinant immunotoxin consisting of two antibody variable domains fused to Pseudomonas exotoxin. Nature. **339**(6223), 394–397 (1989)

Chen, X., Liu, Y.D., Flynn, G.C.: The effect of Fc glycan forms on human IgG2 antibody clearance in humans. Glycobiology. **19**(3), 240–249 (2009)

Chen, L., et al.: In-depth structural characterization of Kadcyla® (ado-trastuzumab emtansine) and its biosimilar candidate. MAbs. **8**(7), 1210–1223 (2016)

Chiang, T.W., et al.: CRISPR-Cas9(D10A) nickase-based genotypic and phenotypic screening to enhance genome editing. Sci. Rep. **6**, 24356 (2016)

Chiu, M.L., et al.: Antibody structure and function: the basis for engineering therapeutics. Antibodies (Basel). **8**(4), 55 (2019)

Cho, E., et al.: Stability of antibody drug conjugate formulations evaluated using solid-state hydrogen-deuterium exchange mass spectrometry. J. Pharm. Sci. **110**(6), 2379–2385 (2021)

Chung, C.H., et al.: Cetuximab-induced anaphylaxis and IgE specific for galactose-alpha-1,3-galactose. N. Engl. J. Med. **358**(11), 1109–1117 (2008)

Conilh, L., et al.: Payload diversification: a key step in the development of antibody-drug conjugates. J. Hematol. Oncol. **16**(1), 3 (2023)

Cosimi, A.B.: OKT3: first-dose safety and success. Nephron. **46**(Suppl 1), 12–18 (1987)

Courtenay-Luck, N.S., et al.: Development of primary and secondary immune responses to mouse monoclonal antibodies used in the diagnosis and therapy of malignant neoplasms. Cancer Res. **46**(12 Pt 1), 6489–6493 (1986)

De Cecco, M., Galbraith, D.N., McDermott, L.L.: What makes a good antibody-drug conjugate? Expert. Opin. Biol. Ther. **21**(7), 841–847 (2021)

de Taeye, S.W., Rispens, T., Vidarsson, G.: The ligands for human IgG and their effector functions. Antibodies (Basel). **8**(2), 30 (2019)

Di, L., et al.: Development and application of high throughput plasma stability assay for drug discovery. Int. J. Pharm. **297**(1–2), 110–119 (2005)

Dorywalska, M., et al.: Molecular basis of valine-citrulline-PABC linker instability in site-specific ADCs and its mitigation by linker design. Mol. Cancer Ther. **15**(5), 958–970 (2016)

Douez, E., et al.: Branched pegylated linker-auristatin to control hydrophobicity for the production of homogeneous minibody-drug conjugate against HER2-positive breast cancer. J. Control. Release. **366**, 567–584 (2024)

Dozier, J.K., Distefano, M.D.: Site-specific PEGylation of therapeutic proteins. Int. J. Mol. Sci. **16**(10), 25831–25864 (2015)

Duerr, C., Friess, W.: Antibody-drug conjugates- stability and formulation. Eur. J. Pharm. Biopharm. **139**, 168–176 (2019)

Egan, P.C., Reagan, J.L.: The return of gemtuzumab ozogamicin: a humanized anti-CD33 monoclonal antibody-drug conjugate for the treatment of newly diagnosed acute myeloid leukemia. Onco. Targets. Ther. **11**, 8265–8272 (2018)

Egorova, V.S., et al.: Smart delivery systems responsive to cathepsin B activity for cancer treatment. Pharmaceutics. **15**(7), 1848 (2023)

Elias, D.J., et al.: Phase I clinical comparative study of monoclonal antibody KS1/4 and KS1/4-methotrexate immunconjugate in patients with non-small cell lung carcinoma. Cancer Res. **50**(13), 4154–4159 (1990)

FDA approves gemtuzumab ozogamicin for CD33-positive AML in pediatric patients (2020) bit.ly/2Y86KRI

Fu, Z., et al.: Antibody drug conjugate: the "biological missile" for targeted cancer therapy. Signal Transduct. Target. Ther. **7**(1), 93 (2022)

Garidel, P., et al.: High-concentration protein formulations: how high is high? Eur. J. Pharm. Biopharm. **119**, 353–360 (2017)

Ghaderi, D., et al.: Implications of the presence of N-glycolylneuraminic acid in recombinant therapeutic glycoproteins. Nat. Biotechnol. **28**(8), 863–867 (2010)

Giddens, J.P., et al.: Site-selective chemoenzymatic glycoengineering of Fab and Fc glycans of a therapeutic antibody. Proc. Natl. Acad. Sci. **115**(47), 12023–12027 (2018)

Gogia, P., et al.: Antibody-drug conjugates: a review of approved drugs and their clinical level of evidence. Cancers (Basel). **15**(15), 3886 (2023)

Golay, J., Andrea, A.E., Cattaneo, I.: Role of Fc core fucosylation in the effector function of IgG1 antibodies. Front. Immunol. **13**, 929895 (2022)

Gómez Román, V.R., Murray, J.C., Weiner, L.M.: Chapter 1 – antibody-dependent cellular cytotoxicity (ADCC). In: Ackerman, M.E., Nimmerjahn, F. (eds.) Antibody Fc, pp. 1–27. Academic, Boston (2014)

GSK: GSK Provides an update on Blenrep (belantamab mafodotin-blmf) US marketing authorisation (2022) [cited 2024 09/15/2024]. Available from: https://www.gsk.com/en-gb/media/press-releases/gsk-provides-update-on-blenrep-us-marketing-authorisation/

Hamblett, K.J., et al.: Effects of drug loading on the antitumor activity of a monoclonal antibody drug conjugate. Clin. Cancer Res. **10**(20), 7063–7070 (2004)

Hammood, M., Craig, A.W., Leyton, J.V.: Impact of endocytosis mechanisms for the receptors targeted by the currently approved antibody-drug conjugates (ADCs)-a necessity for future ADC research and development. Pharmaceuticals (Basel). **14**(7), 674 (2021)

Hedrich, W.D., et al.: Antibody-drug conjugates: pharmacokinetic/pharmacodynamic modeling, preclinical characterization, clinical studies, and lessons learned. Clin. Pharmacokinet. **57**(6), 687–703 (2018)

Heo, Y.A.: Mirvetuximab soravtansine: first approval. Drugs. **83**(3), 265–273 (2023)

Higel, F., et al.: N-glycosylation heterogeneity and the influence on structure, function and pharmacokinetics of monoclonal antibodies and Fc fusion proteins. Eur. J. Pharm. Biopharm. **100**, 94–100 (2016)

Holgate, R.G., Baker, M.P.: Circumventing immunogenicity in the development of therapeutic antibodies. IDrugs. **12**(4), 233–237 (2009)

Hötzel, I., et al.: A strategy for risk mitigation of antibodies with fast clearance. MAbs. **4**(6), 753–760 (2012)

Indini, A., Rijavec, E., Grossi, F.: Trastuzumab deruxtecan: changing the destiny of HER2 expressing solid tumors. Int. J. Mol. Sci. **22**(9), 4774 (2021)

Jackson, D.Y.: Processes for constructing homogeneous antibody drug conjugates. Org. Process Res. Dev. **20**(5), 852–866 (2016)

Jain, N., et al.: Current ADC linker chemistry. Pharm. Res. **32**(11), 3526–3540 (2015)

Jaramillo, M.L., et al.: A glyco-engineering approach for site-specific conjugation to Fab glycans. MAbs. **15**(1), 2149057 (2023)

Jennewein, M.F., Alter, G.: The immunoregulatory roles of antibody glycosylation. Trends Immunol. **38**(5), 358–372 (2017)

Johann, F., Wöll, S., Gieseler, H.: "Negative" impact: the role of payload charge in the physicochemical stability of auristatin antibody-drug conjugates. J. Pharm. Sci. **113**(8), 2433–2442 (2024)

Jones, P.T., et al.: Replacing the complementarity-determining regions in a human antibody with those from a mouse. Nature. **321**(6069), 522–525 (1986)

Junutula, J.R., et al.: Site-specific conjugation of a cytotoxic drug to an antibody improves the therapeutic index. Nat. Biotechnol. **26**(8), 925–932 (2008)

Kalinina, E.: Glutathione-dependent pathways in cancer cells. Int. J. Mol. Sci. **25**(15), 8423 (2024)

Kim, M.T., et al.: Statistical modeling of the drug load distribution on trastuzumab emtansine (Kadcyla), a lysine-linked antibody drug conjugate. Bioconjug. Chem. **25**(7), 1223–1232 (2014)

Ko, S., et al.: An Fc variant with two mutations confers prolonged serum half-life and enhanced effector functions on IgG antibodies. Exp. Mol. Med. **54**(11), 1850–1861 (2022)
Köhler, G., Milstein, C.: Continuous cultures of fused cells secreting antibody of predefined specificity. Nature. **256**(5517), 495–497 (1975)
Kolodych, S., et al.: Development and evaluation of β-galactosidase-sensitive antibody-drug conjugates. Eur. J. Med. Chem. **142**, 376–382 (2017)
Kostova, V., et al.: The chemistry behind ADCs. Pharmaceuticals (Basel). **14**(5), 442 (2021)
Kumar, N., et al.: Mechanistic insights and importance of hydrophobicity in cationic polymers for cancer therapy. J. Mater. Chem. B. **11**(7), 1456–1468 (2023)
Lamb, Y.N.: Inotuzumab ozogamicin: first global approval. Drugs. **77**(14), 1603–1610 (2017)
Lee, A.: Loncastuximab tesirine: first approval. Drugs. **81**(10), 1229–1233 (2021)
Li, Q., et al.: PEG linker improves antitumor efficacy and safety of affibody-based drug conjugates. Int. J. Mol. Sci. **22**(4), 1540 (2021)
Liu, L.: Pharmacokinetics of monoclonal antibodies and fc-fusion proteins. Protein Cell. **9**(1), 15–32 (2018)
Liu, L., et al.: Pharmacokinetics of IgG1 monoclonal antibodies produced in humanized Pichia pastoris with specific glycoforms: a comparative study with CHO produced materials. Biologicals. **39**(4), 205–210 (2011)
Liu, R., Wang, R.E., Wang, F.: Antibody-drug conjugates for non-oncological indications. Expert. Opin. Biol. Ther. **16**(5), 591–593 (2016)
Liu, K., et al.: A review of the clinical efficacy of FDA-approved antibody–drug conjugates in human cancers. Mol. Cancer. **23**(1), 62 (2024)
Lu, J., et al.: Linkers having a crucial role in antibody-drug conjugates. Int. J. Mol. Sci. **17**(4), 561 (2016)
Lu, R.M., et al.: Development of therapeutic antibodies for the treatment of diseases. J. Biomed. Sci. **27**(1), 1 (2020)
Mahmood, I.: Clinical pharmacology of antibody-drug conjugates. Antibodies (Basel). **10**(2), 20 (2021)
Maloney, D.G., et al.: IDEC-C2B8 (Rituximab) anti-CD20 monoclonal antibody therapy in patients with relapsed low-grade non-Hodgkin's lymphoma. Blood. **90**(6), 2188–2195 (1997)
Merck: Patritumab deruxtecan granted priority review in the U.S. for certain patients with previously treated locally advanced or metastatic EGFR-mutated non-small cell lung cancer. FDA approves gemtuzumab ozogamicin for CD33-positive AML in pediatric patients (2023) bit.ly/2Y86KRI
Mills, B.J., et al.: Effect of linker-drug properties and conjugation site on the physical stability of ADCs. J. Pharm. Sci. **109**(5), 1662–1672 (2020)
Morrison, S.L., et al.: Chimeric human antibody molecules: mouse antigen-binding domains with human constant region domains. Proc. Natl. Acad. Sci. USA. **81**(21), 6851–6855 (1984)
Nadkarni, D.V.: Conjugations to endogenous cysteine residues. Methods Mol. Biol. **2078**, 37–49 (2020)
Natsume, A., Niwa, R., Satoh, M.: Improving effector functions of antibodies for cancer treatment: enhancing ADCC and CDC. Drug Des. Devel. Ther. **3**, 7–16 (2009)
Nolting, B.: Linker technologies for antibody-drug conjugates. Methods Mol. Biol. **1045**, 71–100 (2013)
Norsworthy, K.J., et al.: FDA approval summary: mylotarg for treatment of patients with relapsed or refractory CD33-positive acute myeloid leukemia. Oncologist. **23**(9), 1103–1108 (2018)
Pereira, N.A., et al.: The "less-is-more" in therapeutic antibodies: Afucosylated anti-cancer antibodies with enhanced antibody-dependent cellular cytotoxicity. MAbs. **10**(5), 693–711 (2018)
Ponziani, S., et al.: Antibody-drug conjugates: the new frontier of chemotherapy. Int. J. Mol. Sci. **21**(15), 5510 (2020)
Rajvanshi, P., et al.: Hepatic sinusoidal obstruction after gemtuzumab ozogamicin (Mylotarg) therapy. Blood. **99**(7), 2310–2314 (2002)

RemeGen: RemeGen announces US FDA has granted breakthrough therapy designation for disitamab vedotin (RC48) in urothelial cancer. FDA approves gemtuzumab ozogamicin for CD33-positive AML in pediatric patients (2020) bit.ly/2Y86KRI
Reusch, D., Tejada, M.L.: Fc glycans of therapeutic antibodies as critical quality attributes. Glycobiology. **25**(12), 1325–1334 (2015)
Riccardi, F., et al.: A comprehensive overview on antibody-drug conjugates: from the conceptualization to cancer therapy. Front. Pharmacol. **14**, 1274088 (2023)
Ritchie, M., Tchistiakova, L., Scott, N.: Implications of receptor-mediated endocytosis and intracellular trafficking dynamics in the development of antibody drug conjugates. MAbs. **5**(1), 13–21 (2013)
Roopenian, D.C., Akilesh, S.: FcRn: the neonatal Fc receptor comes of age. Nat. Rev. Immunol. **7**(9), 715–725 (2007)
Ruan, D.Y., et al.: Development of antibody-drug conjugates in cancer: overview and prospects. Cancer Commun. (Lond). **44**(1), 3–22 (2024)
Rubinstein, J.D., O'Brien, M.M.: Inotuzumab ozogamicin in B-cell precursor acute lymphoblastic leukemia: efficacy, toxicity, and practical considerations. Front. Immunol. **14**, 1237738 (2023)
Sasso, J.M., et al.: The evolving landscape of antibody–drug conjugates: in depth analysis of recent research progress. Bioconjug. Chem. **34**(11), 1951–2000 (2023)
Schroff, R.W., et al.: Human anti-murine immunoglobulin responses in patients receiving monoclonal antibody therapy. Cancer Res. **45**(2), 879–885 (1985)
Shao, S., et al.: Site-specific and hydrophilic ADCs through disulfide-bridged linker and branched PEG. Bioorg. Med. Chem. Lett. **28**(8), 1363–1370 (2018)
Sharma, P., et al.: Therapeutic antibodies in medicine. Molecules. **28**(18), 6438 (2023)
Sheyi, R., de la Torre, B.G., Albericio, F.: Linkers: an assurance for controlled delivery of antibody-drug conjugate. Pharmaceutics. **14**(2), 396 (2022)
Simmons, J.K., et al.: Reducing the antigen-independent toxicity of antibody-drug conjugates by minimizing their non-specific clearance through PEGylation. Toxicol. Appl. Pharmacol. **392**, 114932 (2020)
Singh, S.K., Luisi, D.L., Pak, R.H.: Antibody-drug conjugates: design, formulation and physicochemical stability. Pharm. Res. **32**(11), 3541–3571 (2015)
Su, D., Zhang, D.: Linker design impacts antibody-drug conjugate pharmacokinetics and efficacy. Front. Pharmacol. **12**, 687926 (2021)
Su, Z., et al.: Antibody-drug conjugates: recent advances in linker chemistry. Acta Pharm. Sin. B. **11**(12), 3889–3907 (2021)
Sun, X., et al.: Effects of drug-antibody ratio on pharmacokinetics, biodistribution, efficacy, and tolerability of antibody-maytansinoid conjugates. Bioconjug. Chem. **28**(5), 1371–1381 (2017)
Surowka, M., Schaefer, W., Klein, C.: Ten years in the making: application of CrossMab technology for the development of therapeutic bispecific antibodies and antibody fusion proteins. MAbs. **13**(1), 1967714 (2021)
Tai, Y.T., Anderson, K.C.: Targeting B-cell maturation antigen in multiple myeloma. Immunotherapy. **7**(11), 1187–1199 (2015)
Tai, Y.T., et al.: Novel anti-B-cell maturation antigen antibody-drug conjugate (GSK2857916) selectively induces killing of multiple myeloma. Blood. **123**(20), 3128–3138 (2014)
Tang, M., et al.: Clinical pharmacology of the antibody-drug conjugate enfortumab vedotin in advanced urothelial carcinoma and other malignant solid tumors. Clin. Pharmacokinet. **63**(4), 423–438 (2024)
Tedeschini, T., et al.: Polyethylene glycol-based linkers as hydrophilicity reservoir for antibody-drug conjugates. J. Control. Release. **337**, 431–447 (2021)
Therapeutics, M.: Mersana therapeutics announces partial clinical hold for XMT-1522 clinical trial. FDA approves gemtuzumab ozogamicin for CD33-positive AML in pediatric patients (2018a) bit.ly/2Y86KRI
Therapeutics, M.: Mersana announces FDA lifts partial clinical hold for XMT-1522. FDA approves gemtuzumab ozogamicin for CD33-positive AML in pediatric patients (2018b) bit.ly/2Y86KRI

Thungappa, C.S., et al.: Comparison of the efficacy, safety, pharmacokinetic and immunogenicity of UJVIRA (ZRC-3256, trastuzumab emtansine) with the kadcyla (trastuzumab emtansine) in the treatment of HER2-positive metastatic breast cancer: a randomized, open-label, multicenter study in India. Clin. Breast Cancer. **22**(4), 300–307 (2022)
Tong, J.T.W., et al.: An insight into FDA approved antibody-drug conjugates for cancer therapy. Molecules. **26**(19), 5847 (2021)
Triantafyllidi, E., Triantafillidis, J.K.: Systematic review on the use of biosimilars of trastuzumab in HER2+ breast cancer. Biomedicines. **10**(8), 2045 (2022)
Tsuchikama, K., An, Z.: Antibody-drug conjugates: recent advances in conjugation and linker chemistries. Protein Cell. **9**(1), 33–46 (2018)
Tzogani, K., et al.: EMA review of belantamab mafodotin (blenrep) for the treatment of adult patients with relapsed/refractory multiple myeloma. Oncologist. **26**(1), 70–76 (2021)
U.S. Food and Drug Administration: F. BLENREP (Belantamab Mafodotin-Blmf): US Prescribing Information (2020) [cited 2024 08/25/2024]. Available from: https://www.accessdata.fda.gov/drugsatfda_docs/label/2020/761158s000lbl.pdf
U.S. Food and Drug Administration, F.: Fast Track, breakthrough therapy, accelerated approval, priority review (2023) [cited 2024 October 25, 2024]. Available from: https://www.fda.gov/patients/learn-about-drug-and-device-approvals/fast-track-breakthrough-therapy-accelerated-approval-priority-review
Valent, P., et al.: Paul Ehrlich (1854–1915) and his contributions to the foundation and birth of translational medicine. J. Innate Immun. **8**(2), 111–120 (2016)
van de Donk, N.W., Dhimolea, E.: Brentuximab vedotin. MAbs. **4**(4), 458–465 (2012)
Vidarsson, G., Dekkers, G., Rispens, T.: IgG subclasses and allotypes: from structure to effector functions. Front. Immunol. **5**, 520 (2014)
Vollmar, B.S., et al.: Calicheamicin antibody-drug conjugates with improved properties. Mol. Cancer Ther. **20**(6), 1112–1120 (2021)
Wakankar, A., et al.: Analytical methods for physicochemical characterization of antibody drug conjugates. MAbs. **3**(2), 161–172 (2011)
Walter, R.B.: Brief overview of antibody–drug conjugate therapy for acute leukemia. Expert. Opin. Biol. Ther. **21**(7), 795–799 (2021)
Wang, Z., et al.: Antibody-drug conjugates: recent advances in payloads. Acta Pharm. Sin. B. **13**(10), 4025–4059 (2023)
Wolf, B., et al.: Therapeutic antibody glycosylation impacts antigen recognition and immunogenicity. Immunology. **166**(3), 380–407 (2022)
Wuhrer, M., et al.: Glycosylation profiling of immunoglobulin G (IgG) subclasses from human serum. Proteomics. **7**(22), 4070–4081 (2007)
Yamazaki, C.M., et al.: Antibody-drug conjugates with dual payloads for combating breast tumor heterogeneity and drug resistance. Nat. Commun. **12**(1), 3528 (2021)
Yip, V., et al.: Preclinical characterization of the distribution, catabolism, and elimination of a polatuzumab vedotin-piiq (POLIVY®) antibody-drug conjugate in Sprague Dawley Rats. J. Clin. Med. **10**(6), 1323 (2021)
You, J., et al.: Cysteine-based coupling: challenges and solutions. Bioconjug. Chem. **32**(8), 1525–1534 (2021)
Yu, J., Song, Y., Tian, W.: How to select IgG subclasses in developing anti-tumor therapeutic antibodies. J. Hematol. Oncol. **13**(1), 45 (2020)
Zidar, M., et al.: High throughput prediction approach for monoclonal antibody aggregation at high concentration. Pharm. Res. **34**(9), 1831–1839 (2017)
Zimmerman, E.S., et al.: Production of site-specific antibody-drug conjugates using optimized non-natural amino acids in a cell-free expression system. Bioconjug. Chem. **25**(2), 351–361 (2014)

Toxicology Considerations on Antibody-Drug Conjugate Discovery and Development

Andrea Rodrigues, Ali Chesney, Amy Kimzey, Brian Welsh, and Damon Demady

Abstract Antibody-drug conjugates (ADCs) link antibodies to cytotoxic payloads to enable targeted therapy. This paper reviews the evolution of ADC toxicology, including manufacturing considerations, health authority perspectives, preclinical testing strategies, and common toxicity findings. Early ADC development faced challenges due to instability and immunogenicity, but improved antibody and linker design has since widened therapeutic index. Currently ADCs are regulated similarly to biologics, requiring less testing of individual components; pivotal studies should evaluate safety of the intact ADC. As ADCs continue to be refined, ongoing dialogue with regulators regarding emerging science is recommended to ensure patient safety.

Keywords Monoclonal antibodies · Cytotoxins · Hematological toxicity · Tumor microenvironment · Tissue distribution · Pharmacodynamics · Drug metabolism · Histopathology · Teratogenicity · Biomarkers

---

A. Rodrigues (✉)
AbbVie, Irvine, CA, USA

A. Chesney
Immunogen, Holmes Beach, FL, USA

A. Kimzey
ToxStrategies LLC, Houston, TX, USA

B. Welsh
MorphoSys, Boston, MA, USA

D. Demady
Aura Biosiences, Boston, MA, USA

H. J. Gukasyan, A. Kulkarni (eds.), *Antibody Drug Conjugates*, AAPS Advances in the Pharmaceutical Sciences Series 52, https://doi.org/10.1007/978-3-031-89319-3_8

A History of Antibody-Drug Conjugate Toxicology

The concept that medicine could specifically target a disease without causing harm to the patient dates back more than a century to Paul Erlich's vision of a "magic bullet" (Smaglo et al. 2014), whereby a medicine could locate exactly where it was needed in the body, avoiding potentially toxic effects in off-target tissues. Antibody-drug conjugates (ADCs) offer a unique way of meeting this challenge by linking the targeted delivery capability of monoclonal antibodies (mAbs) with the potent cytotoxic and therapeutic potential of a small molecule in the hopes of minimizing off-target toxicities (Hoofring et al. 2013). Preclinical data from the first use of ADCs in mouse models with xenografted human tumors demonstrated a high degree of efficacy; however, the models underestimated the degree to which the anti-globulin immune response to non-human immunoglobulins affected ADC performance in a cancer patient (Bodey et al. 1996). Phase I/II clinical trials of these initial drugs resulted in unsatisfactory therapeutic outcomes; therefore, improvements were made that modified the antibody components, replacing the rodent regions (i.e., Fc region and the framework region of the variable domain) with human equivalents (Bodey et al. 1996). Similarly, other initial attempts failed because of inadequate or insufficient linker systems or low potency cytotoxins, issues that were addressed as the drug class expanded and evolved (Roberts et al. 2013). The molecular stability of the chemical antibody-drug linker, which is designed to be relatively stable in circulation, complicated the toxicokinetics of these therapies and played a role in the toxicity and immunogenicity. Acid-cleavable hydrazone linkers were used in early ADCs, which are relatively stable in the bloodstream at neutral pH (pH 7.3–7.5), but are designed to hydrolyze and release their payload in more acidic environments, e.g., within the cellular endosomes (pH 5.0–6.5) and lysosomes (pH 4.5–5.0) (Kaur et al. 2013). Drug release has theoretically become more specific with enzymatically cleavable linkers designed for intracellular targets or linkers that are resistant to enzymatic or chemical cleavage, yet unanticipated drug release may occur in vivo due to chemical or enzymatic activity in plasma. These deconjugation events, whether anticipated or not, led to a complex mixture of drug antibody combinations or other biotransformation products, including adduct formation and generation of peptide fragments or linker drug fragments (Kaur et al. 2013). The failures of the first-generation ADCs have led to better engineered linkers and antibodies that improve the pharmacokinetics (PK) and target disposition (Roberts et al. 2013). Mylotarg (gemtuzumab ozogamicin) was approved via the FDA's accelerated approval program as the first ADC from the Caltech Group and Wyeth-Ayerst Research division (now Pfizer) for the treatment of relapsed acute myeloid leukemia (Niculescu-Duvaz 2000). Mylotarg consisted of a humanized anti-CD33 monoclonal antibody covalently linked to a semisynthetic derivative of calicheamicin, a potent cytotoxic enediyene antibiotic (Bross et al. 2001). The drug was withdrawn voluntarily by Pfizer in the United States after a confirmatory trial failed to show a clinical benefit and noted a higher rate of fatal toxicity compared to chemotherapy; however, Myoltarg remained on the market in Japan and was still available to individual patients through compassionate use programs (Pfizer 2020). During this time, clinicians and investigators continued to evaluate Mylotarg in acute myeloid leukemia patients, which provided more safety and efficacy data, allowing the drug to be reintroduced in 2017 after demonstrating a favorable risk/benefit profile. Low therapeutic indexes (TIs) have been an issue for the cancer-therapy field for decades. For example, maytansine and its derivatives, are potent microtubule-targeted compounds that inhibit proliferation of cells at mitosis but have clinically limiting neuro- and gastrointestinal toxicities (Oroudjev et al. 2010). Due to their specificity, ADCs have the potential to widen the TI of compounds such as these (e.g., ado-trastuzumab emtansine), however, this goal was not fully achieved within the first generation. Early ADCs did not undergo internalization at the target but instead relied on tumor-specific enzymes or physiological conditions within the tumor microenvironment to release the active drug (Peters and Brown 2015). The development of internalized conjugates has increased the therapeutic window by decreasing systemic exposure, thereby reducing side-effects as well as reducing drug resistance (Peters and Brown 2015). This was the hope with the second generation of these drugs, such as Adcetris (brentuximab vedotin) which was approved by the FDA in 2011 for Hodgkin's lymphoma and systemic anaplastic large-cell lymphoma, and Kadcyla (ado-trastuzumab emtansine) approved in the United States in 2013 for HER2-positive metastatic breast cancer (Roberts et al. 2013).

Health Authority Regulation of ADCs

Specialized regulatory knowledge, as it applies to ADCs, is important for understanding ADC development and planning a preclinical strategy. In the United States, the FDA has regulated antibody-drug conjugates as combination products, e.g., drug/biologic (21CFR3.2). Under this framework, the ADC may have more than one mode of action; however, the most important is "the mode of action expected to make the greatest contribution to the overall intended therapeutic effects of the combination product".

Globally, ADCs are regulated under the ICH S6 and S9 guidelines. Until recently, specific mentions of ADCs in these guidelines were limited and often vague. For example, ADCs were discussed specifically only in the original ICH S6 guidance (1997) in a note stating that the presence of an organic linker molecule in a conjugated protein product may necessitate an *in vitro* genotoxicity study (i.e., AMES study), although such a study would not be needed for "normal" biologic products. More recently, additional guidance has been added, and the ICH S6 Addendum (2011) offers the following additional information:

- For studies testing the whole ADC, the ICH S6(R1) species selection principles should be followed

- A novel toxin should be tested in at least one species (usually rodent)
- Toxins that are not novel may not require a toxicology study if sufficient information is available in the literature
- The metabolic stability of the ADC in animals should be compared with the metabolic stability of the ADC in humans

The ICH S9 guidance (2010) simply provided broad outlines and noted that the safety of the actual drug to be administered to humans (i.e., the ADC itself) was the most important consideration. Accordingly, a more limited evaluation of the unconjugated payload and possibly the linker was considered appropriate. "Limited" was not defined, and in practice, it varied by program. In addition, an evaluation of the *in vivo* stability of the ADC and the toxicokinetics (TK) of the conjugate and the unconjugated payload was also requested.

Not surprisingly, because all approved ADCs to date have been oncology products, the ICH S9 Q&A guidance (2018) offers extensive clarification and describes best practices derived from successful programs (Questions 13, 21, and 27–36). For example, studies of the monoclonal antibody alone are not usually warranted, even when the ADC contains a novel antibody. However, the payload, which is the most common source of toxicity observed with ADC molecules, does need to be examined alone if the toxicity is not already known. This can be done either as an arm in a study designed to characterize the toxicity of the whole ADC or in a separate study. If conducting a separate study, generally, a single species (usually the rat) is sufficient. Additional studies may be needed, depending on the results of the initial pilot studies and the nature of the payload; therefore, discussion of the payload evaluation plan with the FDA in a pre-IND (investigational new drug) meeting is advisable.

However, although the strategy for ADC development generally will follow guidance from ICH S6 (or ICH S9 for oncology indications), specifics should be discussed and agreed to in a meeting with regulatory agencies, because individual ADC development programs can differ in specific requirements.

Generally, ADCs have been approved in the United States first, followed by approval in other regulatory regions (Table 1). There appears to be a consensus regarding the overall approach recommended for non-clinical toxicology that was captured by the updates to ICH S6 and ICH S9.

In addition to the ICH guidance, the FDA has published two reviews of oncology ADC submissions (Saber and Leighton 2015; Saber et al. 2019). After reviewing multiple ADC drug candidates, the FDA found that the highest non-severely toxic dose (HNSTD) approach for human dose selection was appropriate for most oncology ADCs, with the exception of immune agonists. Saber and Leighton (2015), noted that the minimum-anticipated-biological-effect level (MABEL) approach would be more appropriate for ADCs that contain an immune agonist, which is also the recommended approach in ICH S9 for immune agonists in general. Saber et al. (2016) and the FDA draft immunotoxicity guideline (2020), provide further guidance on the use of the MABEL approach. Of note, some sponsors are taking a platform approach. These sponsors use the same payload and attach it to different

Table 1 ADC global regulatory approvals (first indication)

Name	Antibody Target/ Payload MoA[a]	United States	European Union	Canada	Australia	Japan
Gemtuzumab ozogamicin (Mylotarg™)	CD33/DNA targeting (N-acetyl gamma calicheamicin)	2000; 2017[b]	2018	2020	2020	2005
Brentuximab vedotin (Adcetris®)	CD30/Antimitotic agent (MMAE)	2011	2012	2013	2013	2014
Trastuzumab emtansine (Kadcyla®)	Her2/Antimitotic agent (DM1)	2013	2013	2013	2013	2013
Inotuzumab ozogamicin (Besponsa®)	CD22/DNA targeting (N-acetyl gamma calicheamicin)	2017	2017	2018	2018	2018
Moxetumomab pasudotox (Lumoxiti®)	CD22/pseudomonas exotoxin (PE38)	2018[c]	2021[c]	–	–	–
Polatuzumab vedotin (Polivy™)	CD79B/Antimitotic agent (MMAE)	2019	2020	2020	2019	2021
Enfortumab vedotin (Padcev™)	Nectin-4/Antimitotic agent (MMAE)	2019	2022	2021	2022	2021
Trastuzumab deruxtecan (Enhertu®)	Her2/Topoisomerase I inhibitor (DxD)	2019	2021	2021	2021	2020
Sacituzumab govitecan (Trodelvy™)	Trop2/Topoisomerase I inhibitor (SN-38)	2020	2021	2021	2021	2024
Belantamab mafodotin (Blenrep®)	BCMA/Antimitotic agent (MMAF)	2020[d]	2020; 2025[e]	2025	–	–
Loncastuximab tesirine (Zynlonta®)	CD19/DNA targeting (PBD SG3199)	2021	2022	2025	–	–
Tisotumab vedotin (Tivdak®)	Tissue Factor/ Antimitotic agent (MMAE)	2021	2025	–	–	2025
Mirvetuximab soravtansine (Elahere™)	FRα/Antimitotic agent (DM4)	2022	2024	2025	–	–
Datopotamab Deruxtecan (Datroway®)	TROP-2/Topoisomerase I inhibitor (DxD)	2025	2025	–	2025	2024
Telisotuzumab vedotin (Emrelis™)	cMet/Antimitotic agent (MMAE)	2025	–	–	–	–

MoA mode of action
[a]All approved ADCs to date are for oncology indications
[b]Original approval 2000; withdrawn 2010; reapproved 2017
[c]Withdrawn from the US market in 2023 due to low clinical use; approved in EU in 2021, EU marketing authorization withdrawn in 2021
[d]Withdrawn in 2022
[e]Originally approved 2020; conditional marketing authorization not renewed in 2024; approved again in 2025

mABs, allowing for the same payload-only studies to be included with each submission (Saber and Leighton 2015).

The toxicity of the ADCs approved to date has been driven largely by the payload component, which has been potent cytotoxins (Saber and Leighton 2015; Saber et al. 2019). However, ADCs are not required to have toxic payloads. The payload of an ADC could be linked to a less toxic oncology therapeutic or could even be used for non-oncology purposes. No therapies of this type have yet been approved, so the regulatory guidance is less clear. Even the ICH S6 Addendum refers to the payload as a "toxin," yet non-oncology uses are being explored by the pharmaceutical industry. For example, Abbvie hoped to enhance the activity of the monoclonal antibody, by combining it with a glucocorticoid receptor modulator payload (ABBV-3373) for the treatment of rheumatoid arthritis (RA). In contrast to the approved oncology therapeutics with cytotoxic payloads, the clinical toxicity of ABBV-3373 resembled that of the antibody component of the molecule, not the payload component (Abbvie 2020).

In summary, the totality of the preclinical development history of ADCs to date suggests that health authorities view ADCs predominantly as targeted small-molecule drug delivery in action; however, they view the required preclinical toxicology dossier as a hybrid, generally closer to a traditional biologic than a small molecule in terms of the types and quantity of studies. Recent updates to regulatory guidance have provided significantly increased clarity regarding cytotoxic ADC development for oncology indications. However, outside of this area, the regulatory needs are still unclear.

Considerations for the Overall Preclinical Toxicology Strategy

Manufacturing and Design

The mechanism of the modern ADC can be broken down into five basic steps (Chau et al. 2019):

1. ADC binding to the antigen
2. receptor-mediated ADC internalization
3. degradation of ADC in lysosomes
4. payload release into the cytosol, and
5. apoptosis of the target cell

The components of ADCs are the antibody, the cytotoxic payload, and the linker. Therefore, the successful preclinical program relies on an appropriate combination of all three components (additional information on the manufacturing and design of ADCs can be found in chapter "Conjugation Chemistry and Chemical Design for the Discovery and Development of Antibody Drug Conjugates"). While chemotherapeutics and mAbs can be efficacious individually, the combination of both can overcome the limitations of each. The mAbs are highly specific, but many mAbs have shown only modest efficacy for most oncology indications when used in combination with a conventional chemotherapy. On the other hand, cytotoxic chemotherapeutics are highly toxic, but nonspecific, and therefore lead to adverse side effects and dose-limiting toxicities. Together, the ADC can keep the majority of the payload bound until it reaches the cancer cell and thereby limit its ability to interact with non-target tissues and diminish non-specific toxicity (Ritter 2012).

The effector (cytotoxicity) function, antigen affinity, biodistribution, and subsequent immune function of the ADC can be influenced by the antibody subclass (e.g., IgG1, IgG2, IgG3, or IgG4) and the function of the Fc region (Hoffmann et al. 2018 as cited in Chau et al. 2019). Using human antibodies as opposed to mouse antibodies is advantageous due to decreased immunogenicity, and increased potency and target selectivity, linker stability, intracellular release (Chau et al. 2019). High antibody specificity for the tumor antigen $K_d < 10$ nM) is essential for efficient uptake into target cells (Panowski et al. 2014). If the antibody cross-reacts with other antigens or displays general non-specific binding, it can lead to high levels of antibody in normal tissues, resulting in both toxicity and/or elimination of the ADC before it can reach the tumor. The ideal antibody in an ADC undergoes rapid internalization to elicit its cell-specific function, while typical therapeutic mAbs tend to focus on effector function or immune-modulating activity (Hinrichs and Dixit 2015). The optimal PK properties of the antibody are a long half-life with slow clearance in plasma (Panowski et al. 2014).

Classic antibody-toxin safety profiles are driven by the structure, distribution, and exposure of the toxin. The cytotoxic payload efficacy overall depends on potency, amenability to conjugation, solubility, and stability (Chau et al. 2019). The general payload classes are nanocarriers, protein toxins, enzymes or targeting DNA, RNA, and tubulin filaments.

Linkers can either be cleavable or non-cleavable. Cleavable linkers use intracellular processes to release the toxin, such as reduction in the cytoplasm, acidic conditions in the lysosome, or cleavage by specific proteases (Ritter 2012). Non-cleavable linkages require catabolic degradation within the cell and are associated with improved safety profiles compared to cleavable linkers (Hinrichs and Dixit 2015). Both types of linkers are designed to not release the cytotoxic payload until after the ADC has reached the interior of the target cell. Choosing the right linker is also important for proper function of the ADC. For example, very different immune responses were observed in cynomolgus monkeys administered trastuzumab antibody–drug conjugates with different linkers (Kaur et al. 2013). During the early development of Mylotarg, the linkages between the mAb and the cytotoxic small molecule were destroyed by endogenous proteases in the blood, causing instability of the ADC in the circulation and premature release of the cytotoxin, resulting in side-effect profiles like that of an unconjugated chemotherapeutic (Ritter 2012). If the linker is not stable in blood, it could result in rapid release of free drug in the circulation, rather than inside the tumor cell, resulting in decreased ADC activity (Panowski et al. 2014) and contributing to off-target toxicity (Hinrichs and Dixit 2015).

Toxicity of the ADC can be influenced by its constituents (i.e., the antibody, drug, or linker), or the other product attributes such as stability, purity, impurities, excipients, formulation, etc. (Roberts et al. 2013). The toxicologist has a responsibility, when possible, to be involved with the team members working on the chemistry and conjugation aspects of the ADC program for an understanding of the process of the manufacturing of the ADC, keeping in mind that the antibody, payload, and linker are all manufactured very differently (Ritter 2012). At early development timepoints, the availability of acceptable material with consistent drug to antibody ratios might be an issue. A low drug-to-antibody ratio (DAR) could result in reduced efficacy and less cytotoxicity, while a high DAR could cause instability and insufficient antibody binding, also reducing efficacy. The report or Certificate of Analysis at these earlier phases of development should, at the minimum, include ADC concentration, free drug concentration, and general impurity levels.

Efficacy Studies and Dose Selection

Efficacy studies can serve as a guide for initial investigative toxicology studies but should be used only for basic measures of toxicity by dose and exposure, because efficacy studies are not set up to identify toxicity as the primary objective. Typically, studies performed in mice can, and usually do, produce quite different results from studies done using rats and monkeys with regard to tolerability. However, performing early discovery studies in monkeys is not particularly common, nor is it a good use of animals.

There are some general approaches to ADC efficacy that can also drive toxicity. One approach is to use a less potent toxin to allow for higher overall ADC doses. A second approach is to focus on more potent toxins but delivered at overall lower ADC doses. A third approach is to focus on high DAR. This approach uses a platform that can conjugate a larger number of payload molecules per antibody molecule. As with the first approach above, the objective is to deliver lower overall doses of ADC but higher overall doses of toxin.

Efficacy studies should include relevant targets for the intended disease indication. An ideal target antigen is one that is expressed at high levels on the cell surface of the tissue of interest, to maximize the amount of ADC bound by the target (e.g., tumor) but is expressed only at low levels elsewhere in the body, to minimize off-target toxicity (Panowski et al. 2014; Ritter 2012). For example, ADCs approved for hematological oncology indications have antigen targets in specific cell lineages for CD22, CD30, and CD33 (Chau et al. 2019) which should be evaluated during initial efficacy studies. CD30 is chosen as a potential target antigen because it is a type II transmembrane protein belonging to the Tumor Necrosis Factor (TNF) family, which is highly and selectively expressed on the surface of Hodgkin's lymphoma (HL) and some T-cell non-Hodgkin's lymphoma cancer cells (Bouchard et al. 2014).

Analytical Considerations

To satisfy dose formulation analysis requirements of the ADC material administered in toxicity studies, ADC concentration, DAR, percent of unconjugated payload, and purity of the test material should be assessed, as well as the determination of ADC stability under the conditions of use (Roberts et al. 2013). An enzyme-linked immunosorbent assay (ELISA) can be used to measure conjugated antibodies but does not provide information on DAR or drug load, and traditional LC-MS assays used to determine drug concentration of small molecules are of limited use for ADCs (Kaur et al. 2013).

Bioanalysis for an ADC poses some unique challenges that are discussed in detail in chapter "Bioanalytical Methodologies for Pharmacokinetic Characterization of Antibody Drug Conjugates". Bioanalysis is complex and is used to characterize biotransformation of the ADC due to catabolism and metabolism (Kaur et al. 2013). A homogeneous ADC will break down into metabolites/catabolites, intact antibodies, adducts, complexes with the antigen, drugs, and linkers and linker/drugs (Kaur et al. 2013). For this reason, the common practice of measuring drug concentration over time and using the therapeutic as the reference standard cannot be applied. Assays should be developed to measure specific analytes for the antibody, total antibody (Tab), payload (and any relevant catabolites), and anti-drug antibodies (ADAs), which is recommended to be performed prior to dose range-finding studies and to be validated prior to GLP study sample analysis.

For bioanalytical analysis, different sampling schemes and more samples are usually needed if the study includes both analysis of the ADC/ Tab and unconjugated payload concentrations, as well as ADAs and pharmacodynamic readouts. For example, a typical 3- to 6-week GLP toxicity study with a 2- to 4-week recovery period will include dosing once every 3 weeks, biweekly, or even weekly. For example, bioanalysis for serum concentrations should be conducted following each dose of a typical 4-week study (usually dosing on Days 1 and 22); however, blood collections for ADA analysis would typically start after 1 month and approximately every 30 days during the recovery phase. Determining immunogenicity by measuring ADA is not required but is useful when changes in the expected exposures or pharmacodynamic effects of the ADC are observed (Roberts et al. 2013), and therefore it is recommended to collect blood for this purpose during the study, even if the analysis for ADAs is put on hold. The presence of ADA can affect exposure when the antibodies are neutralizing (cleared from circulation) and can also cause immune-complex immunotoxicity such as hypersensitivity or infusion reactions (Hoofring et al. 2013), and ADA can be formed against the mAb, the linker, or the cytotoxic payload portion of the biotherapeutic. Anti-ADC antibodies targeting either the mAb or the linker could contribute to additional safety concerns. For example, due to the presence of multiple linker–drug attachments on each ADC, anti-linker–drug antibodies may promote the formation of large crosslinked immune complexes (Hoofring et al. 2013).

Biological stability is typically assessed by incubating the ADC at 37 °C for various time periods, with plasma, serum, or tissue homogenates, and determining the concentrations of the unconjugated payload at various time points, not to exceed the period for which sterility of the incubation can be maintained (Roberts et al. 2013). The half-life is thought to contribute to the overall stability of the conjugate. Compared to the half-life for an antibody (1–3 weeks), the half-life of an ADC may be only 2–5 days, due to the different elimination mechanisms of the antibody vs. the cytotoxic payload (Hinrichs and Dixit 2015). Additional discussion of ADC stability can be found in chapter "Considerations for Antibody-Drug Conjugate Design: Stability and Formulation".

Toxicology Strategy

ADCs entering the clinic for oncology indications should be consistent with ICH S9. The recommended approach to setting a first in human starting dose for an ADC is based on the highest non-severely toxic dose (HNSTD, defined in ICH S9 as the "highest dose level that does not produce evidence of lethality, life-threatening toxicities or irreversible findings,") and/or the dose causing severe toxicity in 10% of animals (STD10), where the highest first clinical dose is one-tenth the STD10 in rodents, or one-sixth the HNSTD in non-rodents (ICH S9), depending on which is the most appropriate or sensitive species. However, consideration should be given with respect to differences in binding affinity between animals and humans and pharmacological properties of the biopharmaceutical. Selecting doses that are too high can result in the need to repeat clinical studies with lower doses, thus leading to program delays. Dose-range finding studies should be used and have flexible designs with the goal of identifying a dose in GLP studies to identify clinical and pathology findings related to ADC administration without causing severe toxicity. A staggered dosing approach is best practice, where each subsequent dose is based on the clinical observations following the preceding dose. Due to the long half-life of most ADCs, dose groups for these staggered studies should be spaced at least two weeks apart, with Days 4–8 expected to start seeing clinical signs of toxicity (such as skin, hematological, and/or gastrointestinal findings, depending on the payload). Potent cytotoxins have strong primary pharmacological activity, so there are usually minimal species differences in the payload-driven toxicity profile (Hinrichs and Dixit 2015). Therefore, one single-dose rodent study for each novel payload, regardless of indication or parent antibody, should be performed. Non-novel payloads can use data from a literature evaluation or a study report from an earlier program.

When planning a toxicology strategy, data should be used to determine whether a two-species or a single-species IND-enabling program will be needed. An appropriate animal species is selected with the expectation that the animal will respond to the ADC in a pharmacologically similar manner (affinity, potency, and/or intended pharmacologic function) as humans (Roberts et al. 2013). As with many other monoclonal antibodies, the ADC may bind to the target antigen only in non-human primates. In vitro binding assays that are within an order of magnitude (e.g., tenfold) are generally considered comparable due to variability between assay systems (Roberts et al. 2013). To support initial clinical trials, the pivotal toxicology study should follow Good Laboratory Practices (GLPs) and mimic the intended clinical dose route and frequency (generally intravenous, once every 3 weeks) (Mecklenburg 2018).

In general, the safety of the ADC should be evaluated as a single component. If the toxicity of the payload or linker-payload has been characterized (e.g., through pilot studies), a GLP study of the payload or linker-payload may not be warranted or could be further abbreviated. If the toxicity of the payload or payload with linker has not been characterized, the payload or payload with linker could be evaluated in one species as a stand-alone study or could be added as an arm into toxicology studies of the ADC. Additional assessments of the payload and its catabolites, such as phototoxicity, genotoxicity, etc. will be required during development. Studies of the unconjugated antibody are generally not warranted. The extent to which the separate components of the ADC should be tested depends on what is already known for each component and considered on a case-by-case basis. For example, the developers of Kadcyla and Enhertu were able to leverage a validated target (HER2) and a well-characterized antibody (traztuzumab) with a known safety and efficacy profile as the starting point for an ADC (Ritter 2012).

Tissue cross-reactivity studies can provide information on the unintended binding of an ADC to normal human and animal tissues and aid in interpretation of findings and their relevance to humans, however TCR studies are not required per the ICH S9 Q&A supplemental guidance. Genotoxicity assays would not be expected for intact ADCs, because the large size of the antibody component is not expected to access DNA (ICH S6[R1]). For novel cytotoxic payloads, genotoxicity should be assessed on the unconjugated payload and, potentially, its catabolites. Safety pharmacology endpoints (respiratory, cardiovascular, and neurological) are usually evaluated as part of the pivotal toxicology studies, similar to standard biologics development programs, and are usually not conducted as separate, stand-alone studies. Evaluation of potassium ion channels using the human ether-a-go-go related gene (hERG) assay is not required for antibodies, but for novel cytotoxins, a stand-alone hERG assay with the unconjugated molecule should be performed if serum levels are detected in toxicology studies. A comprehensive risk assessment with a weight-of-evidence approach should be employed to determine which, if any, carcinogenicity or reproductive and developmental studies are warranted. If a payload is considered genotoxic (mutagenic, clastogenic, or aneugenic), reproductive and developmental toxicity studies are usually waived, but should be agreed upon with regulatory agencies. The approach would consider all known data for the ADC components separately, and as a whole, in addition to the drug class.

Typical Studies Used in ADC Development

Results from efficacy, pilot toxicology, pharmacodynamics, (PD), and pharmacokinetics (PK) studies of the whole ADC product or the drug payload can be used to inform pivotal toxicology studies. If the drug payload has been evaluated previously, ADC development can also leverage toxicity data to inform studies and anticipated dose level selection.

Absorption, Distribution, Metabolism, and Excretion (ADME)

For approved ADCs, absorption, distribution, metabolism, and excretion (ADME) studies are shown in Table 2. At least one single-dose pharmacokinetic and/or pharmacodynamic study is commonly performed during the development program of approved ADCs. In approved ADC products, pharmacokinetics was conducted for the species of interest in the toxicology studies—typically, rats and cynomolgus monkeys—however, rabbits were used in a PK study in the development of Trodelvy™, and tumor-bearing mice were used for a PD/PK study. The pharmacokinetics of the free drug and the ADC were compared either within the same PK study or separately. Tissue distribution or whole-body autoradiography assays are conducted to determine distribution of the free payload or ADC. Potential in vitro metabolites of the released drug are evaluated in studies with cryopreserved human, rat, and/or monkey hepatocytes. Metabolites are profiled and identified in plasma, bile, and urine, usually from rats. In vitro evaluation of the released drug on inhibition and/or induction of cytochrome P450 enzymes were conducted as stand-alone studies (Adcetris®), within metabolism studies (Besponsa®, Kadcyla®, Padcev™), or within clinical studies (Enhertu®). Excretion was evaluated mostly with the free payload, but early ADC development also evaluated excretion of the ADC (see Adcetris®, Kadcyla®, Besponsa®, Polivy®, and Enhertu®). Plasma protein binding is often evaluated in multiple species including human, as well as those non-clinical species the pivotal toxicology studies are conducted with.

While preliminary studies such as pharmacokinetic, pharmacodynamic, and ADME studies are not required to be conducted under compliance with Good Laboratory Practice, the pivotal general toxicology studies, including repeat-dose, tissue cross-reactivity, reproductive toxicology, etc., outlined below, should be GLP-compliant. The analytical expectations for characterization and comparability of the monoclonal intermediate are the same as those for the final drug substance. This includes primary, secondary, and tertiary structure, as well as fragmentations or aggregations, charge, glycosylation or other post-translational modifications, antigen binding, and biological activity (as appropriate).

Table 2 Components evaluated in ADME and other studies in the development of approved ADCs

Drug	PK	Distribution	Metabolism	Excretion	DDI	Plasma stability	Plasma protein binding
Mylotarg™[a]	ADC	–	–	–	–	–	–
Adcetris®	ADC, D	D	D	ADC, D	D	ADC	D
Kadcyla ®	ADC	ADC, mAb	ADC	ADC	D	–	ADC, D
Besponsa®	ADC, mAb	D	ADC, D	ADC, D	–	ADC, D	ADC, D
Lumoxiti®	ADC	–	–	–	–	–	–
Polivy®	ADC, mAb	ADC	ADC	ADC	–	ADC	–
Padcev™	ADC	D	D	D	D	–	D
Enhertu®	ADC	ADC	ADC	ADC	ADC[c]	–	–
Trodelvy™[b]	ADC	–	–	–	–	–	–
Blenrep®	ADC, mAb	ADC, D	ADC, D	D	D	D	ADC
Zynlonta®	ADC, D	D	D	D	D	ADC	D
Tivdak®	ADC, mAb, D	mAb, ADC, D	D	D	–	ADC	D
Elahere™	ADC, D	ADC	ADC	D	D, C	–	D, C
Datroway®	ADC	D	D	D	D	–	D
Emrelis™[d]	ADC	–	–	–	–	–	D

C payload catabolite, *D* drug, *mAb* monoclonal antibody, *ADC* antibody drug conjugate
[a]ADME studies were reviewed previously and not included in the Mylotarg non-clinical review
[b]The BLA submission for Trodelvy and Lumoxiti did not include any metabolism or excretion studies
[c]CYP inhibition/induction was conducted in clinical trials for Enhertu
[d]Emrelis referenced the metabolism, excretion, and DDI studies for MMAE in the BLAs for Adcetris and Tivdak

Safety Pharmacology

In vitro assays for modulation of hERG potassium channels are conducted to support evaluation of the potential cardiac safety liability of the payload. Nearly all approved ADC products used an in vitro assay to evaluate modulation of hERG K+ channels in human embryonic kidney cells or Chinese hamster ovary cells by the released payload. When hERG assays were conducted in the approved ADCs to date, only the drug payload component or the drug payload and linker were assessed in hERG assays. Hemolytic potential, blood compatibility, and the effect on the autonomic nervous system and smooth muscle can also be evaluated in vitro or ex vivo using the ADC drug product, such as was done during the development of Kadcyla.

Stand-alone in vivo safety pharmacology studies were conducted in the development programs of early ADCs, such as for cardiovascular assessment (Adcetris, Besponsa, Enhertu, Kadcyla, and Mylotarg) and respiratory and CNS assessment

(Adcetris, Besponsa). Although standalone safety pharmacology studies may allow greater emphasis on the system being assessed, as well as a dedicated safety pharmacologist to serve as study director, without complicating factors and additional collections that may confound results. These evaluations are now more often included as part of a comprehensive general toxicology study.

Phototoxicity

The guidance in ICH S10 indicates that phototoxicity testing should be evaluated prior to Phase 1 clinical trials if one or more of the following conditions are met: the drug absorbs light within the range of natural sunlight (290–700 nm) (or has a molar extinction coefficient [MEC] >1000 L mol^{-1} cm^{-1} between 290 and 700 nm), generates a reactive species following absorption of UV-visible light, and/or distributes sufficiently to light exposed tissues such as eyes and skin (usually determined in a nonclinical tissue distribution study). This assessment would help determine whether risk minimization steps are necessary in clinical trial protocols to prevent adverse events in patients.

Some payloads are known to be photoreactive, such as the PBD SG3199 that is used in Zynlonta, which has an enhanced ability to crosslink and break DNA when exposed to visible light, UVA, or UVB radiation. This was seen in vivo as a potential contributory mechanism for the epidermal hyperpigmentation and hyperkeratosis in monkeys and skin rash in human patients following administration of Zynlonta, although an in vivo study was not conducted. Dxd, the topoisomerase 1 inhibitor payload used in Enhertu and Datroway, was positive in the in vitro Balb/c 3 T3 mouse fibroblast assay but was negative when assessed for phototoxicity in pigmented rats. The phototoxicity assessments included in the FDA reviews of approved ADCs are shown below in Table 3.

Expected Toxicity/Pathology

ADCs are complex and can cause a variety of histological changes, depending on the type of payload, distribution (as determined by the antibody), time between administrations, and time of tissue preservation for analysis. Most test-article-related microscopic findings increase in severity with dose and are reversible, with many of the findings overlapping, depending on the payloads (Mecklenburg 2018).

Most cytotoxic payloads are associated with myelotoxicity, which may result in cytopenias leading to impaired host resistance and secondary infections. ADCs may also produce histopathological findings in the male and female reproductive organs and the intestine that are associated with their specificity for replicating cells.

Table 3 Phototoxicity evaluations conducted during the development of approved ADCs

Drug	Payload	In vitro Balb/c 3 T3 mouse fibroblasts	In vivo phototox study in pigmented rats
Mylotarg™	N-acetyl gamma calicheamicin	–	–
Adcetris®	MMAE	–	–
Kadcyla®	DM1	–	–
Besponsa®	N-acetyl gamma calicheamicin	–	–
Lumoxiti®	PE38	–	–
Polivy®	MMAE	NA	–
Padcev™	MMAE	NA	–
Enhertu®	DxD	Positive	Negative
Trodelvy™	SN-38	–	–
Blenrep®	MMAF	–	–
Zynlonta®	PBD SG3199	Positive	–
Tivdak®	MMAE	NA	–
Elahere™	DM4	Positive	–
Datroway®	DxD	Positive	Negative
Emrelis™	MMAE	–	–

– not conducted, *L-D* linker-drug payload, *D* drug (payload), *NA* no absorption in the range of 290–700 nm

Dolastatins and maytansine/maytansinoids inhibit polymerization of tubulin, causing arrest of dividing cells and large atypical mitotic figures, which are not found in payloads that directly cause DNA damage, such as alkylating agents, or inhibit enzymes that work to replicate DNA like topoisomerase 1 (Mecklenburg 2018). Maytansines (such as DM1 and DM4) and dolastatins (monomethyl auristatin E, MMAE; and monomethyl auristatin F, MMAF) are associated with peripheral neuropathy and ocular toxicity; topoisomerase 1 inhibitors are associated with lymphohematopoietic and gastrointestinal toxicity; DNA alkylators, such as pyrrolobenzodiazepines have associated liver toxicities, and amanitins are well known to cause findings in liver and kidney with limited or no hematopoietic toxicity. Other target organs have included reproductive organs (epididymides and testes in males; ovary and endometrium in females) eye, intestine, kidney, liver, lymph nodes, mammary glands, ovary, pancreas, skin, testis, thymus, and urinary bladder (Mecklenburg 2018). Findings in nonclinical toxicology species are not always predictive of the adverse events (AEs) seen clinically with ADCs. Dose-limiting toxicities in cynomolgus monkeys, such as adverse skin lesions or gastrointestinal toxicity may be minimal or nonexistent in patients. Likewise, clinical AEs may only appear in patients, sometimes due to species differences, but also due to the more medically complex histories of cancer patients (usually on their second, third, or later line of therapy) compared to naive, healthy animals. Clinical AEs that are not easily predictable in animals include, but are not limited to, infusion-related reactions, veno-occlusive disease, and pneumonitis/interstitial lung disease.

General Toxicology Studies

Selection of Toxicology Species

All approved ADC products used cynomolgus monkeys in general toxicology studies due to pharmacological relevance and/or monkeys as a close phylogenetic species to humans. Most also used rats as a rodent species to evaluate the toxicity of the payload and/or the ADC (although for at least one program [Zynlonta, PBD payload], a dog payload study is also mentioned; however, no details are disclosed). The FDA review of Kadcyla and Trodelvy only listed one species (cynomolgus monkeys) for ADC toxicology studies, however rats were also used to evaluate the toxicity of the DM1 payload in Kadcyla, and a mouse toxicology study for the full ADC (sacituzumab govitecan) was listed in the EMA assessment report for Trodelvy. The FDA review for Trodelvy also mentions that the sponsor submitted data from their own published, peer-reviewed articles, which may have included studies evaluating the toxicity of the payload, SN-38, in a rodent species). Justification for species selection may include an assessment of relevant target sequence homology, binding affinity, and/or functional activity across species. It is generally insufficient to focus a drug development program on one species if there is an option to have two relevant species, one rodent and one nonrodent. However, this should be agreed upon with regulatory bodies prior to starting IND-enabling studies. With that said, and to emphasize an earlier point, a payload study in rodents (usually rats) is typically done as part of the IND-enabling program regardless of any pharmacological relevance to the mAb portion but, rather, to diversify the in vivo characterization of the toxin. When there is no pharmacologically relevant species, such as when the mAb does not exhibit binding to any nonclinical species, usually a toxicology study is one species for the ADC may be sufficient to identify off-target toxicities. However, depending on the components of the ADC, there may be a path to leverage toxicity data from other sources (such as other ADC programs using the same payload and/or payload toxicology studies) and gain agreement with regulatory agencies to start human trials without a toxicology study with the novel ADC. In either case, early regulatory agreement is recommended.

Single-Dose Toxicology Studies

Prior to multiple-dose GLP studies, single-dose pilot or "dose-range finding" toxicology studies are often conducted to inform dose selection for longer studies. Dose range-finding studies are important because the therapeutic index for ADCs is small. The goal is to choose doses that will ultimately facilitate a safe and effective first-in-human (FIH) dose. For example, Kadcyla was administered once to cynomolgus monkeys, with a 3-week recovery period, in two studies. The single administration (at 0, 3, 10, and 30 mg ADC/kg) was used to evaluate acute effects and toxicokinetic parameters after a single dose with or without unconjugated maytansinoid, DM1, the cytotoxic drug component. In the development of Mylotarg and Polivy, single-dose studies in rat and monkey were conducted to determine the toxic dose of the drug component, a calicheamicin derivative or monomethyl auristatin (MMAE), respectively, to inform studies of longer duration.

Repeat-Dose Studies

Dosing frequency should follow the pharmacokinetics of the ADC, as well as the intended frequency in the clinic. Per ICH S9 Q&As, at least two doses of the ADC should be administered to support initial clinical trials of once every 3 or 4 weeks. For approved ADC products, the duration of repeat-dose general toxicology studies varied; however, the duration of studies was at least 4 weeks (or two doses; Mecklenburg 2018), consistent with ICH S6 prior to entering the clinic, and at least 13 weeks duration prior to entering pivotal registrational clinical trials. As with a small molecule or a biologic, the frequency and duration of dosing in nonclinical studies should reflect the anticipated clinical study dosing regimen. Most often, this is once every three weeks; or cyclical dosing like in the development of Trodelvy.

Early ADCs (approved prior to 2020) varied in the approach for evaluation of the toxicity of the ADC, payload, and/or mAb in repeat dose studies. According to the ICH S9 Q&A guidance, studies with the unconjugated antibody are usually not needed, however sometimes this evaluation may be warranted. For example, the unconjugated antibody (traztuzumab) was evaluated in the development of Enhertu, which also conducted standalone nonclinical repeat-dose toxicity studies in rat and monkey for the whole ADC, as well as the drug payload. Traztuzumab, which has well known toxicities as an unconjugated antibody (particularly cardiotoxicities), which likely necessitated the inclusion of these stand-alone evaluations in the nonclinical package (and likely had been done for the evaluation of Herceptin, the unconjugated traztuzumab). As part of the nonclinical development programs for Enhertu and Polivy, standalone repeat-dose toxicology studies were conducted in both rats and monkeys with the drug payload alone. These repeat-dose payload studies were also done for Datroway, although it is assumed the Dxd payload studies were the same as those used for Enhertu. Other programs either leveraged payload studies conducted previously (such as Tivdak referencing the MMAE payload studies used for Adcetris), included the payload as an arm within a study (such as Adcetris and Padcev), or the payload study was a single dose standalone study. Table 4 shows the repeat-dose general toxicology studies conducted to support the nonclinical program. Over time, IND development strategies for ADCs have become increasingly minimal and nimble compared to earlier ADCs.

Table 4 Repeat-dose general toxicology studies in the FDA nonclinical reviews[a] of approved ADCs

Drug	Species	ADC, drug, or mAb	Dose frequency	Number of doses	Dosing phase duration	Recovery duration
Mylotarg™	Rat	ADC, mAb	Weekly	6	6 weeks	4 weeks
	Monkey	ADC	Weekly	6	6 weeks	4 weeks
	Monkey[b]	ADC	Weekly	6	6 weeks	–
Adcetris®	Rat	ADC, D, mAb	Weekly	4	4 weeks	4 weeks
	Monkey	ADC, D	Q3W	4	11 weeks (71 days)	5 weeks
	Monkey	ADC	Q3W	9	6 months	6 weeks
Kadcyla®	Monkey	ADC	Q3W	4	9 weeks (66 days)	3 or 6 weeks
	Monkey	ADC	Q3W	8	21 weeks (151 days)	6 weeks
Besponsa®	Rat	ADC	Weekly	4	4 weeks	4 weeks
	Monkey	ADC	Weekly	4	4 weeks	4 weeks
	Rat	ADC	Weekly	26	26 weeks	–
	Monkey	ADC	Weekly	26	26 weeks	–
Lumoxiti®	Monkey	ADC	Q2-3D	5	11 days	–
	Monkey	ADC	Cyclic dosing[c]	6	19 days	28 days
	Monkey	ADC	TIW	39	13 weeks	6 weeks
Polivy®	Rat	D	Weekly	4	4 weeks	4 weeks
	Monkey	D	Q3W	4	10 weeks	5 weeks
	Rat	ADC	Weekly	4	4 weeks	6 weeks
	Monkey	ADC	Q3W	4	10 weeks	9 weeks
Padcev™	Rats	ADC	Weekly	13	3 months	–
	Rats	ADC	Weekly	4	4 weeks	6 weeks
	Monkey	ADC, D, mAb	Weekly	4	4 weeks	6 weeks
	Monkey	ADC	Weekly	4	4 weeks	6 weeks
Enhertu®[d]	Rat	ADC (1)	Q3W	3	6 weeks	9 weeks
	Monkey	ADC (1)	Q3W	3	6 weeks	6 weeks
	Monkey	ADC (2)	Q3W	3	6 weeks	–
	Monkey	ADC (2)	Q3W	5	3 months	3 months
	Rat	D	Weekly	5	4 weeks	4 weeks
	Monkey	D	Weekly	5	4 weeks	4 weeks
	Rat	mAb	Q3W	3	6 weeks	–
	Monkey	mAb	Q3W	3	6 weeks	–
Trodelvy™	Monkey	ADC	Days 1 and 4	2	11 days	3 weeks
	Monkey	ADC	Cyclic dosing[e]	8	13 weeks	6 weeks
Blenrep®	Rat	ADC	Q3W	4	3 months	12 weeks
	Monkey	ADC	Weekly	13	13 weeks	12 weeks

(continued)

Table 4 (continued)

Drug	Species	ADC, drug, or mAb	Dose frequency	Number of doses	Dosing phase duration	Recovery duration
Zynlonta®	Rat	D	Daily	7	7 days	–
	Rat	D	Daily	28	4 weeks	–
	Dog	D	ND	ND	ND	ND
	Monkey	ADC	Q3W	2	4 weeks	12 weeks
	Monkey	ADC	Q3W	5	13 weeks	12 weeks
Tivdak®	Monkey	ADC	Q3W	5	13 weeks	6 weeks
	Monkey	mAb	Weekly[f]	4	27 days	–
Elahere™	Monkey	ADC	Q3W	4	10 weeks	5 weeks
	Rabbit	ADC	Weekly	4	4 weeks	3 weeks
	Rabbit	ADC	Q3W	4	12 weeks	3 weeks
Datroway®	Rat[g]	ADC	Q3W	5	3 months	2 months
	Monkey	ADC	Q3W	3	6 weeks	–
	Monkey	ADC	Q3W	5	3 months	2 months
	Rat	D	Weekly	5	4 weeks	4 weeks
	Monkey	D	Weekly	5	4 weeks	4 weeks
Emrelis™	Rat	ADC	Weekly	2	2 weeks	–
	Monkey	ADC	Q3W	2	4 weeks	8 weeks
	Monkey	ADC	Q2W	7	13 weeks	8 weeks

– not conducted, *ADC* antibody drug conjugate, *D* drug (payload), *mAb* monoclonal antibody, *ND* not disclosed in FDA nonclinical review

[a]Repeat-dose toxicology studies shown were summarized in pharmacology reviews and may not constitute all studies from the individual development programs

[b]Only males were used in the monkey bridging study between the old and new antibody cell line for Mylotarg

[c]Repeated cycle of Q2Dx3 dosing (doses on days 1, 3, 5 and then again on days 15, 17, and 19)

[d]The production process for Enhertu changed during the development (process number)

[e]Repeated cycle of weekly dosing for 2 weeks then 1 week off; dosing on study days 1, 8, 22, 29, 43, 50, 64, and 71

[f]Intra-animal dose escalation study of vehicle and unconjugated mAb

[g]Rat was not a pharmacologically relevant species and the FDA reviewers noted this in their assessment; a rat toxicology study was likely unnecessary

Tissue Cross-Reactivity

To determine the potential cross-reactivity of an ADC, cryosections obtained as surgical or autopsy specimens from normal human tissues are stained using immunohistochemistry. Most but not all approved ADCs used normal human tissues in a TCR assay; many also determined the ADC cross-reactivity in cynomolgus monkeys and/or rats (see Table 5). Either the antibody alone, or more commonly, the ADC drug product was used in tissue cross-reactivity studies. As clarified in the ICH S9 Q&As, in general, tissue distribution studies of the ADC are not warranted. Emrelis is the first ADC to be approved without a TCR study included in the FDA review.

Table 5 Species used in TCR studies in approved ADC products

Drug	Form evaluated	Human	Cynomolgus monkey	Rat
Mylotarg™	mAb	–	X	X
Adcetris®	mAb	X	X	–
	Biotin-conjugated ADC	X	–	X (liver only)
Kadcyla®	ADC	X	X	–
Besponsa®	mAb	X	–	–
Lumoxiti®	ADC	X	X	–
Polivy®	ADC	X	–	–
Padcev™	ADC	X	X	–
Enhertu®	ADC	X	X	–
Trodelvy™	ADC	X	–	–
	Biotin-conjugated ADC	–	X	–
Blenrep ®	ADC	X	–	–
Zynlonta®	ADC	X	–	–
Tivdak®	ADC	X	X	–
Elahere™	ADC	X	X	–
Datroway®	ADC	X	X	–
Emrelis™	–	–	–	–

– not conducted, *mAb* monoclonal antibody, *ADC* antibody drug conjugate

During the repeat-dose toxicology rat studies in Adcetris, greater hepatobiliary toxicity in the ADC-dosed rats was noted than with the drug payload alone. Although a TCR study with the mAb had already been conducted in human and monkey, a separate TCR study was conducted in human tissues and rat liver to help understand these findings (although there was no liver staining in rats).

Genetic Toxicology

Consistent with the standard core battery of genetic toxicology studies recommended in the ICH S2 guidance, assays used to evaluate the potential for genetic toxicity usually included a test for mutagenicity using the bacterial reverse mutation (Ames) assay and a test for genotoxicity in mammalian cells using either a bone marrow micronucleus assay, a chromosome aberration assay, or a mouse lymphoma assay (all of which are in vitro, or can be done in vitro). In general, in vitro studies have been sufficient without the need for in vivo testing, if results from the in vitro tests are positive. If the in vitro tests are negative, an in vivo micronucleus assay or chromosome aberration assay in rats or mice is warranted. The genetic toxicology studies employed in each ADC drug's development is shown in Table 6.

Reproductive Toxicology Studies

The need for a reproductive toxicology assessment of an ADC should be agreed upon with regulatory bodies prior to initiation. A weight-of-evidence assessment, especially for an ADC with a genotoxic payload component, is likely sufficient to

Table 6 Genetic toxicology assays used in development of approved ADCs

Drug	Payload	Ames	Micronucleus in vitro	Micronucleus in vivo	Chromosomal Aberration	MLA (in vitro)
Mylotarg™	N-acetyl gamma calicheamicin	Positive	Clastogenic	Clastogenic (mice)	–	–
Adcetris®	MMAE	Negative	–	Aneugenic (rat)	–	Negative
Kadcyla®	DM1	Negative	–	Clastogenic (rat)	–	–
Besponsa®	N-acetyl gamma calicheamicin	Positive	Clastogenic	Clastogenic (mice)	–	–
Lumoxiti®	PE38[a]	–	–	–	–	–
Polivy®	MMAE	Negative	–	Aneugenic (rat)	–	Negative
Padcev™	MMAE	Negative	–	Aneugenic (rat)	–	Negative
Enhertu®	DxD	Negative	–	Clastogenic (rat)	Clastogenic	–
Trodelvy™	SN-38	Negative	Clastogenic	–	–	–
Blenrep®	MMAF	Negative	Aneugenic[b]	Negative (rat)	–	–
Zynlonta®	PBD SG3199	–	Clastogenic[c]	–	Clastogenic[a]	–
Tivdak®	MMAE	Negative [c]	–	Aneugenic (rat) [c]	–	Negative [c]
Elahere™	DM4 and S-methyl DM4	Negative	–	Clastogenic (rat)	–	–
Datroway®	DxD	Negative	–	Clastogenic (rat)	Clastogenic	–
Emrelis™	MMAE	Negative [c]	–	Aneugenic (rat) [c]	–	Negative [c]

– not conducted Ames = bacterial reverse mutation test, *MN* micronucleus assay, *MLA* mouse lymphoma assay

[a]PE38 is a truncated form of a pseudomonas exotoxin protein; genetic toxicity was not conducted consistent with ICHS6

[b]The payload, cys-mcMMAF (SGD-1361) used in Blenrep was negative in the in vivo micronucleus assay likely due to not being cell-permeable (due to its charged C-terminal phenylalanine).

[c]Referenced the genetic toxicology studies for MMAE in the BLAs for earlier approved ADC programs

forgo a full or partial reproductive and developmental toxicity program. The target patient population, indication, and known hazards or toxicity of the drug product should be considered when determining the usefulness of a reproductive toxicity study, and if one is needed, the endpoints to include, the gestational period of dosing, the duration of the study, and the species to use. Early ADCs conducted EFD, FEED, or other reproductive studies, however, these studies are generally not considered necessary if genetic toxicity studies are positive (which they are with the payloads in all approved ADCs so far).

In accordance with ICH S9, for drugs intended to treat patients with advanced cancers, fertility and early embryonic development (FEED) and prenatal and postnatal development (PPND) studies are not required. Of the early ADCs approved prior to 2020, Adcetris (in the form of the ADC product and the drug product alone), was evaluated in an embryonic and fetal development (EFD) study in rats and Padcev (as the ADC product) was evaluated in an EFD study in rats. Other ADCs that used MMAE referenced the EFD study conducted with MMAE originally submitted with Adcetris. Besponsa (ADC) was evaluated in a FEED study in rats and two EFD studies in rats and rabbits. An EFD study form the payload (MMAE) of Adcetris was referenced in the Polivy review. Mylotarg conducted both a FEED and an EFD study in rats. Kadcyla was the only approved ADC to reference a non-rodent species, cynomolgus monkeys, in a reproductive toxicology study (EFD study), however, the study was for the antibody component only, trastuzumab, which was shown previously to cause reproductive toxicity clinically and the FDA reviewer agreed adding the genotoxic DM1 payload would likelymake Kadcyla even more cytotoxic to developing fetuses. Additional assessments that were included in some approved ADCs included male fertility assessments (Mylotarg); testicular toxicity assessment (Padcev); and late gestational toxicity, placental transfer, and secretion in milk study, and a menstrual cycle study (fertility evaluation) with the antibody component in Kadcyla.

As ADC development strategy evolved, it became more common to rely on the evidence that genotoxic payloads would have risks for reproduction and the developing fetus.

Challenges to Study Design and Implementation

Toxicity studies are intended to identify effects of the test article in animals prior to administering to humans. Selection of GLP toxicology study doses are based on pharmacokinetics, pharmacodynamics, efficacy studies, and dose-range-finding studies. ADCs are unique, in that the drug component of the test material may have already been tested separately, and some general information may already be known about target or off-target effects and potential toxicities. Although the purpose of an ADC is to direct a potentially toxic molecule to a site of action, toxicities are still present. Additionally, the profiles of active metabolites may be different across preclinical species. Although dose-range-finding studies can help elucidate early effects, severe toxicity or mortality may not be seen at a given dose in the pilot study but may be seen in the GLP study, with little or no exposure relationship between dose and toxicity.

To ensure a robust toxicology program that supports first-in-human clinical administration, it is critical that the test-article chemical characterization is well known, the dose levels have been selected properly in pilot and preliminary studies, and that the study is monitored carefully to identify early signs of toxicity. Prudent nonclinical study strategies may include staggering the doses to allow time for potential dosing adjustments; imposing dosing holidays (periods where animals are not dosed) during pivotal studies to allow for either recovery of toxicities or adjustments to dose levels; closely monitoring clinical observations and clinical chemistry results for developing toxicity, particularly blood (especially alterations in hematology and serum chemistry), liver (increases in AST, ALT, or total bilirubin), and kidney, which are general depositories for ADCs and free payload. Allowing signs of toxicity to continue for too long can result in early euthanasia of animals or groups, resulting in a much smaller body of data being collected. It is worthwhile noting that adverse effects seen in nonclinical toxicity studies and clinical studies do not overlap completely. There are certain clinical AEs that are not very predictable in monkeys and, in contrast, monkeys may overpredict some toxicities that are not dose-limiting in humans.

Conclusions

In a relatively short period of time, ADCs have evolved in terms of nonclinical development strategy, study designs, health authority perspectives, and predicted toxicity. As more ADCs have been developed and approved, modifications to the antibody, payload, and DAR have resulted in improvements to ADR efficacy and toxicity. With dozens of ADCs currently in development, ADC technology is expected to evolve to include identification of biomarkers to optimize patient selection, to capture early response signals and to capture DNA damage (Coats et al. 2019). Continued improvements in stability and specificity of the linkers will result in better outcomes for patients by reducing off-target or unwanted effects. Bispecific ADCs are in development that may improve tumor selectivity without affecting normal cells—for example, by binding to both c-MET and EGFR, which are overexpressed in cancer cells (Mysliwy 2020). To predict safety for patients, each component of ADC chemistry (the payload, the linker, and the antibody) plays a critical role in the action at the target site. Finally, as regulatory agencies are gaining more experience with assessing ADCs, they can provide more direction (e.g., guidance documents, publications, etc.) for nonclinical development strategies. A well-thought-out strategy for designing a development plan consisting of efficacy studies, test-article chemical characterization, accurate dose selection, ADME and safety pharmacology evaluations, and relevant toxicity studies is critical for a successful ADC program.

References

Abbvie Press Release: Novel antibody drug conjugate ABBV-3373 shows improvement in disease activity in phase 2a study of patients with rheumatoid arthritis. https://news.abbvie.com/news/press-releases/novel-antibody-drug-conjugate-abbv-3373-shows-improvement-in-disease-activity-in-phase-2a-study-patients-with-rheumatoid-arthritis.htm (2020)

Bodey, B., Siegel, S.E., Kaiser, H.E.: Human cancer detection and immunotherapy with conjugated and non-conjugated monoclonal antibodies. Anticancer Res. **16**(2), 661–674 (1996)

Bouchard, H., Viskov, C., Garcia-Echeverria, C.: Antibody–drug conjugates—a new wave of cancer drugs. Bioorg. Med. Chem. Lett. **24**(23), 5357–5363 (2014)

Bross, P.F., Beitz, J., Chen, G., Chen, X.H., Duffy, E., Kieffer, L., Roy, S., Sridhara, R., Rahman, A., Williams, G., Pazdur, R.: Approval summary: gemtuzumab ozogamicin in relapsed acute myeloid leukemia. Clin. Cancer Res. **7**(6), 1490–1496 (2001) Erratum in: Clin Cancer Res 2002 (1):300

Chau, C.H., Steeg, P.S., Figg, W.D.: Antibody–drug conjugates for cancer. Lancet. **394**(10200), 793–804 (2019)

Coats, S., Williams, M., Kebble, B., Dixit, R., Tseng, L., Yao, N.S., Tice, D.A., Soria, J.C.: Antibody–drug conjugates: future directions in clinical and translational strategies to improve the therapeutic index. Clin. Cancer Res. **25**(18), 5441–5448 (2019)

FDA: Draft guidance nonclinical safety evaluation of the immunotoxic potential of drugs and biologics. U.S. Department of Health and Human Services, Food and Drug Administration, Center for Drug Evaluation and Research (CDER), Center for Biologics Evaluation and Research (CBER) (2020)

Hinrichs, M.J.M., Dixit, R.: Antibody drug conjugates: nonclinical safety considerations. AAPS J. **17**(5), 1055–1064 (2015)

Hoffmann, R.M., Coumbe, B.G., Josephs, D.H., Mele, S., Ilieva, K.M., Cheung, A., Tutt, A.N., Spicer, J.F., Thurston, D.E., Crescioli, S., Karagiannis, S.N.: Antibody structure and engineering considerations for the design and function of antibody drug conjugates (ADCs). Oncoimmunology. **7**(3), e1395127 (2018)

Hoofring, S.A., Lopez, R., Hock, M.B., Kaliyaperumal, A., Patel, S.K., Swanson, S.J., Chirmule, N., Starcevic, M.: Immunogenicity testing strategy and bioanalytical assays for antibody–drug conjugates. Bioanalysis. **5**(9), 1041–1055 (2013)

ICH S6: Guidance for industry preclinical safety evaluation of biotechnology-derived pharmaceuticals. In: International Conference on Harmonisation of Technical Requirements for Registration of Pharmaceuticals for Human Use. Global (1997)

ICH S6(R1): Guidance for industry preclinical safety evaluation of biotechnology-derived pharmaceuticals. In: International Conference on Harmonisation of Technical Requirements for Registration of Pharmaceuticals for Human Use. Global (2011)

ICH S9: Guidance for industry nonclinical evaluation for anticancer pharmaceuticals. In: International Conference on Harmonisation of Technical Requirements for Registration of Pharmaceuticals for Human Use. Global (2010)

ICH S9 Q&A: Guidance for industry nonclinical evaluation for anticancer pharmaceuticals questions and answers. In: International Conference on Harmonisation of Technical Requirements for Registration of Pharmaceuticals for Human Use. Global (2018)

Kaur, S., Xu, K., Saad, O.M., Dere, R.C., Carrasco-Triguero, M.: Bioanalytical assay strategies for the development of antibody–drug conjugate biotherapeutics. Bioanalysis. **5**(2), 201–226 (2013)

Lode, H.N., Reisfeld, R.A., Handgretinger, R., Nicolaou, K.C., Gaedicke, G., Wrasidlo, W.: Targeted therapy with a novel enediyene antibiotic calicheamicin theta(I)1 effectively suppresses growth and dissemination of liver metastases in a syngeneic model of murine neuroblastoma. Cancer Res. **58**(14), 2925–2928 (1998)

Mecklenburg, L.: A brief introduction to antibody–drug conjugates for toxicologic pathologists. Toxicol. Pathol. **46**(7), 746–752 (2018)

Mysliwy, J.: Advancements in antibody–drug conjugate technology for cancer treatment. Future Drug. Discov. **2**(2), FDD35 (2020)

Niculescu-Duvaz, I.: Technology evaluation: Gemtuzumab ozogamicin, Celltech Group. Curr Opin Mol Ther. **2**(6), 691–696 (2000)

Oroudjev, E., Lopus, M., Wilson, L., Audette, C., Provenzano, C., Erickson, H., Kovtun, Y., Chari, R., Jordan, M.A.: Maytansinoid-antibody conjugates induce mitotic arrest by suppressing microtubule dynamic instability. Mol. Cancer Ther. **9**(10), 2700–2713 (2010)

Panowski, S., Bhakta, S., Raab, H., Polakis, P., Junutula, J.R.: Site-specific antibody drug conjugates for cancer therapy. MAbs. **6**(1), 34–45 (2014). Taylor & Francis

Peters, C., Brown, S.: Antibody-drug conjugates as novel anti-cancer chemotherapeutics. Biosci. Rep. **35**(4), e00225 (2015)

Pfizer: Pfizer receives FDA approval for MYLOTARG™ (gemtuzumab ozogamicin). Press Release (2020). Available online: https://www.pfizer.com/news/press-release/press-release-detail/pfizer_receives_fda_approval_for_mylotarg_gemtuzumab_ozogamicin#:~:text=Pfizer%20Inc.,or%20refractory%20CD33%2Dpositive%20AML

Ritter, A.: Antibody–drug conjugates. Pharm. Technol. **36**(1), 42–47 (2012)

Roberts, S.A., Andrews, P.A., Blanset, D., Flagella, K.M., Gorovits, B., Lynch, C.M., Martin, P.L., Kramer-Stickland, K., Thibault, S., Warner, G.: Considerations for the nonclinical safety evaluation of antibody drug conjugates for oncology. Regul. Toxicol. Pharmacol. **67**(3), 382–391 (2013)

Saber, H., Leighton, J.K.: An FDA oncology analysis of antibody-drug conjugates. Regul. Toxicol. Pharmacol. **71**, 444–452 (2015)

Saber, H., Gudi, R., Manning, M., Wearne, E., Leighton, J.K.: An FDA oncology analysis of immune activating products and first-in-human dose selection. Regul. Toxicol. Pharmacol. **81**, 448–456 (2016)

Saber, H., Simpson, N., Ricks, T.K., Leighton, J.K.: An FDA oncology analysis of toxicities associated with PBD-containing antibody-drug conjugates. Regul. Toxicol. Pharmacol. **107**, 104429 (2019)

Smaglo, B.G., Aldeghaither, D., Weiner, L.M.: The development of immunoconjugates for targeted cancer therapy. Nat. Rev. Clin. Oncol. **11**(11), 637–648 (2014)

Index

H. J. Gukasyan, A. Kulkarni (eds.), *Antibody Drug Conjugates*, AAPS Advances in the Pharmaceutical Sciences Series 52, https://doi.org/10.1007/978-3-031-89319-3

MIX
Papier aus verantwortungsvollen Quellen
Paper from responsible sources
FSC® C105338

If you have any concerns about our products,
you can contact us on
ProductSafety@springernature.com

In case Publisher is established outside the EU,
the EU authorized representative is:
Springer Nature Customer Service Center GmbH
Europaplatz 3, 69115 Heidelberg, Germany

Printed by Libri Plureos GmbH
in Hamburg, Germany